POLYMER THIN FILMS

SERIES IN SOFT CONDENSED MATTER **ISSN: 1793-737X**

Aims & Scope

The study of *Soft Condensed Matter* has stimulated fruitful interactions between physicists, chemists, and engineers, and is now reaching out to biologists. A broad interdisciplinary community involving all these areas of science has emerged over the last 30 years, and with it our knowledge of *Soft Condensed Matter* has grown considerably with the active investigations of polymers, supramolecular assemblies of designed organic molecules, liquid crystals, colloids, lyotropic systems, emulsions, biopolymers, and biomembranes, among others.

The present Book Series, initiated by Pierre-Gilles de Gennes, covers a large number of diverse aspects, both theoretical and experimental, in all areas of *Soft Condensed Matter*. It mainly addresses graduate students and junior researchers as an introduction to new fields, but it should also be useful to experienced people considering a change in their field of research. This Book Series aims to provide a comprehensive and instructive overview of all *Soft Condensed Matter* phenomena.

Published:

Vol. 1 Polymer Thin Films
edited by Ophelia K. C. Tsui and Thomas P. Russell

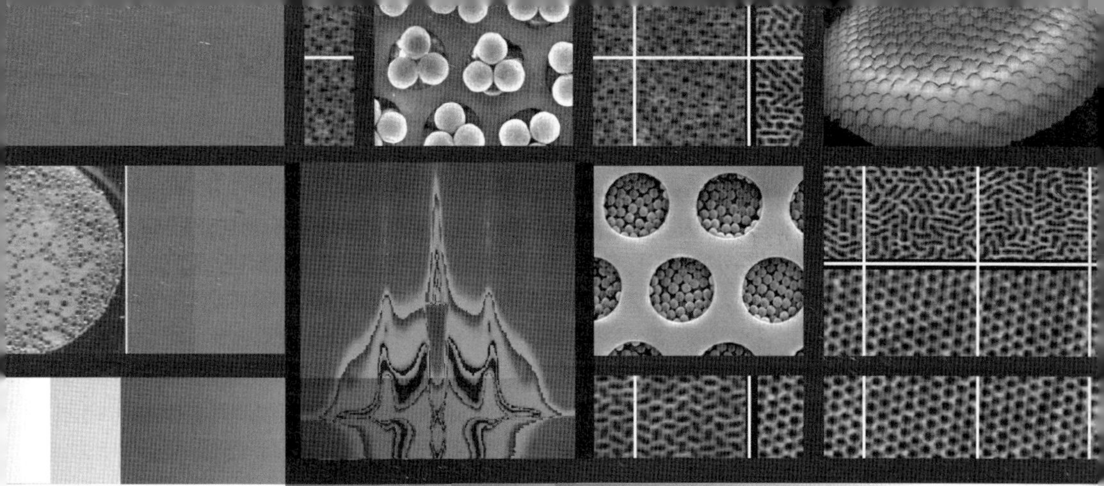

Series in Soft Condensed Matter **Vol. 1**

POLYMER THIN FILMS

Editors

Ophelia K. C. Tsui
Boston University, USA

Thomas P. Russell
University of Massachusetts Amherst, USA

World Scientific

NEW JERSEY • LONDON • SINGAPORE • BEIJING • SHANGHAI • HONG KONG • TAIPEI • CHENNAI

Published by

World Scientific Publishing Co. Pte. Ltd.
5 Toh Tuck Link, Singapore 596224
USA office: 27 Warren Street, Suite 401-402, Hackensack, NJ 07601
UK office: 57 Shelton Street, Covent Garden, London WC2H 9HE

Library of Congress Cataloging-in-Publication Data
Tsui, Ophelia Kwan Chui, 1967-
Polymer thin films / Ophelia K.C. Tsui and Thomas P. Russell.
p. cm.
Includes bibliographical references and index.
ISBN-13: 978-981-281-881-2 (hardcover : alk. paper)
ISBN-10: 981-281-881-2 (hardcover : alk. paper)
1. Thin films. 2. Polymers. 3. Block copolymers. 4. Nanostructured materials.
I. Russell, Thomas P., 1952– II. Title.
TA418.9.T45T78 2008
621.3815'2--dc22

2008033326

British Library Cataloguing-in-Publication Data
A catalogue record for this book is available from the British Library.

Printed in Singapore by Mainland Press Pte Ltd

PREFACE

Everyday we are exposed to a myriad of applications of polymer thin films. This is whether it occurs in food wrapping, the packaging of virtually any item that is sold, or in protective coatings placed on the surface of furniture or glass. More and more often, though, the thickness of the polymer film decreases to dimensions comparable to the dimensions of a single polymer chain, or in the case of block copolymers or mixtures, comparable to the characteristic period of the microphase separated or phase separated morphology. For example, in the case of microelectronic circuits where polymers can be used as a dielectric insulator or as a template to produce a porous oxide layer, the dimensions of and separation distance between the conducting elements is rapidly decreasing to several tens of nanometers. We are forced, therefore, to ask the question as to whether the confinement of the polymer to such small dimensions will change the fundamental characteristics of the polymer. In addition, with such small scale features, the surface to volume ratio increases significantly, so does the abundance of interfacial area affect the nature of the polymer in the confined geometry. In the case of block copolymers, how does such confinement alter the nature or orientation of the microphase-separated morphology? More importantly, if we can understand the influence of such confinement, can we use this to our advantage in designing systems to induce behavior not seen in the bulk or can we control the interactions of the polymer or block copolymer with the interfaces to manipulate the spatial arrangement of the nanoscopic elements in the morphology? If so, then this will open numerous applications of polymers and block copolymers in the burgeoning field of nanotechnology which encompasses applications ranging from ultrahigh density storage media, ultralow dielectric constant materials, high resolution separations media, self-healing and self-corralling nanocomposites, flexible electronics and displays, and photovoltaic devices. Consequently, there has been a significant growth in the amount of research dedicated to understanding the structure and properties of polymer thin films.

The growth in the interest of polymer thin films has also been catalyzed by the increase in the number of techniques available to characterize thin polymer films. While these techniques may have been available for decades, only recently has it been recognized that they could be used to great advantage to characterize polymeric materials. Some of these techniques include scanning probe microscopies, neutron and x-ray reflectivity, grazing incidence x-ray scattering, forward recoil and Rutherford backscattering spectroscopies, dynamic secondary ion mass spectroscopy, x-ray microscopy and electron tomography. The growth in the use of these techniques to characterize polymer thin films, coupled with the use of fairly standard techniques, like ellipsometry, x-ray photoelectron spectroscopy, electron microscopy, x-ray and neutron scattering, dielectric spectroscopy, infrared and Raman spectroscopies, and optical microscopy, has revolutionized our understanding of polymer thin films. At the same time, there have been tremendous advances made in modifying either the chemical nature or the topography of surfaces, by use of photolithography, electron beam lithography, ion beam etching, and surface specific chemistries. This has imparted elegant routes to control the interactions between a polymer and a surface.

Perhaps the most significant development in the characterization of polymer thin films is the interest that a broad scientific community has taken in this area. From the engineering side, there are challenges that are faced in producing perfectly uniform thin films. Yet, this in turn has sparked the interest from the physics community in studies on instabilities in thin films and on the confinement of highly ordered structures. Controlling interfacial interactions has presented numerous challenges to surface chemists, both small molecule chemists and polymer chemists. The influence of confinement on phase transitions like crystallization, phase separation and microphase separation, has piqued the interest of physicists and physical chemists. Producing surfaces with well-defined topographies, chemistries and mechanical properties has enabled the manipulation of the interactions of living cells with surfaces, and as such, has generated interest from the biological and microbiological communities. This convergence of interest from a wide-range of different disciplines has further promoted advances in our understanding of thin polymer films.

An overview encompassing the large variety of areas that have spun off from this vastly developing field is overdue. It is, however, important that this overview be understandable by the novice to the area and appreciated by the experts. This is the rather daunting task that faced the authors of the different chapters in this book. We hope that these chapters will serve as a useful resource for instructors of undergraduate and graduate courses. To reflect the interdisciplinarity, the authors of the different chapters include chemists, engineers, materials scientists and physicists that have made significant contributions in their respective areas.

This book contains eleven chapters that can be categorized into six major areas:

- The design and construction of nanostructures in block copolymer films — the fundamental principles, fabrication methods and applications (Ch. 1–4)
- Alternative methods of fabricating sophisticated nanostructures in polymer thin films (Ch. 5–6)
- Crystallization of polymers confined in nanometer films (Ch. 7)
- Tribology of polymer thin films — friction and adhesion (Ch. 8 and 9, respectively)
- Wetting stability of polymer films supported by a substrate (Ch. 10)
- Novel dynamical properties of polymer nanometer films (Ch. 11)

We are indebted to the contributors of this book who have abided to the cause and taken time and effort in completing the chapters. Despite the crash of a hard disk (Shimomura), laboratory floods (Jacobs) and over-commitment of time (all of the authors), the chapters were completed in a timely manner, are of high-quality, and a pleasure to read. We are grateful to the students and postdoctoral fellows in our groups (Z. Yang, J. Wang, R. Tangirala, W. Chen, J. Xu, J. Chen, D. Chen, L. Li,

P. Dobriyal, J. He and H. Liu) who had helped compile the subject index. Special thanks are due to the anonymous reviewers who had generously contributed their time in reading the chapters and given suggestions for improvements.

We hope that the students, professors and researchers will find this book a useful guide and resource to the field of polymer thin films.

O. K. C. Tsui
T. P. Russell

CONTENTS

CHAPTER 1

BLOCK COPOLYMER THIN FILMS

Jia-Yu Wang, Soojin Park, Thomas P. Russell

Department of Polymer Science and Engineering
University of Massachusetts, Amherst
Amherst, MA 01003, U.S.A.
E-mail: russell@mail.pse.umass.edu

The behavior of amorphous block copolymers (BCPs) in thin films depends on a combination of segmental interactions, interfacial interactions, surface energies and entropy. Commensurability between the film thickness, h, and the natural period, L_0, of the microdomains in the bulk is also of importance. This chapter summarizes recent developments in our understanding of the influence of confinement, surface energies and surface heterogeneities on the morphology of BCP thin films, the use of thin BCP films as scaffolds and templates for the fabrication of nanostructured materials, and the generation BCP arrays in thin films having long-range lateral order for potential addressable media.

1. Introduction

Block copolymers (BCPs) consist of two or more chemically different polymer chains joined covalently at their ends. Due to the positive enthalpy and small entropy of mixing, dissimilar blocks tend to microphase separate into well-ordered arrays of domains, classically termed microdomains. The sizes of these microdomains, due to the connectivity of the blocks, are limited to molecular dimensions and, as such, are tens of nanometers or less. At temperature below an order-to-disorder transition temperature, T_{ODT}, BCPs microphase separate into arrays of spherical, cylindrical, gyroid or lamellar microdomains, depending on the volume fractions of the blocks, *f*, and the degree of microphase separation, χN, where χ is the Flory-Huggins segmental interaction parameter and N is the total number of segments in BCPs. Above the T_{ODT}, BCPs phase mix and are disordered.

The self-assembly of BCPs into well-defined morphologies has opened numerous applications ranging from drug delivery to structural materials. In contrast to the bulk, the morphology of amorphous BCP thin films can be strongly influenced by surface and interfacial energies as well as the commensurability between the film thickness, h, and the period of the microdomain morphology, L_0. With decreasing film thickness these parameters become increasingly important in defining the morphology. By controlling the orientation and lateral ordering the BCP microdomains in thin films, unique opportunities in the use of BCPs in materials science (adhesive properties, lubrication, membranes, and coatings), lithography and microfabrication (addressable memory, magnetic storage, insulating foams) and device technologies (light-emitting diodes, photodiodes, and transistors) are beginning to emerge. In this chapter, two aspects of BCP thin films will be addressed. First, the effect of confinement, surface energies and surface heterogeneities on the morphology of BCP thin films will be discussed on the basis of the simplest and most studied system, namely thin films of compositionally symmetric, amorphous diblock copolymers. This is followed by a review of BCP thin films for nanopatterning with discussion centering on fabrication of long-range ordered nanostructures of BCP thin films by applying various external fields.

2. Morphologies of BCP Thin Films

2.1. *Effect of Confinement*

The presence of a surface or interface can strongly influence the phase behavior, morphology and kinetics of a multicomponent system. Understanding of the influence of boundary surfaces, in particular, commensurability and interfacial interactions, on the morphology of BCP thin films has attracted much attention. Phenomena like symmetric/asymmetric wetting, surface topographies of "islands" or "holes", surface-induced ordering, and commensurability have been extensively studied.

Symmetric/Asymmetric Wetting. Consider a thin film of a block copolymer below T_{ODT}. The connectivity of the block means that the film thickness is not arbitrary but rather defined in terms of L_0.[1-19] For an A-*b*-B BCP, if A or B preferentially locates at both the surface and substrate interface, i.e. symmetric wetting, then the film thickness, h, is defined by nL_0 where n is an integer. If, on the other hand, different blocks segregate to the interfaces, then the film thickness is given by $(n+1/2)L_0$. So, to a first approximation, a thin BCP film is smooth, only if these constraints are satisfied. In all other cases, a surface topography consisting of islands or holes are seen where the step height is L_0. Examples of these are shown in the scanning force micrographs in Figure 1. Optical microscopy can also be used where discrete interference colors are seen, corresponding to the different thickness or optical path lengths.

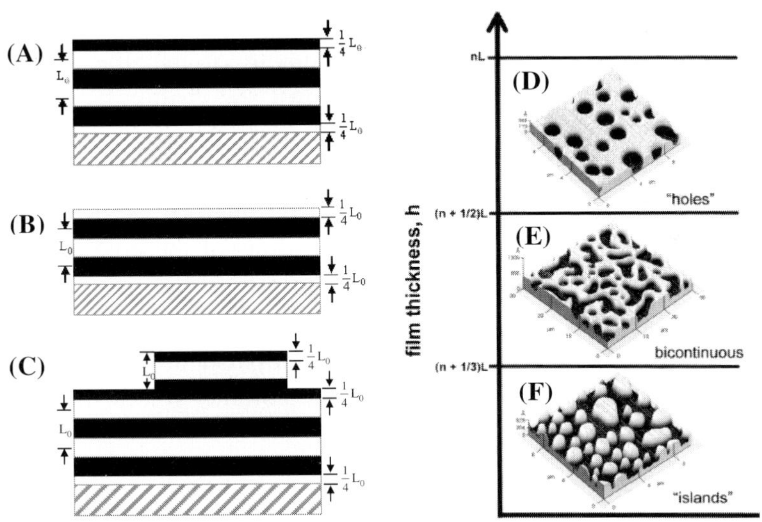

Fig. 1. (A-C) Schematic of three types of film structures. (A) $h = H_n = (n+1/2)L_0$; (B) $h = H_n = nL_0$; (C) ($h = H_n + \Delta h$). (D-E) AFM images of topographical features of (D) holes; (E) bicontinuous and (F) islands. Reproduced with permission from *Advances in Colloid and Interface Science*.[18]

It should be noted that if h is not much different than nL_0 or $(n+1/2)L_0$ for symmetric or asymmetric wetting conditions, respectivel y, then a surface topography may not be seen and the copolymer chains can stretch or compress to accommodate this incommensurability. This, of course, depends on a balance of the energy associated with stretching or compressing the BCP chains and the additional energy arising from the generation of more surface areas.

Frustration. If the film is confined between two parallel, impenetrable walls, having strong interactions with either A or B block, a surface topography cannot form and the incommensurability between L_0 and h causes a frustration, since the hard boundaries prohibit the formation of surface topography and the characteristic period of BCPs must change to accommodate this frustration. Lambooy et al.[19, 20] and, subsequently Koneripalli et al.[6] developed techniques to suppress the formation of surface topography by confining copolymer thin films between two hard walls. Strong interactions of the blocks with the confining walls were sufficient to force the period of BCPs to expand or compress by as much as 50% in some cases. Kickuchi and Binder,[21, 22] Turner[23] and Walton et al.[9] theoretically examined the confinement of lamellar BCPs and found that the frustration imposed on BCPs chains by thickness constraints depended strongly on the strength of the interactions of the blocks with the substrate. In the case of strong interactions, stretched or compressed copolymer multilayers were predicted. However, as observed by Lambooy et al.,[7] as the interactions of the copolymer with the confining walls became weaker, the copolymer domains can change their orientation with respect to the substrate, i.e. orient normal to the wall interface, allowing the period of the copolymer to be L_0 but paying the energetic price associated with unfavorable interaction at the interfaces.

Surface-Induced Instabilities. For $T > T_{ODT}$, if BCPs film is thinner than a characteristic thickness, i.e. $h < h_t$, ($h_t = L_0$ for symmetric wetting and $1/2L_0$ for asymmetric wetting), a "spinodal-like" pattern forms on the surface due to the frustration mentioned above. Prior to the formation of the surface topography, the film is unstable and shows periodic fluctuations.[18, 24-27] These surface patterns form spontaneously across the film surface, similar to a film undergoing dewetting. In the BCPs cases, though, the substrate is the initial copolymer layer that is preferentially located at the substrate. When $h = h_t$, the film is stable and the surface remains smooth. For films of $h > h_t$, a layer of thickness $h - h_t$ becomes unstable and dewets an underlying layer of thickness h_t, forming "islands", "holes" or a "spinodal-like" topography. This type of autophobic dewetting has not been observed in homopolymers.[18, 28-30]

Surface-Induced Ordering. The presence of a surface can induce ordering of a phase-mixed copolymer, i.e. when $T > T_{ODT}$.[1, 7, 16, 26, 31-42] This was first predicted by Fredrickson using mean-field theoretical arguments.[42] Anastasiadia et al. experimentally verified this predication in a neutron reflectivity study on BCPs of polystyrene and poly(methyl methacrylate), denoted PS-*b*-PMMA. By fitting the scattering length density profiles, it was shown that the ordering decays exponentially from the surface with a correlation length very close to the bulk correlation length, in agreement with the arguments of Fredrickson.[42] Subsequently, Menelle et al.[38] studied film thickness dependence of the ordering transition temperature and found that T_{ODT} was significantly elevated by the surface-induced ordering and, in fact, an order-to-disorder transition did not exist in thin films since the order parameter did not decay to zero at any point in the film. However, the films were shown to undergo a transition from a partially to a fully ordered state at a temperature that depended in a power-law manner on the film thickness, as shown in Figure 2.

2.2. *Surface Energy*

The morphology of BCP thin films is strongly influenced by the strength of interfacial interactions.[43-56] Strong preferential interactions of one block with the substrate or a lower surface energy of one component causes a segregation of that block to either the surface of the film or the substrate interface. As a result, the connectivity of the blocks forces a parallel orientation of the microdomain to the substrate. When the surface is neutral, i.e., the interfacial interactions of both blocks are equally favorable or unfavorable, there is no preferential segregation of

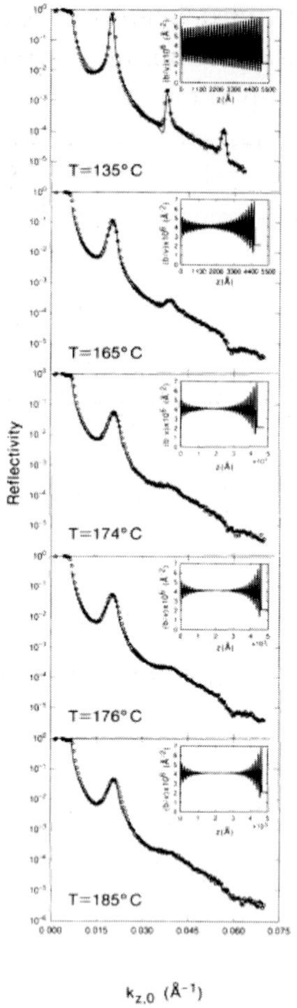

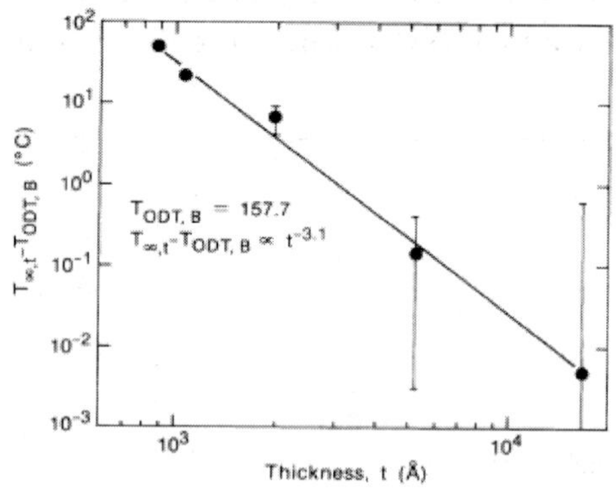

Fig. 2. Left: Neutron reflectivity profiles for a 523.2 nm film of dPS-b-PMMA symmetric diblock copolymer as a function of the neutron momentum normal to the surface at the temperatures indicated. The scattering length density *b/V* profiles as a function of depth z, where z = 0 is the air surface, shown in the insets were used to calculate the reflectivity profiles drawn as the solid lines in the figures. Right: A log-log plot of the difference between $T_{\infty,t}$ and the bulk value of $T_{ODT,B}$ as a function of film thickness using the value of $T_{ODT,B}$ extrapolated from the neutron reflectivity measurements. Reproduced *with permission from Physical Review Letters.*[38]

the components to the interfaces. Any slight incommensurability will cause the microdomains to orient normal to the surface. The interfacial energies of an A-*b*-B BCP with a solid surface can be precisely controlled by anchoring a random copolymer of A and B, A-*r*-B, to the surface, where the volume fraction, *f*, of A monomers in the brush can be varied in the synthesis. As *f* is varied from 0 to 1, the system goes from a condition of preferential wetting of the substrate by A to a preferential wetting by B. However, for one specific value of *f* the interactions of A and B with the substrate are balanced. This was demonstrated with styrene and methyl methacrylate pairs and it was found that a P(S-*r*-MMA) brush surface has equal interfacial energies with PS and PMMA when $f \cong 0.57$ (Figure 3).[43] The use of random

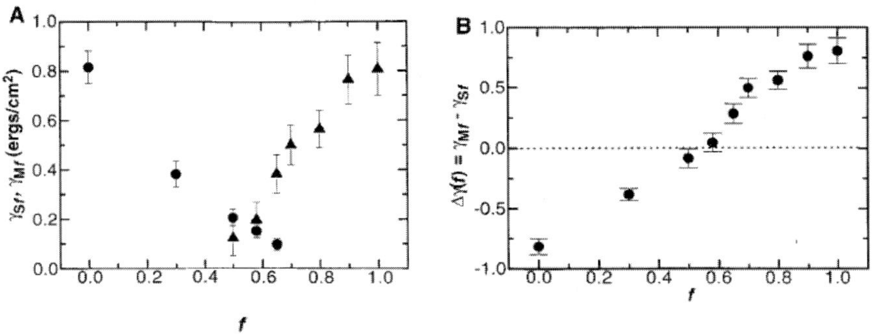

Fig. 3. (A) Interfacial energies γ_{Sf} and γ_{Mf} and (B) $\Delta\gamma(f) = \gamma_{Mf} - \gamma_{Sf}$ for PS (circles) and PMMA (triangles) on a P(S-r-MMA) brush as a function of *f*. Reproduced with permission from *Science*.[43]

copolymers to modify and control interfacial interactions requires a simple synthetic method. Figure 4A shows the synthesis of (PS-*r*-MMA)

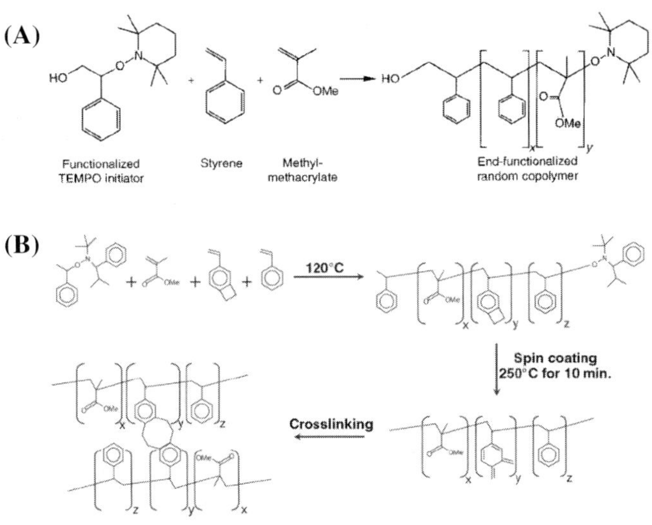

Fig. 4. (A) Schematic synthesis of P(S-r-MMA); (B) Schematic synthesis of P(S-r-BCB-r-MMA) which has 2% reactive benzocyclobutene (BCB) functionality randomly incorporated along the backbone that can be thermally crosslinked. Reproduced with permission from *Science*.[43, 50]

random copolymers by a "living" free-radical polymerization using a unimolecular initiator which has a hydroxyl (OH) group that can be used to end-graft the random copolymer chains to the native silicon oxide surface.[43] However, this covalent attachment of the random copolymer to the surface requires substrate-specific chemistry which limits applications. Hawker and Russell et al.[50] modified the reaction and synthesized a random copolymer containing crosslinkable groups in the chain. Using a simple, highly efficient cross-linking reaction, an insoluble, ultrathin mat of the random copolymer can be formed which is more robust than anchored random copolymer brushes and is independent of the nature of the substrate (Figure 4B) and, as such, can be used to modify virtually any surface on which a thin film could be prepared.

Using anchored and crosslinked random copolymers, Russell and coworkers[44, 46, 49, 52, 53, 56] quantitatively studied the influence of interfacial energies on the orientation of symmetric PS-*b*-PMMA microdomains in thin films. For BCP films placed on random copolymer brushes rich in PS, the PS block segregated to the random copolymer interface, leading to symmetric wetting. (The PS block segregates to the polymer/air interface due to its lower surface energy.) Conversely, PS-*b*-PMMA films on PMMA-rich random copolymers showed asymmetric wetting due to the wetting of the random copolymer layer with the PMMA block. For BCP films cast on random copolymer modified surfaces where interfacial interactions were balanced, preferential segregation of either block did not occur. This condition means that either A or B or both blocks could be located at the interface. Any film thickness that introduces a condition of incommensurability will result in an orientation of the microdomains normal to the interface. It should be noted that by balancing interfacial interactions, commensurability conditions are not the same as when there is a strong preference for either component to the interface. In fact, commensurability can be realized when the film thickness is $n/2L_0$, where n is an integer. This effectively means that an interface can cut a microdomain through the center, i.e. half cylinders or half lamellar layers located at the interface, and still satisfy the constraints imposed by the hard walls. Outside of this condition, if interfacial interactions are balanced, the microdomains are oriented normal to the interface with the orientation persisting through the entire film thickness. Thus, by varying the composition of the random copolymer brushes, four boundary conditions can be achieved as shown in Figure 5.

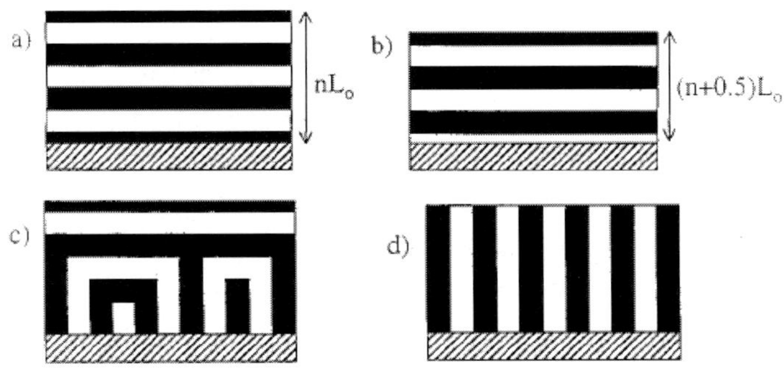

Fig. 5. Block copolymer structures observed for films having (a) symmetric boundary conditions, (b) asymmetric boundary conditions, (c) one nonpreferential and one preferential boundary surface, and (d) nonpreferential boundaries. Reproduced with permission from *Macromolecules*.[46]

Modifying a solid substrate, as described above, can easily be achieved by anchoring a brush to the substrate or by placing a thin crosslinked film on the substrate. However, controlling interactions at the opposite interface, either the air surface or the second hard boundary placed in contact with the free surface, is not as straightforward. This can be achieved by transferring a random copolymer film on top of the BCP film, followed by placing the second hard wall in contact with this transferred layer. Alternatively, a polymer film where interactions between the blocks and this polymer are balanced can be spin-coated on top of the BCP film, as long as a non-solvent for the BCP is used.

2.3. *Heterogeneous Surfaces*

The strength and lateral distribution of interfacial interactions can direct the BCP morphologies in thin films.[57-60] Surfaces where the interfacial interactions were laterally heterogeneous and periodic were used to selectively place blocks at specific locations on a surface. Kramer and coworkers[57] experimentally demonstrated this by designing a surface where CH_3- and HO- terminated thiols were periodically modified on a gold surface to precisely tune the surface potential. This led to a preferential wetting of the components across the surface. Rockford et al.[58] investigated the ordering of BCP ultrathin films on chemically heterogeneous surfaces and found that highly ordered lamellar structures perpendicular to the surface could be obtained on a patterned surface with the period of the surface pattern within 10% of commensuration with the period of BCPs, L_0. When the mismatch is greater than 10%, the lamellar microdomains orient parallel to the surface.[59] By using soft-x-ray-based photolithographic process to make chemically patterned substrates precisely, Nealey and coworkers[61-66] have significantly improved direct assembly of BCP thin films on chemically heterogeneous surfaces Not only have they shown that the patterning can direct the morphology but that the copolymers can be used to improve the line-edge roughness of the features lithographically placed on the surface which is a technologically critical problem in nanolithography.

Heterogeneous surfaces with a gradient in length scales, from the nanoscopic to the microscopic, can also be produced by using mixtures of homopolymers and BCPs to act as templates for surface patterning.[67, 68] In particular, by blade coating heterogeneous surfaces can be prepared, though such surfaces have not been used to control the orientation or ordering of BCPs.

3. Nanopatterning from BCP Thin Films

The self-assembly of BCPs into arrays of nanoscopic elements makes them ideal candidates as templates and scaffolds for the fabrication of nanostructured materials. While self-assembly generates the arrays of elements, it is necessary to harness the self-assembly so as to spatially direct the assembly into usable structures. This may simply mean controlling the orientation of the microdomains into arrays oriented parallel or normal to the surface of the film without any concern over the lateral ordering of the microdomains. Indeed applications ranging from low dielectric constant insulators to filtration membranes for virus particles have already been realized using such a directed self-assembly concept where only the orientation of the microdomains is controlled. However, if BCP thin films are used as templates for addressable media or for polarization applications, it is necessary to control both the orientation and the ordering of the microdomains to a high degree and a biased, directed self-assembly must be used. Here, top-down approaches, like photolithography, which are used to pattern or sector a surface and are coupled with a bottom-up approach, like self-assembly, to obtain multi-length scales control over the ordering and spatial location of nanoscopic elements for device fabrication. Approaches to control the orientation of BCP microdomains include the use of solvent fields,[69-73] electric fields,[74] chemically patterned substrates,[62, 63] epitaxial crystallization,[75] graphoepitaxy,[76] controlled interfacial interactions,[43] thermal gradients,[77] zone casting,[78] optical alignment,[79, 80] and shear.[81-83] This section will discuss long-range ordering of BCP thin films induced by solvent annealing, zone casting and optical alignment approaches that have been used effectively.

3.1. *Solvent Annealing*

The preparation of BCP thin films under various solvent evaporation conditions is an effective way to manipulate the orientation and lateral ordering of BCP microdomains in thin films. The solvent evaporation rate is one of the key factors that control these kinetically trapped nanostructures. For instance, inverted phases consisting of spheres or cylinders of the majority fraction block in a polystyrene-*block*-polybutadiene-*block*-polystyrene (PS-*b*-PB-*b*-PS) copolymer, which were not predicted on the basis of typical thermodynamic considerations for block copolymer melts, were observed by the control over the solvent evaporation rate.[84, 85] Libera and coworkers first reported that solvent evaporation could be used to induce the ordering and orientation of BCP microdomains.[69, 70, 86] Vertically aligned cylindrical PS microdomains could be obtained in a PS-*b*-PB-*b*-PS triblock copolymer thin films with a thickness of ~ 100 nm.[70] The same effect was also observed in a polystyrene-*block*-poly(ethylene oxide) (PS-*b*-PEO) BCP thin films and was attributed to a copolymer/solvent concentration gradient along the direction normal to the film surface giving rise to an ordering front that propagated into the film during solvent evaporation.[87] This orientation was independent of the substrate. However, the lateral ordering of the cylindrical microdomains was poor. Sibener[88] and later Russell et al.[73] showed that evaporation-induced flow, in solvent cast BCP films, produced arrays of nanoscopic cylinders oriented normal to the surface with a high degree of ordering. Recently, Krausch and coworkers[86, 89] demonstrated that solvent annealing could markedly enhance the ordering of BCP microdomains in thin films. By controlling the rate of solvent evaporation and solvent annealing in thin films of PS-*b*-PEO, Russell and coworkers[71, 72] achieved nearly-defect-free arrays of cylindrical microdomains oriented normal to the film surface that spanned the entire films. Moreover, the use of a co-solvent enabled further control over the length scale of lateral ordering. Our most recent results[91] showed that cylindrical microdomains oriented normal to the film surface could be obtained directly by spin-coating polystyrene-*block*-poly(4-vinylpyridine) (PS-*b*-P4VP) BCPs from mixed solvents of toluene and tetrahydrofuran (THF) (Figure 6(a)) and arrays of highly ordered cylindrical microdomains formed over large areas (Figure 6(b)) after exposing the films in the vapor of a toluene/THF mixture for a while. This process was independent of substrates, but strongly dependent on the quality of the solvents for each block and the solvent evaporation rate.

Solvent-induced surface reconstruction has been used to generate nanoporous templates from highly oriented PS-*b*-PMMA thin films.[90] By modifying the interfacial interactions with an anchored random copolymer, arrays of PMMA cylinders oriented normal to the film surface could be produced. By exposing the films to acetic acid, PMMA was solvated, while the glassy PS matrix remained intact. Upon drying, a film reconstruction was observed where pores were opened in the positions of the original PMMA cylinders as the PMMA within the pores was transferred to the surface. This pore generation is completely reversible such that, by heating the film above the glass transition temperature of the PMMA, the original morphology was regenerated. A similar method was also used to generate nanoporous structures in PS-*b*-P4VP BCP thin films by using ethanol to draw the P4VP to the surface.[91]

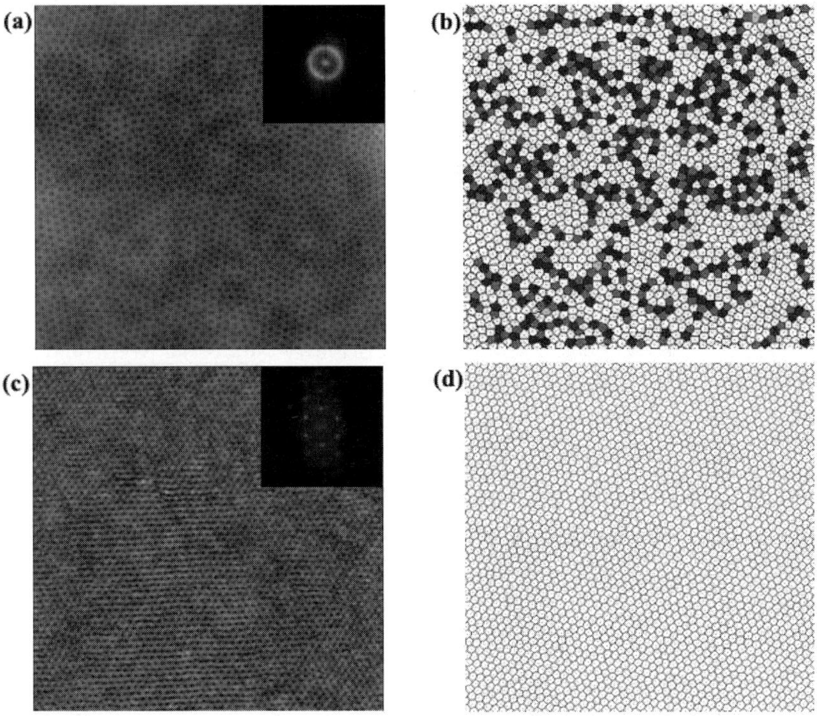

Fig. 6. A PS-b-P4VP thin film obtained by spin-coating (2 μm × 2 μm): (a) SFM height images of an as-spun film; (c) SFM height images of a highly ordered and oriented array of cylindrical microdomains after solvent annealing. Insets are the corresponding Fourier transform spectra. (b, d) the corresponding Voronoi diagrams. Reproduced with permission from *ACS Nano*.[91]

3.2. *Zone Casting*

Zone-casting originally developed for the oriented growth of molecular crystals, has been used to achieve large-scale alignment of nanoscale domains in micro-phase separated BCPs.[79] Kowalewski and coworkers demonstrated a large-scale, long-range ordering of lamellae in thin films of poly(n-butyl acrylate)-*block*-poly(acrylonitrile) (PBA-*b*-PAN) BCPs by using a simple zone-casting technique.[78] Zone-casting was performed by depositing the copolymer solution from *N*, *N*-dimethylformamide (DMF) onto a moving substrate with the aid of a syringe equipped with a flat nozzle. To achieve the desirable solvent evaporation rate, the temperatures of the copolymer solution and of the substrate were controlled. It should also be noted that, by varying the casting condition, the orientation of the microdomains with respect to the casting direction could be controlled.

3.3. *Optical Alignment*

Optical-alignment method on the molecular level is well-established in liquid-crystalline systems.[92, 93] Directionally selective light excitation, using linearly polarized light of photoisomerizable molecules in a liquid-crystalline polymer produced patterned, oriented microdomains in the films. Recent investigations have revealed that such photoexcited collective molecular motions can lead to lateral mass transport over distances of micrometers.[94, 95] Seki and coworkers[96] proposed an optical 3-D (both out-of-plane and in-plane) alignment of nanocylinders of a BCP comprised of a liquid-crystalline photo-responsive block and a PEO block by applying the concept of photo-induced mass migration. The key for out-of-plane alignment (whether the cylinders are oriented normal to or parallel to the substrate surface) is the control of the film thickness, while that for the in-plane alignment is the direction of polarization of the light during the illumination (Figure 7). Moreover, Ikeda and coworkers[80] demonstrated a noncontact-optical method by using polarized light to control a parallel patterning of PEO nanocylinders in an amphiphilic- liquid-crystalline BCP film.

3.4. *Shearing*

BCPs in bulk can easily be aligned by mechanical shear.[81, 82] This technique has been widely used to align lamellar, cylindrical, spherical, and bicontinuous microdomains, but was limited to relatively thick films.

Recently, Register, Chaikin and coworkers[81, 82] have demonstrated that a single layer of cylindrical microdomains[81] and bilayers of spherical microdomains[82] in BCP thin films can be aligned by applying shear. In their experiment, a PDMS pad is placed in contact with a heated BCP film. When subjected to a lateral motion, the stamp elastically distorts and shears the BCP thin film reorienting the microdomains in the direction of the applied shear.

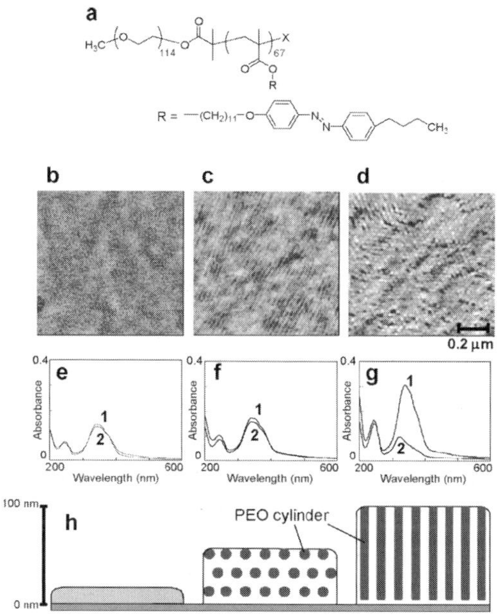

Fig. 7. Thickness dependence of the azobenzene (Az) and cylinder alignment. a) Chemical structure of the diblock copolymer consists of PEO and poly(methacrylate) containing an Az unit. b–d) Phase-mode AFM images (1 µm × 1 µm) of the p(EO_{114}-Az_{67}) film with different film thickness: b) 20 nm, c) 30 nm, and d) 70 nm after annealing and exposure to hexane vapor. e-g) UV-vis absorption spectra of the corresponding p(EO_{114}-Az_{67}) films are shown: e) 20 nm, f) 30 nm, and g) 70 nm thickness for as-cast (1) films and after exposure to hexane vapor (2). h) Schematic illustration of thickness dependence on cylinder alignment. Reproduced with permission from *Advanced Materials.*[96]

4. Applications of Nanopatterned BCP Thin Films

4.1. *Nanoporous Membrane for Filtration of Viruses*

Ultrafiltration membranes with small pore sizes have been used for the separation of viruses.[97, 98] However, they were not very effective, since the virus particles permeates through a small number of abnormally large sized pores.[97] Track-etched polycarbonate (PC) and anodized aluminum oxide (AAO) membranes with uniform pore sizes have also been studied for the separation of viruses. While the pore size distributions are narrow for these membranes, both types of membranes show a very low flux for virus separation.[98] Thus, a new type of membrane, providing both high selectivity and high flux, was needed for this purpose. Recently, Yang et al. introduced a new membrane with an asymmetric film geometry, which shows both high selectivity and high flux.[99] Figure 8 shows schematic diagram of the fabrication of asymmetric nanoporous membranes. This membrane consists of a thin nanoporous layer, prepared from a PS-*b*-PMMA BCP template (~80 nm thick film with cylindrical pores of ~15 nm in diameter and a narrow pore size distribution), and a supported membrane that provides mechanical strength. This asymmetric membrane showed ultrahigh selectivity while maintaining a high flux for the separation of human rhinovirus type 14 (HRV14), which has a diameter of ~ 30 nm.[103] This virus is a major pathogen for the common cold in humans. Since the pore diameter in the top layer can be tuned from 10 to 40 nm by changing the molecular weights of BCP or by adding a homopolymer miscible with the minor component of the BCP,[100, 101] the cutoff size of the membrane filter could be precisely controlled.

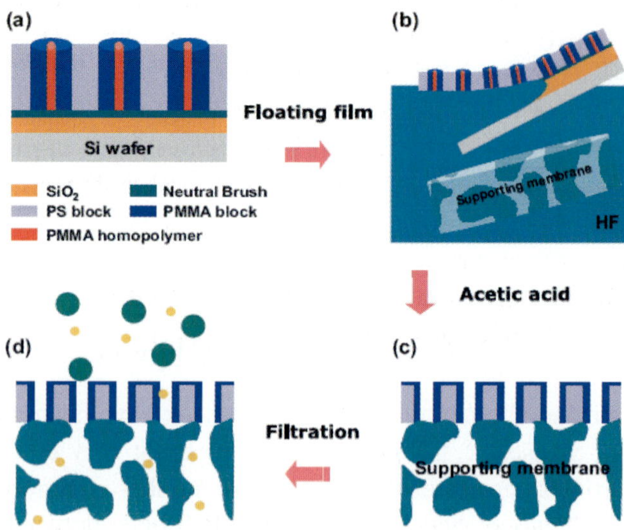

Fig. 8. Schematic depiction of the procedure for the fabrication of asymmetric nanoporous membranes. Reproduced with permission from *Advanced Materials*.[99]

With these pore sizes, these asymmetric membranes allow biomolecules like proteins, present in the unfiltered solution, to pass through the membrane, while only the viruses are screened. The unique characteristic of this new membrane filter eliminates the risk of contamination from viruses while processing biotherapeutic proteins such as vaccines and hormones. Therefore, this new membrane can be used to develop new types of blood filtering systems, such as a haemodialysis membrane that is free of the risk of viral infection.

4.2. *Incorporation of Nanocrystals and Nanoreactors*

Quantum size effects and large surface to volume ratios contribute to the unusual properties of inorganic nanoparticles and studies of these effects have sparked great interests in novel fabrication processes of metal or metal oxide nanoparticles. The synthesis of nanoparticles from BCP microdomains is an effective route to control the size distributions, shapes, and spatial placements of nanoparticles. In the simplest rendition, one may use BCPs where one block has a higher affinity to the inorganics. For instance, Jaeger and coworkers demonstrated that metals deposited directly to the surface of a film would segregate into the underlying microdomains.[102] The formation of ordered patterns is also possible by assembling BCP micelles upon casting, which has evoked a major interest for potential applications. Micellar cores offer a unique microenvironment, namely a nanoreactor, where inorganic precursors can be loaded and then processed by wet chemical methods to produce nanoparticles with a narrow size distribution, in a similar way as it is done with microemulsions.[103] The ordered deposition of gold and silver nanoclusters from micellar PS-*b*-P4VP[104, 105] and PS-*b*-PAA[106] have been reported.

4.3. *Planar Optical Waveguide*

While polymer-based optical waveguide materials have been widely discussed,[107] little attention has been paid to BCP systems in regard to their applications in the field of optical elements. Kim et al.[108]

demonstrated that thin films of BCPs with controlled orientation of microdomains can be used as planar optical waveguides and they investigated the waveguiding properties using optical waveguide spectroscopy (OWS).[109, 110] Thin films of mixtures of PS-*b*-PMMA and PMMA homopolymers, with cylindrical PMMA microdomains oriented normal to the film plane, were used as optical waveguides with light being coupled into the film at different modes. The nanofabrication processes occurring inside the layer were monitored by OWS. The confinement of the PMMA homopolymer to the microdomains markedly enhances the aspect ratio of the microdomains by ~10 times over that seen in the bulk period,[111] which makes the film suitable for waveguiding applications. Resonance coupling between surface plasmons (or plasmon surface polaritons) and incident photons can occur at a metal/dielectric interface, especially in the experimental setup known as the Kretschmann configuration (Figure 9).[109, 112] If the thickness of the dielectric layer is

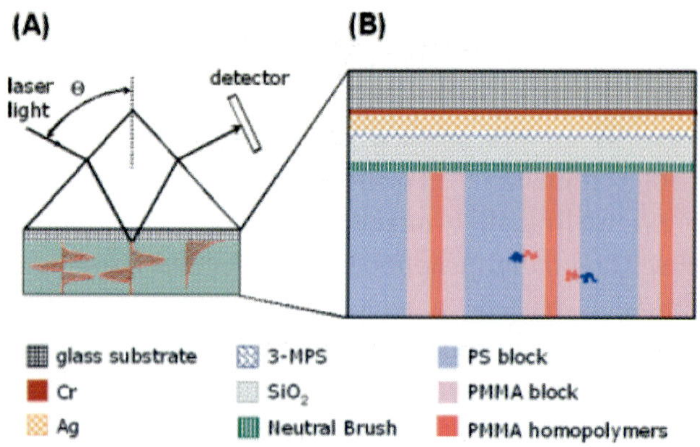

Fig. 9. A) Schematic diagram of the OWS setup based on the Kretschmann configuration, and of the idealized field distributions of several guided modes in the waveguiding layer. B) Schematic diagram of a thin film of PS-b-PMMA/PMMA homopolymer mixture with PMMA microdomains aligned normal to the substrate surface fabricated onto glass substrate. PS: polystyrene; PMMA: poly(methyl methacrylate); 3-MPS: 3-mercaptopropyl trimethoxysilane. Reproduced with permission from *Advanced Materials.*[108]

further increased (in organic polymeric layers typically thicker than ~ 200 nm), waveguide optical modes, in addition to surface plasmon resonance, can be observed.[109, 110]

4.4. *Templated Growth of Nanowires*

An advantage of the cylindrical microdomains in BCPs is the high aspect ratio and non-connectivity of the microdomains. For a 20-μm-thick film containing cylindrical microdomains that are 20 nm in diameter, an aspect ratio of 1000:1 is obtained. This advantage has been used where, under an applied electric field, the cylindrical microdomains of PMMA were oriented normal to the surface in films with thicknesses of ~30 μm. Thurn-Albrecht et al.[74] oriented the PMMA cylindrical microdomains in a PS-*b*-PMMA BCP by applying an electric field across the polymer film (Figure 10). Upon removal of the PMMA microdomains, metals, including cobalt, lead, and iron were placed

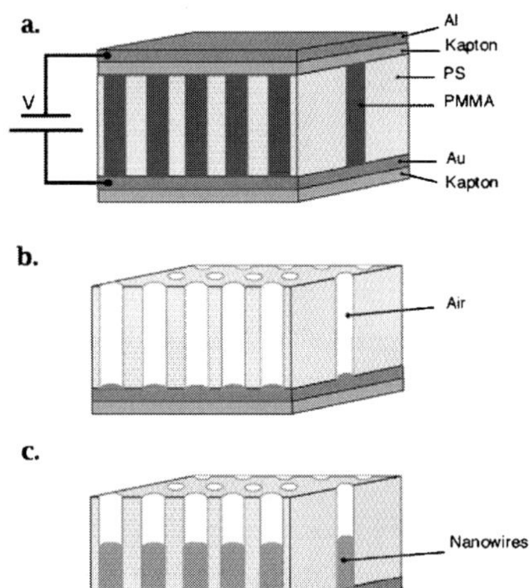

Fig. 10. Left: A schematic representation of high-density nanowire fabrication in a polymer matrix. (a) An asymmetric diblock copolymer annealed above the glass transition temperature of the copolymer between two electrodes under an applied electric field, forming a hexagonal array of cylinders oriented normal to the film surface. (b) After removal of the minor component, a nanoporous film is formed. (c) By electrodeposition, nanowires can be grown in the porous template, forming an array of nanowires in a polymer matrix. Reproduced with permission from *Science*.[74]

within the nanopores by electrochemical deposition. The cobalt nanowires showed a single magnetic domain behavior[74] and enhanced coercivities[113] in comparison to continuous cobalt films, and hold promise for use in ultrahigh-density magnetic storage devices (approaching one terabit per square inch). The effectiveness of nanoporous PS-*b*-PMMA thin films in generating such structures has motivated in-depth work in optimizing the fidelity of the structures and controlling the feature sizes.[114]

In addition to above mentioned applications, BCP thin films have received attention for use in the nanolithography process. The periodic nanostructure patterns of BCP films can be subsequently transferred to various kinds of substrates by a standard lithographic ion-etching process, due to the different etching sensitivities of the two building blocks. The regular nanopattern achieved can be used for a variety of applications, like high-density magnetic recording media, and quantum dot arrays.[115,116] Another application of BCP thin films is in photonic materials. This is especially the case of BCP thin films that self-assemble into well-defined alternating layered structures.[117]

5. Summary and Outlook

Recent research efforts have demonstrated the utility of BCP thin films as components in a new generation of optical and electronic devices. Lithographic and micro-fabrication techniques using BCP thin films offer unprecedented feature dimensions and densities. Moreover, elegant methods of harnessing the microphase separation to produce ordered dispersions of inorganic nanoparticles are being developed. However, for many potential applications, e.g., addressable memory devices, limitations arise from imperfections in the long-range lateral order and the restricted choice of motifs. Significant advances have been made, though, by exploiting interfacial interactions and confinement effects, the use of external fields, and the use of chemically and topographically patterned substrates. Such morphological control, necessary in any applications, evolved directly from research on the fundamental physics, chemistry and engineering of BCP films. Further control in generating hierarchical assemblies with 3-D structures can easily be envisioned with BCPs where control over structures and assembly on multiple length scales will be possible.

References

1. S. H. Anastasiadis, T. P. Russell, S. K. Satija, C. F. Majkrzak, *Physical Review Letters* 62, 1852 (1989).
2. S. H. Anastasiadis, T. P. Russell, S. K. Satija, C. F. Majkrzak, *The Journal of Chemical Physics* 92, 5677 (1990).
3. G. Coulon, T. P. Russell, V. R. Deline, P. F. Green, *Macromolecules* 22, 2581 (1989).
4. C. S. Henkee, L. T. Edwin, L. J. Fetters, *Journal of Materials Science* 23, 1685 (1988).
5. T. P. Russell, A. Menelle, S. H. Anastasiadis, S. K. Satija, C. F. Majkrzak, *Macromolecules* 24, 6263 (1991).
6. N. Koneripalli, N. Singh, R. Levicky, F. S. Bates, P. D. Gallagher, S. Satija, *Macromolecules* 28, 2897 (1995).
7. G. J. Kellogg, D. G. Walton, A. M. Mayes, P. Lambooy, T. P. Russell, P. D. Gallagher, S. K. Satija, *Physical Review Letters* 76, 2503 (1996).
8. A. M. Mayes, G. J. Kellogg, D. G. Walton, P. Lambooy, T. P. Russell, *Polymeric Materials Science and Engineering* 71, 282 (1994).
9. D. G. Walton, G. J. Kellogg, A. M. Mayes, P. Lambooy, T. P. Russell, *Macromolecules* 27, 6225 (1994).
10. P. Mansky, T. P. Russell, *Macromolecules* 28, 8092 (1995).
11. A. M. Mayes, S. K. Kumar, *MRS Bulletin* 22, 43 (1997).
12. T. P. Russell, A. M. Mayes, P. Bassereau, *Physica A: Statistical Mechanics and Its Applications* 200, 713 (1993).
13. T. P. Russell, P. Lambooy, G. J. Kellogg, A. M. Mayes, *Physica B: Condensed Matter* 213&214, 22 (1995).
14. T. P. Russell, P. Lambooy, G. J. Kellogg, A. M. Mayes, *Macromolecules* 28, 787 (1995).
15. A. M. Mayes, T. P. Russell, V. R. Deline, S. K. Satija, C. F. Majkrzak, *Macromolecules* 27, 7447 (1994).
16. A. M. Mayes, T. P. Russell, P. Bassereau, S. M. Baker, G. S. Smith, *Macromolecules* 27, 749 (1994).
17. M. J. Fasolka, A. M. Mayes, *Annual Review of Materials Research* 31, 323 (2001).
18. P. F. Green, R. Limary, *Advances in Colloid and Interface Science* 94, 53 (2001).
19. P. Lambooy, T. P. Russell, G. J. Kellogg, A. M. Mayers, P. D. Gallagher, S. K., Satija, *Physical Review Letters* 72, 2899 (1994).
20. P. Lambooy, J. R. Salem, T. P. Russell, *Thin Solid Films* 252, 75 (1994).
21. M. Kikuchi, K. Binder, *Europhysics Letters* 21, 427 (1993).
22. M. Kikuchi, K. Binder, *The Journal of Chemical Physics* 101, 3367 (1994).
23. M. S. Turner, *Physical Review Letters* 69, 1788 (1992).
24. P. F. Green, T. M. Christensen, T. P. Russell, R. Jerome, *The Journal of Chemical Physics* 92, 1478 (1990).

25. R. Limary, P. F. Green, *Macromolecules* 32, 8167 (1999).
26. P. F. Green, T. M. Christensen, T. P. Russell, *Macromolecules* 24, 252 (1991).
27. P. F. Green, T. M. Christensen, T. P. Russell, R. Jerome, *Macromolecules* 22, 2189 (1989).
28. F. Brochard-Wyart, P. Martin, C. Redon, *Langmuir* 9, 3682 (1993).
29. A. R. Segalman, P. F. Green, *Macromolecules*, 801 (1990).
30. J. L. Masson, P. F. Green, *The Journal of Chemical Physics* 112, 349 (2000).
31. G. Brown, A. Chakrabarti, *The Journal of Chemical Physiscs* 101, 3310 (1994).
32. Y. Liu, W. Zhao, X. Zheng, A. King, A. Singh, M. H. Rafailovich, J. Sokolov, K. H. Dai, E. J. Kramer, *Macromolecules* 27, 4000 (1994).
33. K. R. Shull, E. J. Kramer, F. S. Bates, J. H. Rosedale, *Macromolecules* 24, 1383 (1991).
34. T. P. Russell, A. Menelle, S. H. Anastasiadis, S. K. Satija, C. F. Majkrzak, *Progress in Colloid & Polymer Science* 91, 97 (1993).
35. B. J. Factor, T. P. Russell, M. F. Toney, *Physical Review Letters* 66, 1181 (1991).
36. T. P. Russell, G. Coulon, V. R. Deline, D. C. Miller, *Macromolecules* 22, 4600 (1989).
37. T. P. Russell, A. Menelle, S. H. Anastasiadis, S. K. Satija, C. F. Majkrzak, *Makromolekulare Chemie, Macromolecular Symposia* 62, 157 (1992).
38. A. Menelle, T. P. Russell, S. H. Anastasiadis, S. K. Satija, C. F. Majkrzak, *Physical Review Letters* 68, 67 (1992).
39. C. S. Henkee, E. L. Thomas, L. J. Fetters, *Journal of Materials Science* 23, 1685 (1988).
40. L. T. J. Korley, B. D. Pate, E. L. Thomas, P. T. Hammond, *Polymer* 47, 3073 (2006).
41. E. L. Thomas, E. J. Roche, *Polymer* 20, 1413 (1979).
42. H. Fredrickson Glenn, *Macromolecules* 20, 2535 (1987).
43. P. Mansky, Y. Liu, E. Huang, T. P. Russell, C. Hawker, *Science* 275, 1458 (1997).
44. E. Huang, P. Mansky, T. P. Russell, C. Harrison, P. M. Chaikin, R. A. Register, C. J. Hawker, J. Mays, *Macromolecules* 33, 80 (2000).
45. M. Husseman, E. E. Malmstroem, M. McNamara, M. Mate, D. Mecerreyes, D. G. Benoit, J. L. Hedrick, P. Mansky, E. Huang, T. P. Russell, C. J. Hawker, *Macromolecules* 32, 1424 (1999).
46. E. Huang, S. Pruzinsky, T. P. Russell, J. Mays, C. J. Hawker, *Macromolecules* 32, 5299 (1999).
47. T. P. Russell, T. Thurn-Albrecht, M. Tuominen, E. Huang, C. J. Hawker, *Macromolecular Symposia* 159, 77 (2000).
48. Z. H. Cai, K. Huang, P. A. Montano, T. P. Russell, J. M. Bai, G. W. Zajac, *The Journal of Chemical Physics* 98, 2376 (1993).
49. E. Huang, T. P. Russell, C. Harrison, P. M. Chaikin, R. A.Register, C. J. Hawker, J. Mays, *Macromolecules* 31, 7641 (1998).
50. D. Y. Ryu, K. Shin, E. Drockenmuller, C. J. Hawker, T. P. Russell, *Science* 308, 236 (2005).
51. C. J. Hawker, T. P. Russell, *MRS Bulletin* 30, 952 (2005).
52. T. Xu, C. J. Hawker, T. P. Russell, *Macromolecules* 38, 2802 (2005).
53. P. Mansky, T. P. Russell, C. J. Hawker, M. Pitsikalis, J. Mays, *Macromolecules* 30, 6810 (1997).

54. C. J. Hawker, E. Elce, J. Dao, W. Volksen, T. P. Russell, G. G. Barclay, *Macromolecules* 29, 2686 (1996).
55. P. Mansky, T. P. Russell, C. J. Hawker, J. Mays, D. C. Cook, S. K. Satija, *Physical Review Letters* 79, 237 (1997).
56. P. Mansky, O. K. C. Tsui, T. P. Russell, Y. Gallot, *Macromolecules* 32, 4832 (1999).
57. J. Heier, Kramer E. J., S. Walheim, Krausch G., *Macromolecules*, 6610 (1997).
58. L. Rockford, Y. Liu, P. Mansky, T. P. Russell, M. Yoon, S. G. J. Mochrie, *Physical Review Letters* 82, 2602 (1999).
59. L. Rockford, S. G. J. Mochrie, T. P. Russell, *Macromolecules* 34, 1487 (2001).
60. M. Böltau, S. Walheim, J. Mlynek, G. Krausch, U. Steiner, *Nature* 391, 877 (1998).
61. S. Xiao, X. Yang, E. Edwards, Y.-H. La, P. F. Nealey, *Nanotechnology* 16, S324 (2005).
62. S. O. Kim, H. H. Solak, M. P. Stoykovich, N. J. Ferrier, J. J. de Pablo, P. F. Nealey, *Nature* 424, 411 (2003).
63. M. P. Stoykovich, M. Mueller, S. O. Kim, H. H. Solak, E. W. Edwards, J. J. de Pablo, P. F. Nealey, *Science* 308, 1442 (2005).
64. E. W. Edwards, M. F. Montague, H. H. Solak, C. J. Hawker, P. F. Nealey, *Advanced Materials* 16, 1315 (2004).
65. S.-M. Park, M. P. Stoykovich, R.Ruiz, Y. Zhang, C. T. Black, P. F. Nealey, *Advanced Materials* 19, 607 (2007).
66. M. P. Stoykovich, E. W. Edwards, H. H. Solak, P. F. Nealey, *Physical Review Letters* 97, 147802/1 (2006).
67. I. Y. Tsai, M. Kimura, T. P. Russell, *Langmuir* 20, 5952 (2004).
68. I. Y. Tsai, J. A. Green, M. Kimura, B. Jacobson, T. P. Russell, *Journal of Biomedical Materials Research part A*, 509 (2006).
69. G. Kim, M. Libera, *Macromolecules* 31, 2670 (1998).
70. G. Kim, M. Libera, *Macromolecules* 31, 2569 (1998).
71. S. H. Kim, M. J. Misner, T. P. Russell, *Advanced Materials* 16, 2119 (2004).
72. S. H. Kim, M. J. Misner, T. Xu, M. Kimura, T. P. Russell, *Advanced Materials* 16, 226 (2004).
73. M. Kimura, M. J. Misner, T. Xu, S. H. Kim, T. P. Russell, *Langmuir* 19, 9910 (2003).
74. T. Thurn-Albrecht, J. Schotter, G. A. Kastle, N. Emley, T. Shibauchi, L. Krusin-Elbaum, K. Guarini, C. T. Black, M. T. T uominen, T. P. Russell, *Science* 290, 2126 (2000).
75. J. Y. Cheng, A. M. Mayes, C. A. Ross, *Nature Materials* 3, 823 (2004).
76. R. A. Segalman, H. Yokoyama, E. J. Kramer, *Advanced Materials* 13, 1152 (2001).
77. J. Bodycomb, Y. Funaki, K. Kimishima, T. Hashimoto, *Macromolecules* 32, 2075 (1999).
78. C. Tang, A. Tracz, M. Kruk, R. Zhang, D.-M. Smilgies, K. Matyjaszewski, T. Kowalewski, *Journal of the American Chemical Society* 127, 6918 (2005).
79. A. Tracz, J. K. Jeszka, M. D. Watson, W. Pisula, K.Muellen, T. Pakula, *Journal of the American Chemical Society* 125, 1682 (2003).
80. H. Yu, T. Iyoda, T. Ikeda, *Journal of the American Chemical Society* 128, 11010 (2006).
81. D. E. Angelescu, J. H. Waller, D. H. Adamson, P. Deshpande, S. Y. Chou, R. A. Register, P. M. Chaikin, *Advanced Materials* 16, 1736 (2004).

82. D. E. Angelescu, J. H. Waller, R. A. Register, P. M. Chaikin, *Advanced Materials* 17, 1878 (2005).
83. M. A. Villar, D. R. Rueda, F. Ania, E. L. Thomas, *Polymer* 43, 5139 (2002).
84. Q. Zhang, O. K. C. Tsui, B. Du, F. Zhang, T Tang, T. He, *Macromolecules*, 33, 9561 (2000).
85. H. Huang, F. Zhang, Z. Hu, B. Du, T. He, *Macromolecules*, 36, 4084 (2003).
86. K. Fukunaga, H. Elbs, R. Magerle, G. Krausch, *Macromolecules* 33, 947 (2000).
87. Z. Q. Lin, D. H. Kim, X. D. Wu, L.Boosahda, D. Stone, L. LaRose, T. P. Russell, *Advanced Materials* 14, 1373 (2002).
88. J. Hahm, S. J. Sibener, *Langmuir* 16 4766 (2000).
89. K. Fukunaga, T. Hashimoto, H. Elbs, G. Krausch, *Macromolecules* 35, 4406 (2002).
90. T. Xu, J. Stevens, J. A. Villa, J. T. Goldbach, K. W. Guarini, C. T. Black, C. J. Hawker, T. P. Russell , *Advanced Functional Materials* 13, 698 (2003).
91. S. Park, J.-Y. Wang, B. Kim, J. Xu and T. P. Russell, *ACS Nano* 2, 766 (2008).
92. K. Ichimura, *Chemical Reviews* 100, 1847 (2000).
93. T. Ikeda, *Journal of Materials Chemistry* 13, 2037 (2003).
94. R. H. Berg, S. Hvilsted, P. S. Ramanujam, *Nature* 383, 505 (1996).
95. A. Natansohn, P. Rochon, *Chemical Reviews* 102, 4139 (2002).
96. Y. Morikawa, S.Nagano, K. Watanabe, K. Kamata, T. Iyoda, T. Seki, *Advanced Materials* 18, 883 (2006).
97. T. Urase, K. Yamamoto, S. Ohgaki, *Water Science and Technology* 30, 199 (1994).
98. T. Urase, K. Yamamoto, S. Ohgaki, *Journal of Membrane Science* 115, 21 (1996).
99. S. Y. Yang, I. Ryu, H. Y. Kim, J. K. Kim, S. K. Jang, T. P. Russell, *Advanced Materials* 18, 709 (2006).
100. U. Jeong, D. Y. Ryu, D. H. Kho, D. H. Lee, J. K. Kim, T. P. Russell, *Macromolecules* 36, 3626 (2003).
101. U. Jeong, D. Y. Ryu, J. K Kim, D. H. Kim, T. P. Russell, C. J. Hawker, *Advanced Materials* 15, 1247 (2003).
102. W. A. Lopes, H. M. Jaeger, *Nature* 414, 735 (2001).
103. M. Breulmann, S. Forster, M. Antonietti, *Macromolecular Chemistry and Physics* 201, 204 (2000).
104. J. P. Spatz, S. Mössmer, C. Hartmann, M. Möller, T. Herzog, M. Krieger, H.-G. Boyen, P. Ziemann, B. Kabius, *Langmuir* 16, 407 (2000).
105. S.-H. Yun, S. I. Yoo, J. C. Jung, W.-C. Zin, B.-H. Sohn, *Chemistry of Materials* 18, 5646 (2006).
106. Y. Boontongkong, R. E. Cohen, *Macromolecules* 35, 3647 (2002).
107. H. Ma, A. K. Y. Jen, L. R. Dalton, *Advanced Materials* 14, 1339 (2002).
108. D. H. Kim, K. H. A. Lau, J. W. F. Robertson, O.-J. Lee, U. Jeong, J. I. Lee, C. J. Hawker, Russell, T. P. J. K .Kim, W. Knoll, *Advanced Materials* 17, 2442 (2005).
109. W. Knoll, *Annual Review of Physical Chemistry* 49, 569 (1998).
110. J. D. Swalen, *The Journal of Physical Chemistry* 83, 1438 (1979).
111. U. Jeong, D. Y. Ryu, D. H. Kho, J. K. Kim, J. T. Goldbach, D. H. Kim, T. P. Russell, *Advanced Materials* 16, 533 (2004).
112. E. Kretschmann, T. L. Ferrell, J. C. Ashley, *Physical Review Letters* 42, 1312 (1979).
113. T. Shibauchi, L. Krusin-Elbaum, L. Gignac, C. T. Black, T. Thurn-Albrecht, T. P. Russell, J. Schotter, G. A. Kastle, N. Emley, M. T. Tuominen, *Journal of Magnetism and Magnetic Materials* 226-230, 1553 (2001).

114. T. Xu, H.-C. Kim, J. DeRouchey, C. Seney, C. Levesque, P. Martin, C. M. Stafford, T. P. Russell, *Polymer* 42, 9091 (2001).
115. C. T. Black, R. Ruiz, G. Breyta, J. Y. Cheng, M. E. Colburn, K. W. Guarini, H.-C. Kim, Y. Zhang, *IBM Journal of Research & Development* 51, 606 (2007)
116. M. Park, C. Harrison, P. M. Chaikin, R. A. Register, D. H. Adamson, *Science* 276, 1401, *Science* 276, 1401 (1997).
117. Y. Fink, J. N. Winn, S. Fan, C. Chen, ,M. Jurgen, J. D. Joannopoulos, E. L.Thomas, *Science* 282, 1679 (1998).

CHAPTER 2

EQUILIBRATION OF BLOCK COPOLYMER FILMS ON CHEMICALLY PATTERNED SURFACES

Gordon S. W. Craig, Huiman Kang and Paul F. Nealey

Department of Chemical and Biological Engineering
University of Wisconsin-Madison
Madison, Wisconsin, 53706
E-mail: nealey@engr.wisc.edu

We provide a summary of the assembly of block copolymer films as they equilibrate on substrates patterned with regions of different chemical functionality at the approximate length scale of the domain dimensions of the copolymer. Important characteristics of the chemically patterned surface are shown to be the interfacial energy contrast of alternating regions of the pattern, and the pattern layout. As the interfacial energy contrast with respect to the two blocks of the copolymer increases, the selective wetting of the alternating regions of the pattern by different blocks of the copolymer also increases. As the contribution of interfacial energy is increased, it plays a larger role in the equilibration of the copolymer, and drives the assembly towards vertical orientation of domains away from the surface, and alignment and registration of the domains with the underlying chemical pattern. In some cases, the interfacial energy of the chemical pattern can eliminate structural defects, elongate or compress the configuration of copolymer chains, and create morphologies that do not naturally occur in the bulk. The forced registration of the copolymer domains to the chemical pattern automatically causes low defect densities in periodic and nonregular structures registered to the chemical pattern. Research on this subject has benefited from a combined theoretical and experimental approach.

1. Introduction

One of the fundamental characteristics of block copolymers is that they can self-assemble into discrete domains in a process governed by thermodynamics.[1] The size and configuration of the domains depends upon the number and relative lengths of the blocks in the copolymer, the total degree of polymerization, and the Flory-Huggins interaction parameter, χ, of the disparate blocks. In the case of thin films, added to these intrinsic characteristics that govern self-assembly are external factors of the copolymer film system, such as film thickness[2] and surface and interfacial interactions.[3] The resulting domain structures have length scales on the order of 5-100 nm, which span current and future integrated circuit element dimensions.[4] As a result, interest has grown in the use of thin film block copolymer domain structures as templates for device fabrication[5-12] or as precursors for the fabrication of electronic elements.[13-15] However, for many template applications, it is necessary to control the orientation and ordering of the domains,[16,17] and desirable to create patterns beyond what one would obtain with the predominantly occurring morphologies of block copolymers, e.g. series of parallel lines or hexagonal arrays of spots.[18,19]

One straightforward approach to control domain orientation relies upon tailoring of surface and interfacial interactions. Surfaces that are wet preferentially by one of the blocks can force lamellar[20,21] or cylindrical[2,5,22] structures to lie parallel to the surface. In contrast, a surface that is not preferentially wet by either block of the copolymer will lead to perpendicularly oriented lamellae[3,23] or cylinders,[7,24,25] although these structures usually lack any lateral long-range order.[3,26,27]

Several methods have been developed to create long-range lateral order of block copolymer domains in thin films, including solvent annealing,[28-31] electric field alignment,[32,33] shear alignment,[34] and directed assembly on topographically[10,35-40] or chemically[3,41-43] patterned substrates. In this chapter we will focus on the process of directed assembly on chemically patterned substrates, in which alternating regions of the chemical pattern are spaced to match approximately the domain spacing of the block copolymer, and also have a significant contrast in their interfacial energies with respect to the two blocks of the copolymer, resulting in selective wetting of the alternating regions by the different blocks. The selective wetting of the chemical pattern regions drives the domain assembly process, to which we refer in this chapter simply as "directed assembly."

The effect that the contrast in interfacial energy of neighboring regions of the chemical pattern has on the block copolymer morphology as it equilibrates on the pattern can be understood in terms of the thermodynamic free energy of the total system. The energy of the total system, comprised of the block copolymer film on the chemical nanopattern, can be divided into the primary factors that contribute to the energy of the system, and then analyzed to determine which factors dominate the free energy of the system. In the case of a thin block copolymer film on a chemically patterned substrate the primary factors are assumed to be the following: the polymer chain configuration, the interface between domains, and the contrast of interfacial energy between the polymer and neighboring regions of the chemical pattern.[3] The energy of these factors can be estimated with equations in the literature. For example, one can estimate the free energy per chain to change the width of the domains in a thin block copolymer film on a chemically neutral surface by using equations for the energy associated with stretching or compressing chains and for the interfacial energy per chain at the domain interface.[3,44] In an example system of a 40-nm-thick film of symmetric polystyrene-*block*-poly(methyl methacrylate) (P(S-*b*-MMA)), with a M_n of ~100 kg/mol, the estimated free energy per chain to change the chain configurational entropy such that the domain period is decreased by 5% would be 0.011, in units of kT, which includes a free energy change of -0.141 kT for the compression of the chain and 0.152 kT for the change of the S-MMA domain interface. It is the balance between chain configuration and interfacial energy that sets the bulk lamellar period, L_o.[1] These free energy values can be compared to the free energy per chain provided by moving one block of the example system from a neutral surface chemistry to a surface chemistry with great affinity for the block, which equals 0.064 kT, based on equations[3] and surface energy values[23] in the literature. The interfacial energy contribution to the film by the chemical pattern is a function of film thickness, such that the surface would contribute 0.129 *kT* per chain to a 20-nm-thick film, but only 0.026 *kT* to a 100-nm-thick film. Thus, in the example system as the block copolymer morphology equilibrates in the presence of the chemical pattern, it will be thermodynamically driven to register to the pattern, even for film thicknesses of 200 nm. In the case of morphologies that do not match the chemical pattern, the chemical pattern can direct the copolymer to completely reconfigure its morphology, again over significant film thicknesses.

Knowledge of the relative energy contributions in the block copolymer film on a chemical nanopattern can shed light on both the lateral long-range order and defect reduction achieved with the directed assembly of block copolymer films. In terms of lateral long-range order, the energy imparted by the chemical nanopattern to the block copolymer film drives the registration and orientation of the domains to the pattern,[41] such that the long-range, lateral domain order matches that of the underlying chemical nanopattern. In terms of defect reduction, the energy associated with the various defects in a block copolymer domain structure has not been calculated, but one could consider the 5% width change mentioned in the previous paragraph as a type of defect. Similarly, one could consider a curved lamellar domain as another type of defect. We can estimate the free energy per chain required to deform a lamellar structure about a curve[45] with a radius equal to 5 times L_o to be approximately 0.012 *kT* in the example system. These free energy values are all fractions of a *kT* unit, so there is not a large free energy penalty for the existence of a defect in the lamellar structure. Additionally, large activation energy barriers exist to prevent the block copolymer from removing the defect on its own.[27] However, the energy associated with these example "defects" is significantly less than the interfacial energy imparted to the film by the chemical pattern. Thus, in the example case given in the preceding paragraph, the interfacial energy can overwhelm the relatively small free energy values of defect structures, orient and position domains, and even force the copolymer into morphologies that do not exist in the bulk.[46,47]

2. Directed Assembly of Block Copolymers

The general directed assembly process, shown in Fig. 1, starts with the creation of a chemical surface pattern. A thin (≤ 100 nm) film of block copolymer is coated onto the chemical pattern, and subsequently annealed. The different blocks of the copolymer selectively wet alternating regions of the pattern, causing the block copolymer domains to register to the pattern.[41,42] For technological applications, one of the blocks can be selectively removed, leaving the remaining block to serve as a template. Unlike the formation of the block copolymer morphology in the bulk, the domain orientation and ordering is driven primarily by the chemical pattern, and less strongly by neighboring domains.

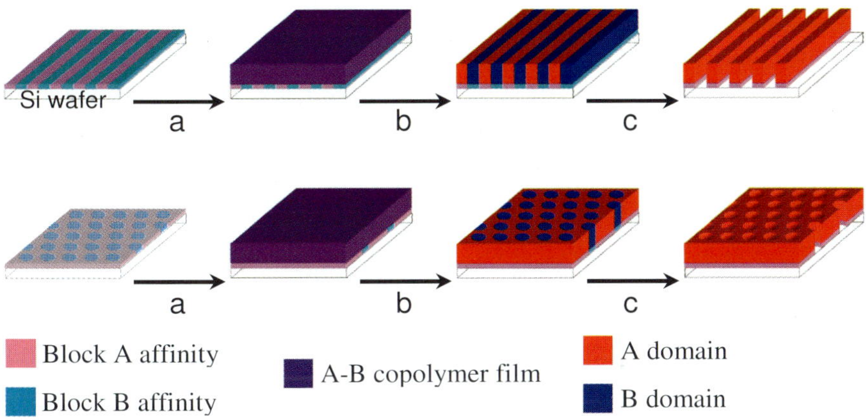

Fig. 1. Directed assembly of a thin film of A-B block copolymer on a patterned surface. The top and bottom rows depict the directed assembly of lamellar and cylindrical morphologies, respectively. The directed assembly process starts with a chemical nanopattern on a substrate. In step (a), the block copolymer is spin coated onto the chemical nanopattern. Annealing in step (b) leads to self-assembly of the copolymer domains, directed by the underlying chemical nanopattern. In step (c), one of the domains is selectively removed to leave a template consisting of the other domain.

Assuming the pattern period L_s is commensurate with L_o, the block copolymer domains will be oriented perpendicular to the substrate, and the morphology will have the same lateral long-range order as the chemical pattern at the copolymer/pattern interface. In this case, the only question concerning long-range order is how far the morphology at the interface propagates through the thickness of the film toward its surface. As the distance from the interface increases, the influence of the chemical pattern decreases, and for sufficiently thick films, the copolymer will revert to a bulk-like morphology at its free surface. As L_s and L_o become less commensurate, the thickness over which the chemical pattern can direct the polymer morphology will decrease. Finally, in the case when $L_s >> L_o$, the domains will assume a parallel orientation at the substrate interfaces.[48]

2.1. *Importance of Commensurability*

Initial research by Rockford *et al.* demonstrated that one could use a chemical nanopattern to influence the orientation and order of a block copolymer's morphology in a thin film.[49] They created a chemical surface pattern by evaporating Au at an oblique angle on the crystal lattice steps of an etched, mis-cut silicon wafer, resulting in a substrate that had alternating Au and silicon oxide stripes with L_s of 60 nm. They then used symmetric P(S-*b*-MMA) of varying molecular weights to investigate the effect that the commensurability of L_o and L_s had on copolymer self-assembly on their chemically patterned surface. They defined the degree of commensurability δ as the ratio of L_o to L_s, and used a two-dimensional order parameter f, calculated from the Fourier transform of the image data, to analyze the assembled morphologies as a function of δ. The resulting P(S-*b*-MMA) thin films had vertically oriented domains for δ values of 0.78, 0.99, and 1.40, as shown in Fig. 2. When δ was 0.99, the most ordered domain structure was obtained. However, even when commensurate, many defects in the domain structure were apparent. Such defects could have been caused by a number of factors, such as imperfections in the underlying chemical pattern, insufficient interfacial energy contrast between the alternating stripes of the chemical pattern to overcome defect formation in the film, or intrinsic defects in the block copolymer self-assembly. The statistical analysis done by Rockford *et al.* showed that the Au/oxide pattern had approximately the same f as the sample with $\delta = 0.99$ (see inset in Fig. 2), but unfortunately, no direct mapping of the block copolymer morphology to the underlying chemical pattern was done. As a result, it was impossible to discern whether the defects in the assembled structure were due to replication of defects in the underlying chemical pattern, or to a fundamental problem with the directed assembly process. In hindsight, we believe that the commensurate samples were the only samples with sufficient commensurability to have a high degree of registration of the domains to the underlying chemical pattern, and that the defects observed in the commensurate samples were likely due to

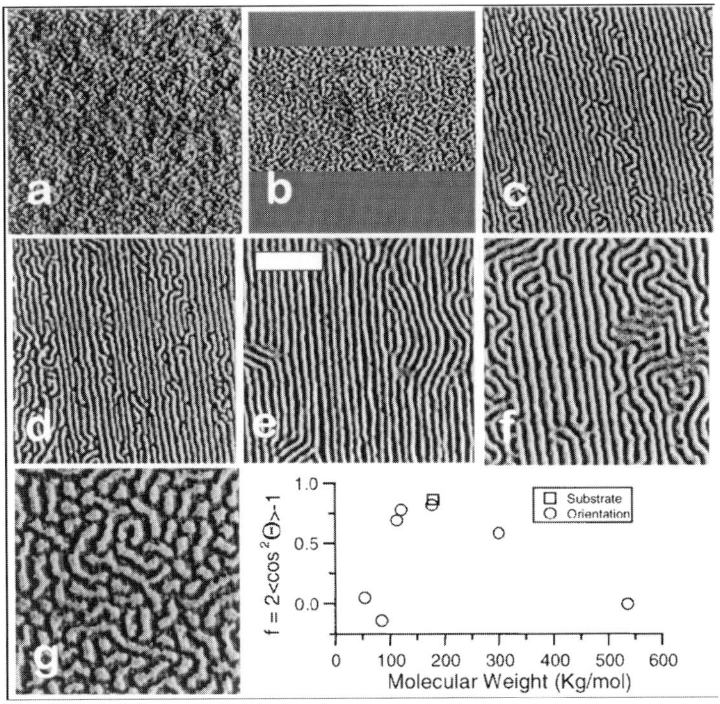

Fig. 2. AFM tapping mode phase images of the free surface of P(S-*b*-MMA) films of a range of molecular weights, solution cast on a 60 nm heterogeneous substrate. Molecular weights (x 103 g/mol) [degree of commensurability δ], are as follows: (a) 57 [0.48]; (b) 84 [0.61]; (c) 113 [0.74]; (d) 121.3 [0.78]; (e) 177 [0.99], (f) 300 [1.4]; (g) 535 [2.05]. The scale bar shown in (e) is 0.5 μm; same scale applies to all images. Inset is the calculated 2D orientation function for each sample. (Reprinted from Ref. 49 with permission of authors and the American Physical Society. Copyright 1999 by the American Physical Society.)

registration of the P(S-*b*-MMA) domains to defects in the chemical pattern.

In subsequent work, Yang *et al.*[48] assembled P(S-*b*-MMA) on chemically nanopatterned substrates made from self-assembled monolayers (SAM) patterned with the interference fringes produced by a Lloyd's mirror.[50] They were able to generate substrates with a variety of L_s values, and studied systems with δ ranging from 1 to 25. When $L_s >> L_o$, the lamellar domains oriented parallel to the substrate, similar to earlier studies of block copolymer thin films.[51,52] However, similar to the work of Rockford *et al.*, when $L_s \approx L_o$, the lamellar domains oriented vertically to the substrate. But the work of Yang *et al.* suffered from the same uncertainty as that of Rockford *et al.* in that it was impossible to verify whether defects were due to problems with the directed assembly process, or to defects in the underlying structure that were replicated in the P(S-*b*-MMA) by the directed assembly process.

Along with the significance of commensurability, the work of the Rockford and Yang teams pointed to the importance of quantitatively mapping the chemical pattern, including whatever defects it may have, to the final copolymer structure. Such mapping is necessary to discern whether the directed assembly process is limited by defects that occur in the copolymer self-assembly, or by defects in the chemical pattern. If the process is limited by defects in the copolymer itself, then even a perfect chemical pattern could not always induce a perfect structure in the copolymer, and the technological applications of directed assembly would be severely limited. By contrast, if the quantitative mapping showed that the directed assembly process was limited by defects in the chemical pattern, such that defects in the copolymer structure were only due to defects in the underlying chemical pattern, then it could be inferred that the copolymer domains were perfectly registered to the underlying chemical pattern. In such a situation, registration of the copolymer to a perfect chemical nanopattern would induce a defect-free block copolymer morphology. Thus, the ability to quantitatively map the chemical pattern to the final, equilibrated copolymer structure addresses a singularly important question because it is ties directly to the efficacy of the directed assembly process to generate defect-free domain structures that are oriented and have long-range order.

Kim *et al.*[41] developed a process that enabled the quantitative mapping needed to analyze the efficacy of the directed assembly process. They used extreme ultraviolet interference lithography (EUV-IL) with a transmission grating[53] to pattern a photoresist layer on top of a SAM. The use of EUV-IL with a transmission grating was a key technological advancement in the study of directed assembly because it enabled the creation of defect-free, fine-scale patterns over the entire patterned area. The pattern formed in the photoresist was transferred to the underlying SAM layer through an oxygen plasma process. An essential feature of this process was that, although mapping the chemical pattern directly was not feasible, one could quantitatively map the chemical pattern by mapping the resist pattern used to create it. Importantly, the pattern also had sufficient interfacial energy contrast to direct the assembly of the copolymer. The combination of high interfacial energy contrast, commensurability, and chemical surface pattern quality led to an assembled morphology that covered a large area with a high degree of perfection, as shown in Fig. 3.[41] The perfection of the structure shown in Fig. 3 proved that it is possible to put enough energy into the block

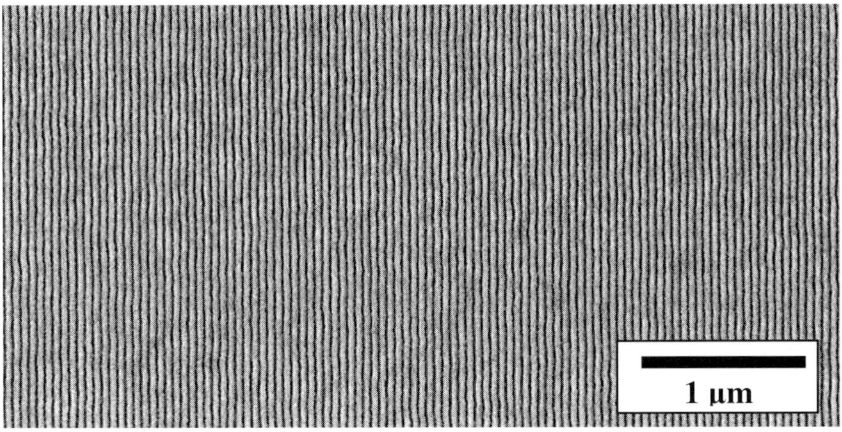

Fig. 3. Top-down SEM image of the perfect ordering of a P(S-*b*-MMA) block copolymer patterned by an underlying SAM chemical pattern. The chemical pattern period was commensurate with the copolymer bulk pattern period. The light and dark regions were PS and PMMA domains, respectively.

copolymer film through the interfacial energy contrast of the chemical nanopattern to overcome defect structures. However, Kim *et al.* were only able to achieve this structure when L_s and L_o were commensurate. When L_s was 2.5 nm larger or smaller than L_o, such a perfect structure was not obtained.

An example of the quantitative mapping of the chemical pattern to the block copolymer structure performed by Kim *et al.* is shown in Fig. 4.[41] They imaged a section of photoresist that had a range of exposures during EUV-IL (Fig. 4a), ranging from unexposed to overexposed. They then used this section of chemical nanopattern to direct the assembly of a thin film of P(S-*b*-MMA) (Fig. 4b). The unpatterned or underexposed regions of PMMA caused the SAM to be untreated chemically, resulting in the formation of perpendicular lamellae without long-range order in the P(S-*b*-MMA) film that covered these regions. By contrast, the overexposed region resulted in the oxidation of the underlying SAM during the chemical treatment. The oxygenated SAM had affinity for the PMMA in the P(S-*b*-MMA), resulting in a morphology that was parallel to the substrate.

2.2. *Importance of Pattern Interfacial Energy Contrast*

At the core of the directed assembly process is the effect that the chemical pattern has on the thermodynamics of the block copolymer thin film system. As alluded to in the introduction, the total free energy of the block copolymer film can be viewed as a balance of thermodynamic factors intrinsic to the bulk copolymer, such as the energy of the S-MMA interface and the energy of stretching or compressing a polymer chain configuration, and thermodynamic factors extrinsic to the bulk copolymer such as the surface and interfacial free energies.[3] On a neutral, unpatterned substrate, there is minimal contribution to the free energy of the system by the extrinsic factors, and the intrinsic factors dominate, resulting in perpendicular lamellae that have precise dimensions but that are also relatively disordered. When the substrate is chemically patterned, the interfacial energy can add to the free energy of the system. As the amount of interfacial energy is increased, it can affect the ordering and orientation of the domains of the film, as shown above.

To achieve perfect structures like that shown in Fig. 3, there needs to be sufficient free energy to overcome the formation of defect structures. However, when L_s and L_o are commensurate, the amount of free energy

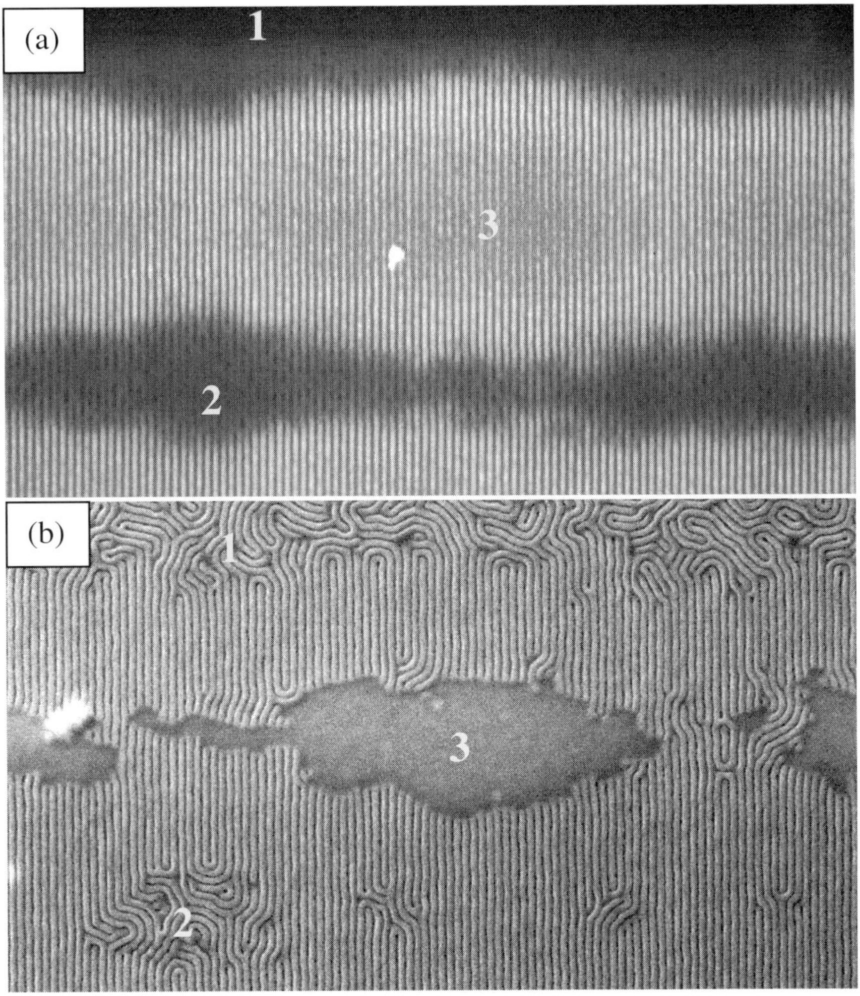

Fig. 4. Comparison of SEM images of chemical nanopattern and block copolymer morphology. (a) Unpatterned (1), underexposed (2), and overexposed (3) regions of photoresist used to form chemical nanopattern. (b) Resulting directed assembly morphology, with the regions corresponding to (1), (2), and (3) denoted in (a).

required to overcome the formation of defect structures can be achieved without supplying the additional energy that would be required to stretch or compress the polymer chains away from their bulk configuration. The fact that additional energy is required to stretch or compress polymer chains is why the directed assembly work of Kim *et al.*[41] was only completely successful when L_s and L_o were commensurate.

Based on this phenomenological argument, one can discern two key attributes of the chemical nanopattern that affect the ability of the chemical pattern to direct the assembly of the copolymer morphology: commensurability of L_s and L_o, as referenced above in the work of Rockford[49] and Yang,[48] and the strength of the interfacial energy of the alternating sections of the chemical nanopattern. Edwards *et al.* were able to demonstrate the thermodynamic balance of interfacial energy contrast and commensurability by testing directed assembly on a series of chemical patterns with increasing interfacial energy contrast.[3] Edwards *et al.* built on the seminal work of Mansky *et al.* on neutral wetting surfaces, which reported interfacial free energies of a series of PS-PMMA random copolymer (P(S-*r*-MMA)) brushes with varying PS content.[23] Edwards *et al.* used EUV-IL to pattern a similar series of P(S-*r*-MMA) brushes with varying PS content. The pattern was transferred from the photoresist used in EUV-IL to the underlying P(S-*r*-MMA), resulting in oxidized regions that were preferential to PMMA blocks, and unoxidized regions that were preferential to PS blocks. The least energy contrast was achieved with a patterned, oxidized 50:50 P(S-*r*-MMA) brush, and the greatest energy contrast was achieved with a patterned, oxidized PS homopolymer brush. They studied the range of commensurability over which they could achieve perfect directed assembly, as shown in Fig. 5. As they increased the interfacial energy contrast, they were able to move away from the requirement of commensurability. In the case of the patterned, oxidized PS homopolymer brush, they were able to assemble perfect structures when L_s was as much as 10% different from L_o. The key point of their work was that they demonstrated that the contribution of interfacial energy to the thermodynamics of the system could overcome other thermodynamic factors, such as the energy required to stretch or compress copolymer chains. The final structures of the copolymers in their work were truly equilibrated, taking into account both the intrinsic and extrinsic factors described above. With such strong interfacial energy contrast of the alternating regions of the pattern with respect to the different blocks of the copolymer, the alternating regions of the chemical pattern would be wet by the different blocks regardless of the value of L_s. When the copolymer and the pattern were approximately commensurate, the

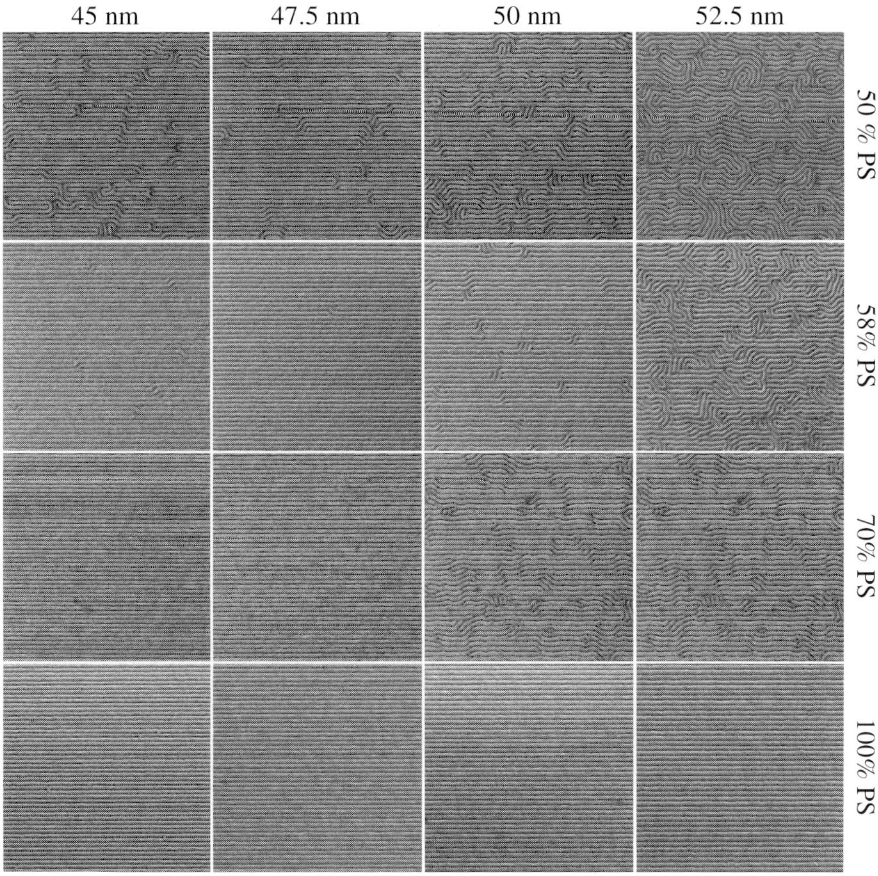

Fig. 5. Top-down SEM images of the directed assembly morphologies of P(S-*b*-MMA) on chemically patterned surfaces of P(S-*r*-MMA) brushes as a function of pattern L_s (columns) and brush PS content (rows). Each micrograph shows a 2um x 2um area.

impact of the patterned interface extended throughout an equilibrated, 50-nm-thick film.

Similar research has shown that the chemical pattern could supply sufficient energy to overcome other thermodynamic barriers, such as the energy required to bend lamellar domains, which increases as the radius of curvature increases. For example, Wilmes *et al.* used a chemically patterned PS homopolymer brush to direct the assembly of a lamellae-forming P(S-*b*-MMA) copolymer into a series of concentric rings, as shown in Fig. 6.[43] The results of Wilmes *et al.* underscore one of the basic principles of directed assembly, which is that the chemical pattern will direct the assembly of the block copolymer as long as the energy supplied by the chemical pattern is greater than the energy required for the copolymer to replicate the pattern.

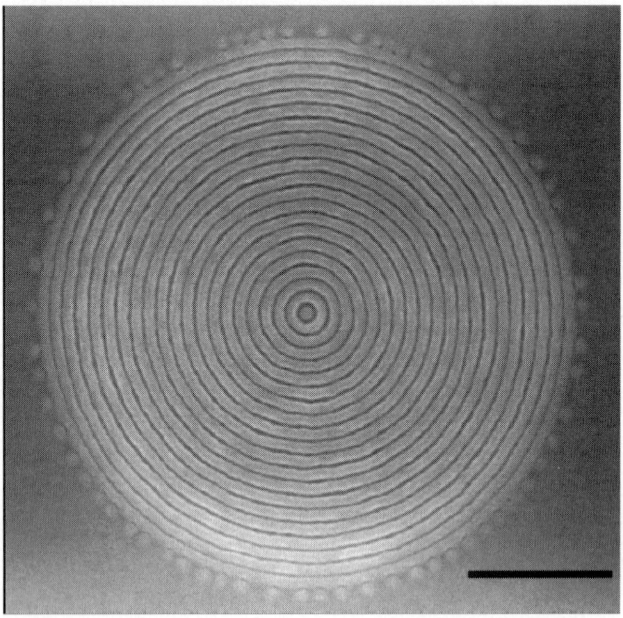

Fig. 6. SEM micrograph of P(S-*b*-MMA) on a silicon wafer with concentric circles chemically patterned into a PS brush. Scale bar is 500 nm. (Figure reprinted from Ref. 43 with permission of the American Chemical Society.)

3. Directed Assembly Kinetics and Mechanism

While the preceding investigations have focused on the thermodynamics of directed assembly, and shown the importance of the interfacial energy contrast, it is also important to understand the kinetics and the mechanism of the process. The mechanism of the directed assembly process was experimentally investigated by examining the top surface of a thin film of lamellae-forming P(S-*b*-MMA) on a chemically patterned surface at regular intervals during the course of annealing.[54] Immediately after spin-coating, the P(S-*b*-MMA) is typically assumed to be in a disordered state. After a short period of annealing, the copolymer formed hexagonally-close-packed (HCP) styrene domains at the surface. Additional annealing caused the styrene domains to coalesce to form a mixed morphology of spheres and linear styrene domains. With continued annealing, the linear styrene domains broke and reformed until lamellae were formed. Several elements of this process can be observed in Fig. 7, including the HCP styrene domains, vertical lamellar structures, and disclination defects as the lamellar structures reform to register with the underlying chemical pattern. In keeping with the importance of the interfacial energy contrast, it is worth noting that the P(S-*b*-MMA) film shown in Fig. 7 was on a chemical pattern with relatively weak interfacial energy contrast, such that more defects were apparent, and a longer time was required to anneal the sample to perfection compared to other samples that were made with chemical nanopatterns with greater interfacial energy contrast.[54]

Single-chain-in-mean-field (SCMF) simulations provided significant insight into the experimental observations by delineating the interfacial as well as the three-dimensional structure of the film as the copolymer equilibrates in the presence of the chemical pattern (see Fig. 8). The simulations suggested that first the PS and PMMA blocks wet their preferred regions of the chemical nanopattern, forming an ordered layer at the substrate that is registered with the underlying chemical surface pattern. As the simulated annealing proceeds, the order at the interface propagates upward through the film to the free surface. The unregistered linear domains near the free surface break and reform new conformations of linear domains until defects are removed.

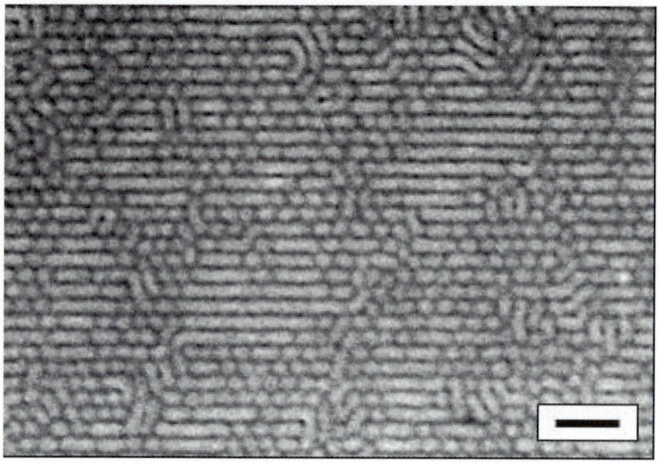

Fig. 7. Plan-view scanning electron micrograph of hexagonally-close-packed styrene domains of P(S-*b*-MMA) transitioning into linear domains over a chemically patterned neutral brush with L_s and L_o commensurate. The chemically nanopatterned lines run from left to right in the image. The black scale bar represents 200 nm.

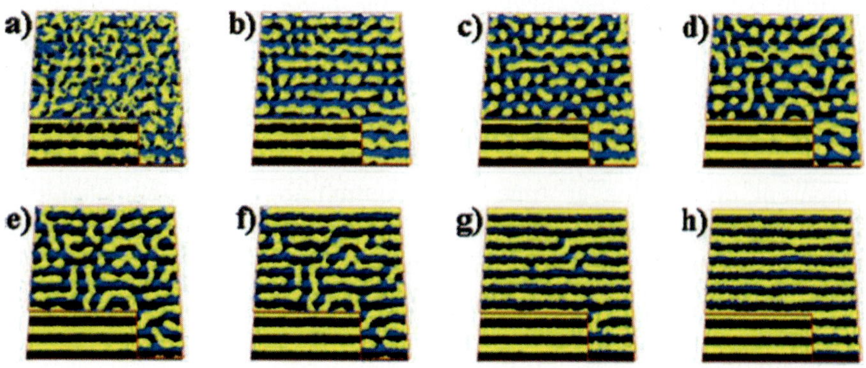

Fig. 8. Snapshot images of the three dimensional behavior of P(S-*b*-MMA) on chemically patterned substrates with $L_s = 46$ nm and $L_o = 47.5$ nm after various time steps in SCMF simulations. PS domains are in yellow; PMMA domains have been removed from the images. Blue areas represent an interface between PS and PMMA domains. In the bottom left of each image, the top 75% of the film has been removed to reveal the behavior of the block copolymer at the substrate. The number of Monte Carlo steps corresponding to each image is (a) 100, (b) 1000, (c) 3000, (d) 5000, (e) 7500, (f) 10,000, (g) 15,000, and (h) 21,000. (Reprinted from Ref. 54 with permission of authors and John Wiley & Sons, Inc. Copyright © 2005.)

There are important implications of the simulations and experiments on the mechanism of directed assembly. First, at the interface the PS and PMMA blocks will rapidly and uniformly wet their preferred regions on the chemical pattern. Second, only after long annealing times can defects at the surface be assumed to be equilibrium structures. The SCMF simulations reveal that the morphology observed in the SEM in Fig. 7 represents surface structures, and is likely not indicative of the three-dimensional structure. Third, the kinetics of non-equilibrium defect annihilation leading to the equilibrium domain structure are governed primarily by the thermodynamics of the chemical pattern/copolymer interface, which is markedly different from defect annihilation in self-assembled block copolymers on chemically homogeneous surfaces. In the latter case, fully developed domains are formed, and the defects in those domains must diffuse until they are in close proximity to other defects prior to annihilation.[55] The effect of the interfacial thermodynamics on the kinetics of defect annihilation in directed assembly was readily apparent in the longer annealing times required to achieve defect-free block copolymer domains on chemical surface patterns that were either incommensurate with the block copolymer or had lower interfacial energy contrast.

The mechanistic studies of directed assembly suggest that the time frame for assembly relates to polymer mobility. The chains must have the opportunity to diffuse over a characteristic length scale on the order of the film thickness or L_o. Indeed, annealing studies[56] have revealed that the time required for assembly decays exponentially with increasing temperature, indicative of a thermally activated process. These annealing studies showed that for the P(S-*b*-MMA) system, the annealing time to reach perfect assembly can be shortened significantly by increasing the annealing temperature, as shown in Fig. 9.[56] Because of the weak temperature dependence of χ on T for P(S-*b*-MMA),[57] even at elevated temperatures the P(S-*b*-MMA) remains below the ODT. In the case of a commensurate chemical pattern with a high interfacial energy contrast, by increasing the temperature to 250 °C, a perfect structure was achieved within 1 minute.

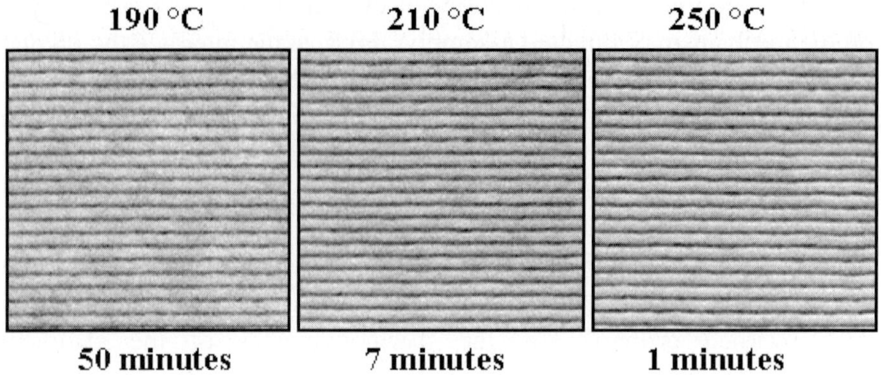

Fig. 9. SEM micrographs of the morphologies achieved from directed assembly of thin films of P(S-*b*-MMA) on chemically patterned surfaces. The annealing temperature and time to reach perfection for each image is shown above and below the image, respectively. (All micrographs show a 1 μm x 1 μm area.)

4. Role of Block Copolymer Thermodynamics

Under certain conditions, the assembly process is not particularly sensitive to the detailed geometry of the underlying chemical pattern. In this sense, the directed assembly process is not "epitaxial", as it has been sometimes labeled.[41,58] For example, irrespective of the width *W* of adjacent stripes in the pattern, for a chemical pattern that has a large interfacial energy contrast and has L_s commensurate with L_o, directed assembly occurs with domain periods equal to L_s and the domain widths equal to ½ L_s. In other words, L_s matters more than *W*. This dependence on L_s, and relative independence from *W*, occurs because the fundamental thermodynamic properties of the block copolymer still play a significant role.

In a demonstration of a block copolymer's ability to correct for variation in *W*, Edwards *et al.* recently reported on the directed assembly of P(S-*b*-MMA) on a series of chemical patterns with fixed L_s and varying *W*.[59] Through experimentation and modeling, they demonstrated that a chemically patterned PS brush could direct the assembly of P(S-*b*-MMA) domains over a range of *W*, for a given L_s, as shown in Fig. 10.

Similar to previous work on directed assembly, they found that perfect directed assembly could be achieved over a wider range of W values when L_s and L_o were closer to commensurability. Edwards *et al.* also performed SCMF simulations to analyze the cross-sections of the assembled lamellar domains as W was varied (see Fig. 10(c)). On chemical surface patterns with a value of W/L_S of approximately 0.50, the block copolymer structures oriented perpendicularly to the substrate throughout the film. As W/L_s diverged from 0.50, the domains created trapezoidal structures to accommodate the underlying chemical surface pattern. The width of the PS domains at half height, as determined from

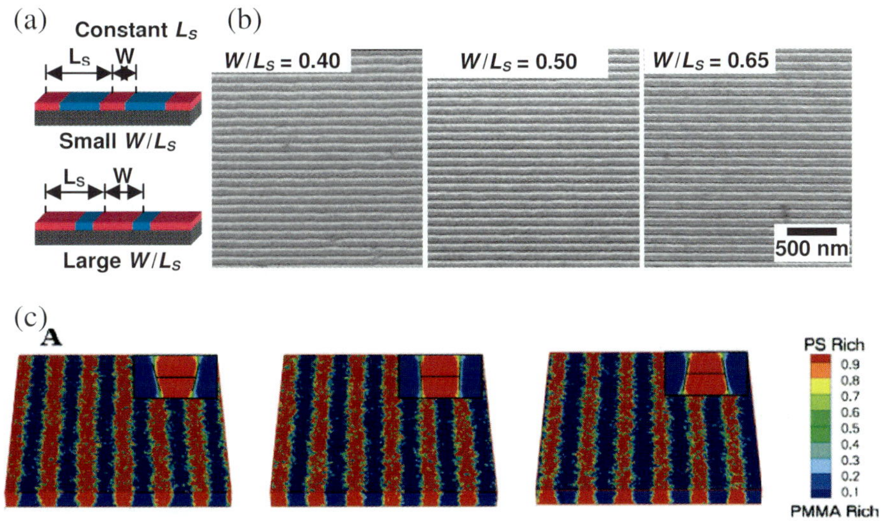

Fig. 10. (a) Schematic showing variation of line-width W while maintaining constant pattern period L_s. (b) Top-down SEM images showing uniformity of copolymer domain width over constant L_s chemical nanopatterns with variations of linewidth W ranging from 20% smaller to 30% larger than the median value of ½ L_s. (c) Results of SCMF simulations of block copolymer thin films on chemically patterned substrates with varying W. Three-dimensional snapshot (large) and two-dimensional contour plots (inset) of SCMF simulations with W/L_S = 0.30, 0.45, and 0.65, from left to right, and $L_S = L_O$ = 80 nm. Red domains are PS and blue domains are PMMA.

the SCMF simulations, was equal to $L_s/2$, regardless of W, thereby correcting for pattern regions where $W \neq L_s/2$.

5. Achievable Structures with Directed Assembly

Based on the understanding that the interfacial energy of the chemical pattern can have a significant impact on the morphology of the block copolymer film as it equilibrates, such that chain configurations that are unfavorable in the bulk can be stabilized to minimize the overall free energy of the system, one would expect that a variety of interesting and potentially useful structures could be obtained. Recent work by Park *et al.* demonstrated the variety of chemical pattern features that can be replicated by a block copolymer.[26] Park *et al.* chemically modified the disordered fingerprint pattern that is generated when a lamellae-forming block copolymer is placed on a neutral surface. The chemically modified fingerprint structure served as a chemical pattern for a second layer of the same block copolymer. Thus, L_s and L_o were automatically commensurate. As shown in Fig. 11, the fingerprint chemical pattern directed the assembly of the block copolymer for a variety of pattern features, including line terminations, sharp bends, small radii of curvature, and junctions. Also, although the domains in the copolymer shown in Fig. 11 had different lengths, they all had the same block copolymer period.

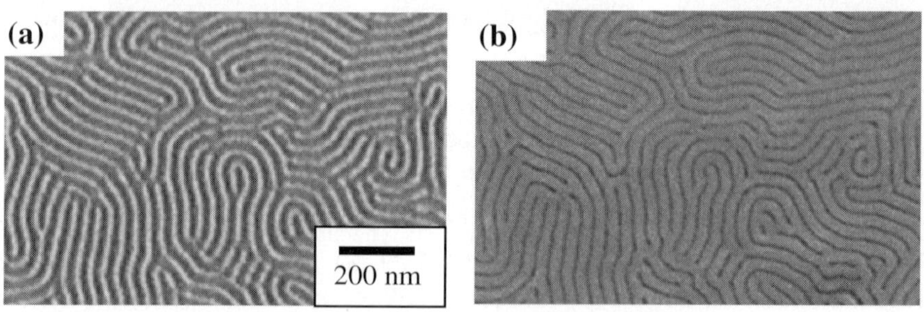

Fig. 11. SEM images of (a) fingerprint chemical nanopattern and (b) corresponding block copolymer morphology that was assembled on top of pattern in (a).

One of the applications of directed assembly is the formation of templates for integrated circuit manufacturing.[58,60,61] While the pattern in Fig. 11 was generated randomly, elements of the fingerprint pattern, such as T-junctions and line terminations, are regular fabric architectures necessary for integrated circuit design. Advanced lithographic techniques such as EUV-IL and electron beam lithography can be used to make periodic patterns of these fabric architectures. Curved lines were shown in Fig. 6. Other useful fabric structures are shown below in Fig. 12.[19]

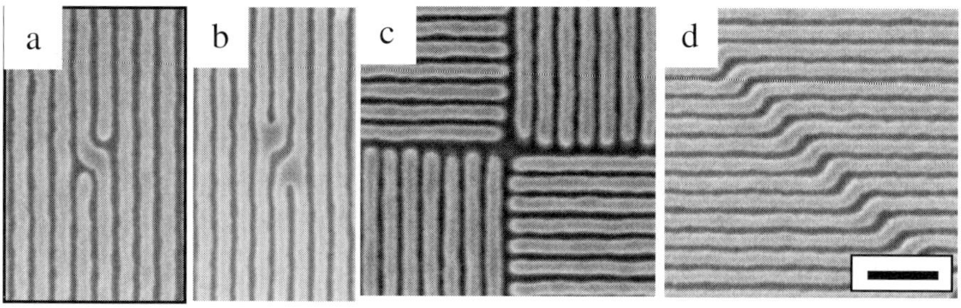

Fig. 12. SEM images of P(S-*b*-MMA) structures after directed assembly on a variety of regular fabric architectures generated by electron beam lithography: (a) line terminations, (b) junctions, (c) T-junctions, and (d) jogs. Scale bar represents 250 nm.

It would also be beneficial to have the ability to generate structures in which the spacing between domains could be varied. Stoykovich *et al.* demonstrated this possibility in the formation of angled domains, as shown in Fig. 13(a).[42] The presence of an angle in a chemical pattern creates two different pattern periods, as shown in Fig. 13(b). In addition to the period L_s away from the corner, there is the period L_c at the corners. In the case of a 90° bend, L_c is 41% larger than L_s. Comparison with the commensurability results presented above suggests that for a block copolymer with $L_o \approx L_s$, L_c of $1.41 * L_s$ is a much less favorable geometry than defect structures at L_o. The addition of PS and PMMA homopolymer assisted the domain formation across L_s and L_c. SCMF simulation of the blend formation showed that the PS preferentially gathers at the angle of the cornered structure, as shown in Fig. 13(c). It is significant to note that in this case, the interfacial energy supplied by the chemical pattern assembled a non-bulk period, and also stabilized a homopolymer concentration gradient in the P(S-*b*-MMA)/PS/PMMA blend, which would not normally exist in the bulk.

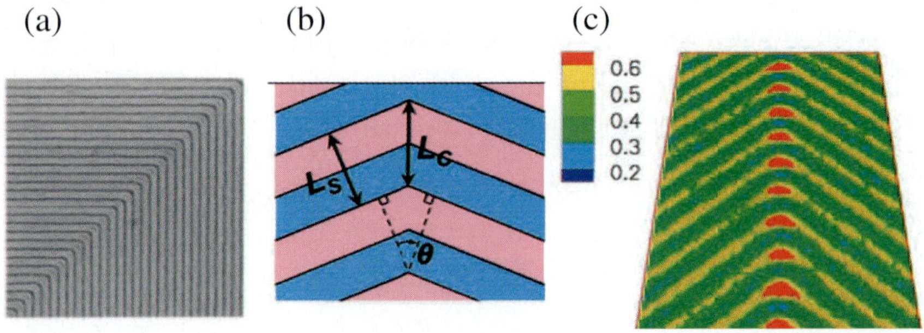

Fig. 13. Angled domains of a ternary P(S-*b*-MMA)/PS/PMMA blend assembled on an angled chemical nanopattern. (a) Top-down SEM of a 2μm x 2μm section of the ternary blend directed to assemble on an angled chemical pattern with L_s of 70 nm and a 90° bend. (b) Schematic of the increased lamellar period at the corners of the bends. The linear structures have a lamellar period of L_s, whereas the corner-to-corner lamellar period is L_c such that $L_c > L_s$. (c) Contour plot of the total homopolymer concentration obtained from SCMF simulations for the blend directed to assemble on the same pattern shown in (a), with the bulk period of the blend equal to 70 nm. The periodic red areas are enriched in PS homopolymer.

6. Conclusion

Several key points can be drawn from the investigations of the directed assembly of block copolymers on chemically nanopatterned substrates. First, the final film morphologies represent thermodynamic minima of the block copolymer equilibrating in the presence of the chemical nanopattern. The contribution of the total free energy of the system from the interfacial interactions between the film and the chemical pattern can be a substantial fraction of the system free energy, and can drive the orientation and order of block copolymer domains. The result is block copolymer structures with low defect densities that are registered to the underlying chemical pattern, and that can overcome propensity of the system to form defects. Second, the directed assembly process is most readily accomplished when there is a large pattern interfacial energy contrast and L_s is commensurate with L_o. However, with sufficiently large interfacial energy contrast, the pattern can drive the block copolymer into structures that have a period equal to L_s, even if $L_s \neq L_o$, or into structures that do not naturally appear in the bulk. In the case of symmetric block copolymers, the domain width will equal ½ L_s, and not the width of the corresponding pattern region. Third, because of the interfacial energy contrast with respect to the different blocks of the copolymer, the correct block will always wet the correct portion of the chemical pattern. The morphology of the block copolymer will propagate from the interface through the film if the pattern and copolymer are approximately commensurate. Finally, the kinetics of the process are thermally activated, with assembly time decaying exponentially as temperature is increased.

These fundamental findings have garnered interest in using directed assembly of block copolymers in advanced lithographic applications.[5,60] Directed assembly can satisfy the essential lithographic process attributes of being able to pattern a variety of semiconductor fabric architectures with low defect densities over areas limited only by the lithographic equipment used. The block copolymer is registered to the pattern, and insomuch as one can control the formation of the chemical pattern, one can control the placement of the block copolymer domains equally well. The process is thermally activated, so it is possible to achieve the desired block copolymer structures in under 1 minute, meeting one of the goals of the International Technology Roadmap for Semiconductors.[4]

The directed assembly process has the potential to improve advanced lithography in the arena of process control, because the assembled copolymer structure is at a thermodynamic minimum. As a result, it is not subject to change due to minor process variations, unlike kinetically-limited processes. Additionally, the length scale of the structure that is obtained depends on the period of the chemical pattern, and not the width of a particular chemical pattern element. Fluctuations in linewidth that might occur during the initial patterning of the chemical pattern would be reduced or eliminated by the equilibrated block copolymer structure overlying the chemical pattern. As a result, directed assembly of block copolymers on chemically patterned surfaces offers the potential to reduce line edge roughness and improve critical dimension control in advanced lithographic processes.

References

1. F. S. Bates and G. H. Fredrickson, *Annu. Rev. Phys. Chem.* **41**, 525 (1990).
2. A. Knoll, A. Horvat, K. S. Lyakhova, G. Krausch, G. J. A. Sevink, A. V. Zvelindovsky, and R. Magerle, *Phys. Rev. Lett.* **89**, 035501 (2002).
3. E. W. Edwards, M. F. Montague, H. H. Solak, C. J. Hawker, and P. F. Nealey, *Adv. Mater.* **16**, 1315 (2004).
4. "International Technology Roadmap for Semiconductors — Lithography." International SEMATECH, Austin, TX, 2005.
5. M. Park, C. Harrison, P. M. Chaikin, R. A. Register, and D. H. Adamson, *Science* **276**, 1401 (1997).
6. P. Mansky, P. Chaikin, and E. L. Thomas, *J. Mater. Sci.* **30**, 1987 (1995).
7. T. Thurn-Albrecht, R. Steiner, J. DeRouchey, C. M. Stafford, E. Huang, M. Bal, M. Tuominen, C. J. Hawker, and T. P. Russell, *Adv. Mater.* **12**, 1138 (2000).
8. K. W. Guarini, C. T. Black, Y. Zhang, I. V. Babich, E. M. Sikorski, and L. M. Gignac, *in* "IEEE International Electron Device Meeting", p. 541, 2003.
9. R. Olayo-Valles, M. S. Lund, C. Leighton, and M. A. Hillmyer, *J. Mater. Chem.* **14**, 2729 (2004).
10. X. M. Yang, S. G. Xiao, C. Liu, K. Pelhos, and K. Minor, *Journal of Vacuum Science & Technology B* **22**, 3331 (2004).
11. R. D. Peters, X. M. Yang, Q. Wang, J. J. de Pablo, and P. F. Nealey, *Journal of Vacuum Science & Technology B* **18**, 3530 (2000).
12. L.-W. Chang and H.-S. P. Wong, *Proc. of SPIE* **6156**, 615611 (2006).
13. T. Thurn-Albrecht, J. Schotter, C. A. Kastle, N. Emley, T. Shibauchi, L. Krusin-Elbaum, K. Guarini, C. T. Black, M. T. Tuominen, and T. P. Russell, *Science* **290**, 2126 (2000).
14. C. T. Black, *Appl. Phys. Lett.* **87**, 163116 (2005).
15. C. T. Black, K. W. Guarini, K. R. Milkove, S. M. Baker, T. P. Russell, and M. T. Tuominen, *Appl. Phys. Lett.* **79**, 409 (2001).
16. D. J. C. Herr, *in* "Future Fab International" (B. Dustrud, ed.), Vol. 20, p. 82. Montgomery Research Incorporated, San Francisco, 2006.
17. M. Harada, *J. Chem. Eng. Jpn.* **37**, 577 (2004).
18. *In* "Directed Self Assembly of Materials for Patterning Workshop". Semiconductor Research Corporation / National Nanotechnology Initiative, Madison, WI.
19. M. P. Stoykovich, H. Kang, K. C. Daoulas, G. Liu, C.-C. Liu, J. J. de Pablo, xfc, M. ller, and P. F. Nealey, *ACS Nano* **1**, 168 (2007).

20. S. H. Anastasiadis, T. P. Russell, S. K. Satija, and C. F. Majkrzak, *Phys. Rev. Lett.* **62**, 1852 (1989).
21. P. Mansky, T. P. Russell, C. J. Hawker, M. Pitsikalis, and J. Mays, *Macromolecules* **30**, 6810 (1997).
22. Y. Liu, W. Zhao, X. Zheng, A. King, A. Singh, M. H. Rafailovich, J. Sokolov, K. H. Dai, E. J. Kramer, and et al., *Macromolecules* **27**, 4000 (1994).
23. P. Mansky, Y. Liu, E. Huang, T. P. Russell, and C. Hawker, *Science* **275**, 1458 (1997).
24. A. Knoll, R. Magerle, and G. Krausch, *J. Chem. Phys.* **120**, 1105 (2004).
25. D. Y. Ryu, K. Shin, E. Drockenmuller, C. J. Hawker, and T. P. Russell, *Science* **308**, 236 (2005).
26. S. M. Park, Y. H. La, P. Ravindran, G. S. W. Craig, N. J. Ferrier, and P. F. Nealey, *Langmuir* **23**, 9037 (2007).
27. R. Ruiz, R. L. Sandstrom, and C. T. Black, *Adv. Mater.* **19**, 587 (2007).
28. J. Bang, S. H. Kim, E. Drockenmuller, M. J. Misner, T. P. Russell, and C. J. Hawker, *J. Am. Chem. Soc.* **128**, 7622 (2006).
29. S. H. Kim, M. J. Misner, and T. P. Russell, *Adv. Mater.* **16**, 2119 (2004).
30. S. H. Kim, M. J. Misner, T. Xu, M. Kimura, and T. P. Russell, *Adv. Mater.* **16**, 226 (2004).
31. G. Kim and M. Libera, *Macromolecules* **31**, 2569 (1998).
32. P. Mansky, J. DeRouchey, T. P. Russell, J. Mays, M. Pitsikalis, T. Morkved, and H. Jaeger, *Macromolecules* **31**, 4399 (1998).
33. T. L. Morkved, M. Lu, A. M. Urbas, E. E. Ehrichs, H. M. Jaeger, P. Mansky, and T. P. Russell, *Science* **273**, 931 (1996).
34. D. E. Angelescu, J. H. Waller, D. H. Adamson, P. Deshpande, S. Y. Chou, R. A. Register, and P. M. Chaikin, *Adv. Mater.* **16**, 1736 (2004).
35. R. A. Segalman, H. Yokoyama, and E. J. Kramer, *Adv. Mater.* **13**, 1152 (2001).
36. D. Sundrani, S. B. Darling, and S. J. Sibener, *Nano Letters* **4**, 273 (2004).
37. D. Sundrani, S. B. Darling, and S. J. Sibener, *Langmuir* **20**, 5091 (2004).
38. J. Y. Cheng, A. M. Mayes, and C. A. Ross, *Nature Materials* **3**, 823 (2004).
39. C. T. Black and O. Bezencenet, *IEEE Transactions on Nanotechnology* **3**, 412 (2004).
40. D. Sundrani and S. J. Sibener, *Macromolecules* **35**, 8531 (2002).
41. S. O. Kim, H. H. Solak, M. P. Stoykovich, N. J. Ferrier, J. J. de Pablo, and P. F. Nealey, *Nature* **424**, 411 (2003).
42. M. P. Stoykovich, M. Muller, S. O. Kim, H. H. Solak, E. W. Edwards, J. J. de Pablo, and P. F. Nealey, *Science* **308**, 1442 (2005).
43. G. M. Wilmes, D. A. Durkee, N. P. Balsara, and J. A. Liddle, *Macromolecules* **39**, 2435 (2006).
44. Q. Wang, S. K. Nath, M. D. Graham, P. F. Nealey, and J. J. de Pablo, *J. Chem. Phys.* **112**, 9996 (2000).
45. Z. G. Wang, *J. Chem. Phys.* **100**, 2298 (1994).

46. K. C. Daoulas, M. Muller, M. P. Stoykovich, S. M. Park, Y. J. Papakonstantopoulos, J. J. de Pablo, P. F. Nealey, and H. H. Solak, *Phys. Rev. Lett.* **96**, 036104 (2006).
47. S. M. Park, G. S. W. Craig, Y. H. La, H. H. Solak, and P. F. Nealey, *Macromolecules* **40**, 5084 (2007).
48. X. M. Yang, R. D. Peters, P. F. Nealey, H. H. Solak, and F. Cerrina, *Macromolecules* **33**, 9575 (2000).
49. L. Rockford, Y. Liu, P. Mansky, T. P. Russell, M. Yoon, and S. G. J. Mochrie, *Phys. Rev. Lett.* **82**, 2602 (1999).
50. H. H. Solak, D. He, W. Li, S. Singh-Gasson, F. Cerrina, B. H. Sohn, X. M. Yang, and P. Nealey, *Appl. Phys. Lett.* **75**, 2328 (1999).
51. B. Collin, D. Chatenay, G. Coulon, D. Ausserre, and Y. Gallot, *Macromolecules* **25**, 1621 (1992).
52. G. Coulon, B. Collin, D. Ausserre, D. Chatenay, and T. P. Russell, *Journal De Physique* **51**, 2801 (1990).
53. H. H. Solak, C. David, J. Gobrecht, V. Golovkina, F. Cerrina, S. O. Kim, and P. F. Nealey, *Microelectron. Eng.* **67-8**, 56 (2003).
54. E. W. Edwards, M. P. Stoykovich, M. Muller, H. H. Solak, J. J. De Pablo, and P. F. Nealey, *Journal of Polymer Science Part B-Polymer Physics* **43**, 3444 (2005).
55. C. Harrison, D. H. Adamson, Z. D. Cheng, J. M. Sebastian, S. Sethuraman, D. A. Huse, R. A. Register, and P. M. Chaikin, *Science* **290**, 1558 (2000).
56. A. M. Welander, K. O. Stuen, H. Kang, H. H. Solak, M. Müller, J. J. De Pablo, and P. F. Nealey, *Macromolecules* **41**, 2759 (2008).
57. T. P. Russell, R. P. Hjelm, and P. A. Seeger, *Macromolecules* **23**, 890 (1990).
58. J. Y. Cheng, C. A. Ross, H. I. Smith, and E. L. Thomas, *Adv. Mater.* **18**, 2505 (2006).
59. E. W. Edwards, M. Muller, M. P. Stoykovich, H. H. Solak, J. J. de Pablo, and P. F. Nealey, *Macromolecules* **40**, 90 (2007).
60. M. P. Stoykovich and P. F. Nealey, *Materials Today* **9**, 20 (2006).
61. C. J. Hawker and T. P. Russell, *MRS Bulletin* **30**, 952 (2005).

CHAPTER 3

STRUCTURE FORMATION AND EVOLUTION IN CONFINED CYLINDER-FORMING BLOCK COPOLYMERS

G.J.A. Sevink and J.G.E.M. Fraaije

Soft Condensed Matter Group, Leiden Institute of Chemistry, Leiden University, PO Box 9502, 2300 RA Leiden, The Netherlands

We review the main results of an experimental/computational collaborative effort. This research aimed at determining the fundamental mechanisms of structure formation in block copolymer thin films. Although the experimental system was a cylinder-forming SBS block copolymer thin film that forms terraces upon annealing, the results are general and can be useful for the understanding of phase behavior also in other confined systems. In this review, we focus on the computational part of this study: we formulate the dynamic density functional method (DDFT), that was developed earlier in our group, and show several applications. We discuss in detail the important factors in thin films with both symmetric and asymmetric surface wetting conditions, and relate these findings to experimental observations in annealed thin films. Finally, we show how our computational method and an advanced dynamic scanning force microscopy (SFM) method can be exploited in a synergetic fashion to extract information about the elementary steps in structural transitions at the mesoscopic level. In particular, the experiments validate the dynamic DDFT method, and the DDFT calculations rationalize the experimental surface measurements away from the surface.

1. Introduction

Block copolymers (BCP) are attractive materials, as they are lightweight, corrosion resistant, easy to process at low temperatures, and generally inexpensive. Moreover, due to the different chemical nature of the blocks, (rational) design strategies, key to the philosophy of modern 'soft' nanotechnology, are within reach. For example, styrene-butadiene (PS-*block*-PB) BCP can be devised that serve either as rigid, tough and transparent thermoplastics or as soft, flexible and thermoplastic elastomers, by appropriate control of the chain architecture and PS/PB ratio.[1] They are part of a fascinating class of materials of ordered fluids, that combines crystal-like order on mesoscopic length scales with liquid-like disorder on microscopic scales, and tend to self-assemble into patterns with characteristic length scales determined by the molecular size, i.e. in the 10-100 nm range. In bulk materials, the design manifests itself in the observation that desired macroscopic properties, such as mechanical strength and opacity, depend on the way that the polymer blocks are organised in space: in particular, the structure on the mesoscale. In 'soft' nanotechnology, for instance in the design of delivery systems for hydrophobic medicines or scaffolds for other nanostructures, the macroscopic bulk properties are of lesser importance, but there is a desire to assembly structures with specific mesoscopic structure and/or behavior.

In the absence of external factors, the (equilibrium) patterns in BCP are dictated by the interaction between the segments that comprise the copolymer, the volume fraction of the blocks, and the molecular architecture of the BCP. The patterns can, depending on these variables, be very complex.[2] In reality, patterns are often also affected by kinetic factors, such as the slow thermodynamic relaxation towards equilibrium and external fields (for instance, shear fields) during their preparation, resulting in irregular and/or metastable structures. Understanding the fundamental relation between molecular properties, processing conditions and kinetic factors on one hand, and supra-molecular structures (or patterns) on the other hand, is therefore key to any rational design strategy. A very important condition is *confinement*, since molecular conformations and assembly are strongly influenced by confinement. Interfacial interactions, symmetry breaking, structural frustration and confinement-induced entropy loss play a determining role and may even lead to structures that differ from the ones found in bulk. Especially the interfacial interactions give rise to an additional driving force, because typically one component/block has a lower interfacial energy than the other(s). This phenomenon belongs to a class of interfaces of modulated phases.[3] In more general terms, confinement can be employed to induce global orientation of microdomains, whereas in the bulk, grains of ordered microdomains are randomly oriented. Recently, phase separation in confinement has been the subject of extensive theoretical and experimental work due to the intriguing prospects from a technological viewpoint. Here we consider pattern formation in block copolymers in one-dimensional confinement. This type of confinement is present in thin block copolymer films, which can either be *free standing*, bounded by air at two sides, *supported*, bounded by air and a solid substrate, or *slits*, confined between two solid interfaces. Pattern formation in two- and three-dimensional confinement, nanopores and nanospheres respectively, have been considered as well recently, but are outside the scope of this chapter. Surface induced ordering was first identified in lamellae-forming BCP thin supported films. The first detailed experimental work, including a simple model, dates back to the eighties of last century.[4,5] The first theoretical investigation used mean-field theory to study the 1-D equilibrium composition profile for near-symmetric BCP in the presence of a confining wall.[6] Later studies identified two major effects in lamellae-forming BCP.[7–9] The preferential segregation of one type of blocks to the surface (the surface field) breaks the symmetry and causes lamellae to align parallel to the confining interface. Moreover, the film forms terraces, islands or holes, where the film thickness deviates from the favored thickness, quantized with the bulk lamellar domain distance L (the commensurability effect). This favored thickness can be either nL or $(n+1/2)L$, with n an integer, depending on whether the surface field promotes the same block (symmetric conditions) or different blocks (asymmetric conditions) close to the surface. Slits are a special case of confinement, as terrace formation is prohibited, and frustration due to a mismatch between the favored and the actual film thickness can give rise to perpendicular lamellae (with respect to the surface).

Several authors noticed that the situation for lamellae is special. The symmetry of the bulk structure and the planar surface is the same, and consequently the formation of parallel structures does not involve additional elastic chain deformation. The situation for systems forming spheres, cylinders or gyroid in the bulk (the other equilibrium morphologies in $A-B$ systems) confined between planar interfaces is different, and, as a result, the phase behavior is expected to be much more complex. In these systems, a planar surface, regardless of its orientation, always breaks the symmetry of the bulk structure, and the microdomain structure has to adjust. Here, we focus on thin films of cylinder-forming BCP. Experimentally, a variety of deviations from the bulk cylinder structures was observed near surfaces and in thin films, such as a wetting layer,[10] spherical microdomains,[11] a perforated lamellae,[11] cylinders with necks,[12] and more complicated structures.[13,14] A number of early modelling studies showed qualitative agreement, and only in parts.[15] In particular, it remained unclear which of the reported phenomena were specific to the particular system and/or route of film preparation, and which were general behavior.

As systems with non-lamellar bulk structures play an significant role in technological applications,[16] several groups have focussed on an unifying description of the role of confinement, including a determination of key factors for different bulk structures. Here, we primarily focus on the combined computational/experimental contribution of our group to the development of this field, in collaboration with leading experimentalists from Bayreuth University. This approach is truly synergetic: the measurements validated the computational model, and the computational model allowed us to rationalize the thin film behavior away for the surface at which the measurements were performed. Several phenomena in thin films of cylinder-forming BCP were considered: phase behavior in symmetric conditions, in asymmetric conditions, the dynamics of a particular phase transition, and the dynamic of terrace formation. We review the most important computational results and conclusions of these works. The general aim is to give the reader an understanding of static and dynamic aspects of pattern formation in thin films of BCP in general. First, we introduce the particular experimental system for completeness and consider the computational model in more detail.

2. Experimental System and Measurements

The experimental model system is a thin film of cylinder-forming polystyrene-*block*-polybutadiene-*block*-polystyrenen (SBS) triblock copolymer swollen in chloroform vapor, with molecular weights $M_{w,PS} = 14k$, $M_{w,PB} = 73k$ and $M_{w,PS} = 15k$, respectively (PS is polystyrene, PB is polybutadiene). More information about this system, that forms PS cylinders in a PB matrix in the bulk, can be found in Ref. 17; only relevant information is shortly reviewed here. Thin SBS films were spun cast from toluene solution onto polished silicon substrates. The films were annealed by exposing them for 7 h to a controlled partial pressure p of chloroform vapor. The resulting microdomain structures were quenched via fast solvent removal. The nonselective solvent acts as a plasticizer during annealing, and merely induces chain mobility. Within the studied range of $\tilde{p} = p/p_S$ (p_S is the partial pressure of saturnated chloroform vapor at annealing temperature) the lateral spacing between two neighboring PS cylinders, a_0, decreases with increasing $\tilde{p}$ from $a_0 \approx 41$ to 39 nm. Initially the film is flat. Terrace formation takes place during vapor annealing. The mesostructure formation is monitored by TappingMode scanning force microscopy (TM-SFM). Phase images were recorded along with height images to map the

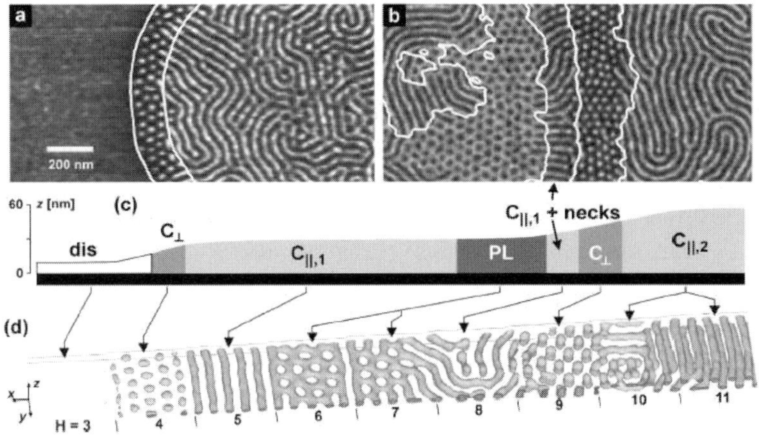

Fig. 1. (a,b)TM-SFM phase images of thin SBS films on Si substrates after annealing in chloroform vapor at $\tilde{p} = 0.62$. The surface is everywhere covered with an ≈ 10 nm thick PB layer. Bright (dark) corresponds to PS (PB) microdomains below this top PB layer. Contour lines calculated from the corresponding height images are superimposed. (c) Schematic height profile of the phase images shown in (a,b). (d) Simulation in *one* simulation box of $352 \times 32 \times H(x)$ dimension, with increasing film thickness $H(x)$, $\varepsilon_{AB} = 6.5$ and $\xi = 6$. The isodensity surface $\theta_A = 0.5$ is shown. Reprinted from[28] with permission. Copyright APS (2002)

lateral distribution of PS (bright) and PB (black) near the film surface, via the difference in modulus.[18] Figures 1(a)-(b) show phase images of two annealed SBS films for $\tilde{p} = 0.62$ with different initial film thickness. Both films have formed regions of well-defined film thickness (terraces) as indicated by the height profile shown in Fig. 1(c). At the same time, well-defined microdomain patterns have formed, systematically changing with the gradually changing film thickness. In particular, boundaries between different structures correspond to height contour lines. Anticipating on the computational results, we denote the structures by a shorthand symbol. Bright stripes, indicative of PS cylinders parallel to the film surface ($C_{||}$), are the majority phase on the two terraces. Earlier work showed that the parallel orientation on the higher terrace continues though the depth of the film.[19,20] The thinnest region of the film does not display any lateral structure at all, indicative of either a disordered (dis) phase or a lamellar wetting (W) layer. Thickness measurement suggest the presence of a wetting layer everywhere in the film, so we conclude that a wetting layer is most probable.[21] On the slopes towards the first and second terrace, hexagonal patterns of bright dots are found, indicative of PS cylinders oriented perpendicular to the surface ($C_\perp$). Hexagonally ordered black dots indicate the formation of a non-bulk structure of PB microdomains in an otherwise continuous PS layer (a perforated PS lamellae, *PL*). Besides dots, and perforated lamellae, stripes are also found on this slope. Since these stripes are found to coexist with dots for a particular film thickness (in between contour lines), we conclude that this structure displays features associated with both parallel and perpendicular cylinders. Modulated cylinders, ie parallel cylinders with undulations or perpendicular necks, possess this property and '$C_{||}$+necks' was therefore assigned to this structure. We refer to separate publications for a detailed discussion of this type of cylindrical shape modulations[39] and the role of these interconnected structures in structure transitions.[46] Since all these phases appear in a single system and under identical experimental conditions, we conclude that local film thickness is an important control parameter. The experiments were repeated for a range of

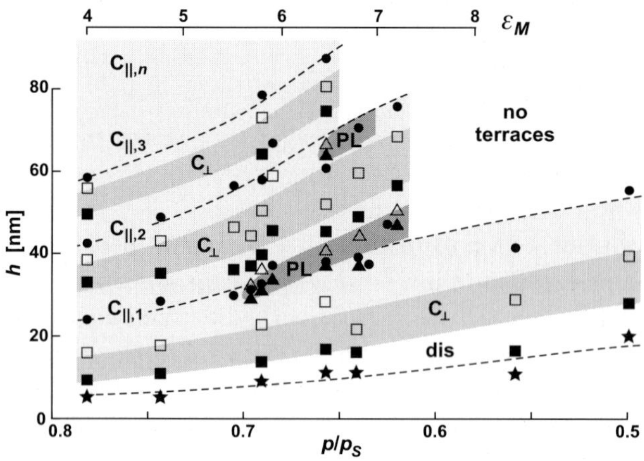

Fig. 2. Structure diagram of thin SBS block copolymer films on Si substrate after annealing in chloroform vapor. Data are given for equilibrium film thicknesses of $C_{||,n}$ (circles) and 'dis' (star), and for the upper and lower bounds, open or closed symbols respectively, of $C_\perp$ (squares) and *PL* (triangles). The latter correspond to contour lines such as those shown in Figs 1(a,b). All lines and areas are drawn to guide the eye. Reprinted from[28] with permission. Copyright APS (2002).

$\tilde{p}$ during the annealing process, and surface morphologies were determined for varying film thickness (after drying, see Fig. 2). With increasing $\tilde{p}$, the total polymer concentration in the film decreases, which has the effect of reducing the strength of the surface field (for a discussion, see the end of next section). From Fig. 2, one can observe that the non-bulk *PL* phase is predominantly observed near the lower terrace, nested between $C_{||}$. With increasing $\tilde{p}$ the total area of the PL phase decreases, and for $\tilde{p} > 0.71$ no PL is observed. For larger thickness, only a small fraction of PL is formed at particular film thicknesses.

3. Computational Model

The dynamic density functional theory (DDFT[22,23]) is used to describe pattern formation on the mesoscale. This method belongs to the class of field-theoretic computer simulations (FTS) methods; for a recent overview, see Ref. 24. This type of approaches were devised to overcome the drawbacks of many microscopic particle-based methods. Except in rare instances, particle-based methods are incapable of equilibrating sufficiently large systems of polymers at realistic densities to extract meaningful information about structure and thermodynamics. The strategy in field approaches involve a number of steps: (1) representation of the BCP by a suitable particle-based model, leading to a reduction of the degrees of freedom to particle coordinates, (2) conversion to a field theory, by replacing the variables in the partition function from particle coordinates to one or more fluctuating potential fields, (3) discretization on a computational grid, and (4) sampling with the proper statistical weights. The Gaussian chain representation in DDFT, step (1), is known to be a sufficiently good model for most (semi)flexible BCP. It is the same coarse-grained molecular model as used in the well-known self-consistent field (SCF) theory. The goal in these methods, the determination of the equilibrium solutions (or fields), requires the localization of (inhomogeneous) saddle points. An powerful approach to solve the SCFT equations (and find the equilibrium structures) are spectral methods that act in the Fourrier domain.[26] In contrast, the DDFT method is a real space strategy. Compared to the spectral methods, it has the advantage that no symmetry is required in advance. At the same time, it has the disadvantage of being rather inefficient for the determination of equilibrium solutions. If the DDFT algorithm reaches equilibrium (the situation where the driving forces vanish, see below for details), the structure represents a solution of the SCF theory. In general, there are a finite number of saddle points. The great advantage of DDFT over SCF theory is that it follows the structural evolution in inhomogeneous BCP systems towards equilibrium based upon a physical model. It is increasingly recognized that many experimental structures are metastable (see, for instance, Ref. 25), and that an understanding of dynamic pathways is crucial. In contrast to locating the saddle point based upon mathematical techniques, the dynamic equations in DDFT describe the collective diffusive dynamics of the BCP system, and are therefore expected to relate directly to the dynamic pathways and saddle points that are relevant in experimental systems. Although one can argue that some physical phenomena, like viscoelastic and hydrodynamic effects and chain entanglements, are not properly captured in this model, we will show that the computational pathways capture the dynamics in experimental thin film system very well (see section 6).

The dynamic equations, describing the time evolution of the relevant field variables $\rho_I(\mathbf{r})$, are given by[27]

$$\frac{\partial \rho_I}{\partial t} = M\nabla \cdot \rho_I \nabla \mu_I + \eta_I \,. \tag{1}$$

Here, the type I denotes different bead types on the coarse-grained chain. Each block copolymer molecule is represented by a set of connected harmonic springs, or a Gaussian chain, with N the total number of beads in the chain, Gaussian bond length a and bead volume ν. We choose $A_3B_{12}A_3$ as a coarse-grained chain representation for the SBS triblock copolymer.[28] The different bead types (A or B) relate to difference in chemical nature in the actual chain; these beads typically represent a number of monomers.[22] The fields ρ_I are the concentration of blocks of type I, which have dimensions of inverse volume; μ_I is chemical potential, M is constant mobility and η_I is thermal noise, which is related to M via the fluctuation-dissipation theorem.[22] We note that the noise term is subject to some debate in the literature[29] and does not extend the method beyond the mean-field approximation. However, fluctuations do not play a large role in the systems considered in this chapter. In the absence of the noise term, the free energy F (see below) is monotonically decreasing; the role of the noise is primarily to cross small barriers in the free energy landscape. We use a fairly simple kinetic model; an extensive description of this and other dynamical models can be found in Refs. 30 and 31. The chemical potential $\mu_I = \delta F/\delta \rho_I$ can be found from the free energy[22,23]

$$\begin{aligned} F[\rho] = &-kT \ln \frac{\Phi^n}{n!} - \sum_I \int_V U_I(\mathbf{r})\rho_I(\mathbf{r})d\mathbf{r} \\ &+ \frac{1}{2}\sum_{I,J} \int_{V^2} \varepsilon_{IJ}(|\mathbf{r}-\mathbf{r}'|)\rho_I(\mathbf{r})\rho_J(\mathbf{r}')d\mathbf{r}d\mathbf{r}' \\ &+ \frac{\kappa\nu^2}{2}\int_V \left(\sum_I \rho_I(\mathbf{r}) - \rho_I^0\right)^2 d\mathbf{r} \,, \end{aligned} \tag{2}$$

for a system with mean-field interactions, occupying a simulation volume V, in the absence of any external fields. This free energy functional does not explicitly depend on time. The time evolution of the system as described by Eq. 1 assumes that the system is in quasi-equilibrium, so that there exists a free energy functional, Eq. 2. On every time instance of the systems evolution, all intra-molecular degrees of freedom are assumed to be equilibrated. This is a good approximation in case that the internal dynamics of a single chain is faster than the collective dynamics of the ensemble of chains. Although the free energy description, Eq. 2, incorporates spatial resolution on the level of beads of chains, the time evolution described by Eq. 1 is coarse grained. The bead-bead interaction potential $\varepsilon_{IJ}(\mathbf{r}-\mathbf{r}')$ is chosen Gaussian and given by

$$\varepsilon_{IJ}(\mathbf{r}-\mathbf{r}') = \varepsilon^0_{IJ}\left(\frac{3}{2\pi a^2}\right)^{\frac{3}{2}} \exp\left[-\frac{3}{2a^2}(\mathbf{r}-\mathbf{r}')^2\right] . \tag{3}$$

Furthermore, in equation (2) k denotes the Boltzmann constant, T is the temperature in Kelvin, n is the number of polymer molecules in the volume V, and Φ is the intra-molecular partition function for ideal polymer chains. The last term is added to keep the system (nearly) incompressible. Exact incompressibility would lead to an extra constraint for the chemical potentials, and slow convergence due to the stiffness of the set of equations (1). Hence, the parameter κ determines the compressibility of the system and ρ^0_I is the mean concentration of the I-block (where the average is taken over the sample volume V). The external potentials U_I and the concentration fields ρ_I are related via the density functional.[22] Confinement is incorporated as stationary 'mask' fields, $\nu\rho_M(\mathbf{r}) = 1$ for $\mathbf{r} \in V^0$, in the simulation volume V ($V^0 \subset V$). The concentration and external potential fields (and integration of these fields) are now restricted to V/V^0. The presence of the walls is accounted for by rigid wall boundary conditions in equation (1), $\mathbf{n} \cdot \nabla\mu_I = 0$, where $\mathbf{n}$ is the normal towards the surface.[23] In the other directions, the standard periodic boundary conditions apply. Moreover, the energetic interaction of the BCP with the wall is accounted for by an extra term in the free energy

$$\sum_I \int_{V^2} \varepsilon_{IM}(|\mathbf{r}-\mathbf{r}'|)\rho_I(\mathbf{r})\rho_M(\mathbf{r}')d\mathbf{r}d\mathbf{r} , \tag{4}$$

similar to the third term for the cohesive interaction in equation (2).

The (continuous) dynamic equations are numerically integrated on a regular lattice by a Crank-Nicholson scheme, and specially developed finite-difference stencils.[32] The spatial resolution is limited to the grid distance Δx, which is the base unit for all distances in the remainder. Prior to the implementation, all variables were made dimensionless, and only the most important choices are briefly reviewed here. From numerical considerations, the time step $\Delta\tau = kTM\Delta t\Delta x^{-2}$ should be taken equal or smaller than 0.5 to ensure stability of the numerical scheme.[22] We chose a slightly larger timestep, $\Delta\tau = 0.73$, but still small enough for stability. Since t and M enter in the simulation only as a product, all simulation can be done without specifying either of them explicitly. As a result, the timescale for the structure evolution can be determined from the experimentally measurable mobility. Here, we focus on an alternative route, were the time scales are derived from a direct comparison of pronounced stages in the experimental and computational pathways. We make the relevant variables, the concentration fields, dimensionless by multiplication with the smallest volume ν, and follow the evolution of $\theta_I(\mathbf{r}) = \nu\rho_I(\mathbf{r})$. In 3-D, the microstructure can be determined from isodensity surfaces, which are computed by connecting all positions in space with the same dimensionless concentration (isodensity) value. The energetic mean-field interactions are determined by the weights ε^0_{IJ} (bead-bead) and ε^0_{IM}(bead-surface) in the Gaussian kernels in Eqs. 2 and 4. The dimension of these parameters is kJ/mol, as in all our previous works,[23] and we omit zero-superscripts in the following for simplicity. These parameters are directly related to the more familiar dimensionless Flory-Huggins (FH) interaction parameter $\chi = \varepsilon/(N_A kT)$, where ε in J/mol and N_A is Avogadro number.[22] Both the BCP architecture and length ($A_3B_{12}A_3$, $N = 18$) and this FH interaction parameter are important, as they determine the equilibrium morphology. We previously discussed in detail how these parameters can be obtained for a three-component system of Pluronics 64 in an aqueous environment.[22] Although solvent is present in the experimental SBS system, it is non-selective. Theoretical considerations show that the solvent can be incorporated only implicitly,[33] by considering effective interaction parameters $\varepsilon = \Phi\tilde{\varepsilon}$ (where Φ is a function of the polymer concentration, and tilde denotes the value in the melt). In our earlier study,[34] this effective $A - B$ interaction parameter for an A_3B_6 system was obtained as $\chi = 2$ ($T = 413K$), by fitting the order-disorder transition temperature of an experimental SBS triblock copolymer.[35] For the particular experimental SBS triblock copolymer used here, we determined this effective interaction to be $\varepsilon_{AB} = 6.5$ kJ/mol or $\chi = 1.9$ ($T = 413$). This value was determined by matching the phase sequences for increasing thickness, and used in all later studies.[28] It gives rise to a hexagonally packed cylindrical phase in the bulk.[36] The slit calculations have been performed in volumes V of dimension $L \times L \times (H + 1)$, with H the thickness of the film. The volume is always chosen large enough to avoid strong boundary effects on the microstructure and small enough to ensure numerical efficiency, needed to do systematic analysis. The slit surfaces were represented as planes of one layer thickness, one of which is positioned parallel to the $L \times L$ faces of the volume, at the volume boundary (the other one is automatic, due to the periodic boundary conditions). Only in the computational system of Fig. 1(d), with dimensions $352 \times 32 \times H(x)$, the film thickness H was not taken constant throughout the simulation volume. Instead, we imposed a laterally varying film thickness $H(x)$ similar (but not equal) to the profile observed in the experimental terrace-forming thin film (see Fig. 1(c)). In particular, the profile in the computational system is stepwise, and the film thickness is incremented (from $H = 3$ to $H = 11$, see Fig. 1(d)) at regular intervals along the x-direction (each 44 gridpoints). The surface field strength is quantized by the value of ε_{AM} and ε_{BM}. We consider the difference $\xi = \varepsilon_{AM} - \varepsilon_{BM}$, since only the difference counts in the calculation of the chemical potential.[37] For all simulations we set $\varepsilon_{BM} = 0$. The sign of ξ shows the preferential attraction of A-beads (negative) or B-beads (positive) to the surface. Finally we note that the experimental structures for a decreased polymer concentration (increasing $\tilde{p}$) should be compared to computational structures for a lower effective surface interaction ξ (or weaker surface field), following the reasoning above. This experimental condition will also affect the effective ε_{AB}, the bead-bead interactions, that was determined for $\tilde{p} = 0.62$. However, we will disregard this effect for a small range around this value, since the experimental diagram in Fig. 2 suggests that we remain in the cylinder-forming regime (see Ref. 36 for a more extensive discussion).

4. Symmetric Conditions

Here, we consider the computational systems where the surface field, ξ, is equal at both surfaces. Focussing on the experimental thin film behavior (Fig. 1(a)-(c)) once more, we note that the thickness gradient is actually rather modest: the height of the step between the first and second terrace is approximately 30 nm, while these terraces are laterally 1-2 μm apart.

Consequently, the film can be considered almost flat on the scale of the microdomain structures. Combined with the fact that the interfaces between different structures coincide very well with the contour lines derived from height measurements, our computational focus lies on slits with fixed thickness H (and periodic boundary conditions in other directions, $L = 32$), and the relation between microstructure and two important system parameters in particular: the slit thickness H and surface interaction ξ. We have systematically varied these parameters in a broad range. The computational H and experimentally measured film thickness can be directly related by matching the microdomain distances c_0, which is the relevant length scale in the systems (see for a matching of ξ, lateron in this section). To be complete, we have also studied the role of the interfaces between different structures on the thermodynamic behavior. Starting with the latter, we first consider the result of DDFT in a simulation volume where the surface interactions are appropriately chosen, and a thickness gradient is imposed (wedge-shaped simulation volume V, Fig. 1(d), $\xi = 6$). We note that we have chosen a *single* wedge shaped simulation volume and that the microstructure is not compiled from structures obtained by separate simulations; more details about the simulation setup can be found in the previous paragraph. With varying film thickness H, one can clearly observe the same sequence of phases as in the experimental data (Fig. 1(a)-(b)): a disordered layer ($H = 3$), short upright cylinders $C_\perp$ ($H = 4$), parallel cylinders $C_{||,1}$ ($H = 5$), perforated lamellae PL ($H = 6$), coexistence of perforated lamellae and parallel cylinders $PL/C_{||}$ ($H = 7$), cylinders with necks $C_{||}C_\perp$ ($H = 8$), $C_\perp$ ($H = 9$) and two layers of parallel cylinders $C_{||,2}$ ($H = 10 - 12$). First, we note that the film thickness associated with different structures is correctly predicted. Also the spacing of the perforation in PL is 1.15 times larger than the spacing a_0 ($= 7.0 \pm 0.5$) between cylinders, similar to the experiments. The calculated structure diagram in Fig. 3, showing final microstructure types in simulation volumes with constant ξ and H, is not necessarily a phase diagram. In contrast to SCF theory, structures calculated with DDFT do not only represent equilibrium states; they may be kinetically trapped metastable states, like in experiments. For the determination of the final structure, we have used the requirement that the free energy (2) does not vary with time any longer. For a quantitative comparison of the diagrams in Figs. 2 and 3, we have to take into account that the film thickness on the vertical axis in Fig. 2 was measured in the dried state (hereafter denoted by h^{dry}), whereas structure formation takes place in the swollen state. We can estimate the thickness h^{wet} in the swollen film

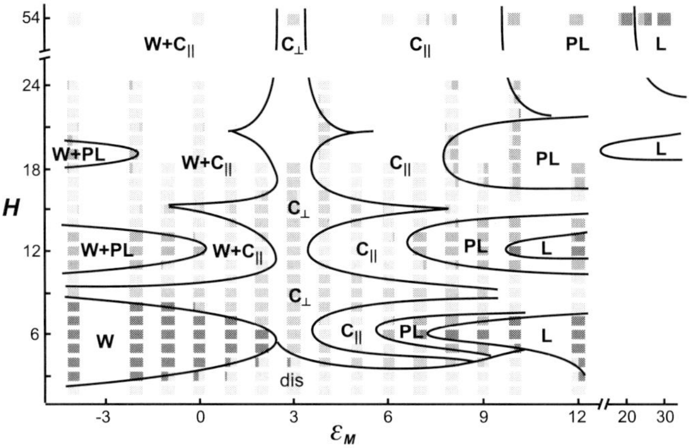

Fig. 3. Diagram of surface structures of a $A_3B_{12}A_3$ block copolymer thin film calculated by DDFT for $\varepsilon_{AB} = 6.5$. The shades of grey in the boxes indicate that two phases were found to coexist after finite simulation time. Smooth phase boundaries have been drawn to guide the eye. Reproduced from[28] with permission. Copyright APS (2002).

from the particular observed film structure by simple geometrical means (for perfect $C_{||,2}$, the thickness $h^{wet} = a_0\sqrt{3}$). This procedure has the effect of lining up the phase boundaries to the (almost) horizontal position. For the horizontal axis, we previously assumed a simple qualitative relation between vapor pressure $\tilde{p}$ and surface field strength ξ via the polymer concentration Φ. Since the exact relation is unknown, we map both diagrams, by fixing $\xi = 6$ to $\tilde{p} = 0.68$ and using linear scaling around this value (see axis in Fig. 2). We conclude that there is a striking agreement between measured and calculated diagrams of intriguing complexity. In particular, several important features, such as the onset of the PL phase for different thickness, are well reproduced. In the remainder of this section, we therefore use the computational model to distinguish between the effects of two constraints that are simultaneously present in thin films: the surface field and the film thickness. For very thick films ($H = 54$, see Fig. 3) the effect of the surface field is most prominent. While the structure in the bulk of the film is (parallel) cylindrical, independent of the surface field, this field is sufficient to induce considerable rearrangements of microdomains near the surface, i.e. *surface reconstructions.* When there is a preferential attraction of the A-block to the surface ($\xi < 2$), a thin A-wetting layer (W) forms, which may be seen as half a lamella ($L^{\frac{1}{2}}$). In contrast to confined lamella-forming systems, a A-wetting layer is also formed for a vanishing and slightly repulsive surface field. The cause of this behavior is the asymmetry of the molecule; chain conformations with the shortest part of the chain towards the surface are entropically favored. The cross-over point where the entropic contributions and A repulsion are (nearly) balanced depends on the asymmetry, and for this chain found at $\xi \approx 3$. The result is $C_{\perp}$ (chain conformations parallel to the surface). For larger ξ, the repulsion of the A-block dominates, and the B block is preferential at the surface. In the experimental thin film, the difference in interfacial energy between PS and PB is like in this situation, and dictates PB surface-coverage. Next to the B-wetting layer, cylinders of A align parallel to the surface ($C_{||}$). When ξ is further increased, surface reconstructions with noncylindrical microdomains are induced: first as PL, then a lamella (L). In these structures, the averaged mean curvature is gradually decreased to adopt to the planar symmetry of the surface. On a chain level, the formation of these surface reconstructions requires an increased elastic chain deformation (in the layer closest to the surface) with increasing surface field. Since only the layer next to the surface forms surface reconstructions, the *range* of the surface field is limited. Our results indicate that surface fields extend into the bulk with a decay length of about one microdomain spacing, and that they are additive. The additivity explains why weaker surface fields are sufficient to form a PL (or L) for thin films, compared to thick films. It also explains the formation of PL in combination with a wetting layer ($W+PL$). Similar to the situation in lamellae-forming systems, *confinement effects* modulate the stability regions of the parallel phases ($C_{||}$, PL and L). The natural domain distance for hexagonally packed cylinders, $c_0 = a_0\sqrt{3}/2 \approx 6$, dictates these commensurability effects, and a thickness $H = nc_0$ (n an integer) is energetically favored. This causes easier deformable cylindrical phases to occur at intermediate film thicknesses. For very small H ($\leq c_0$) and weak surface fields, confinement prevents microphase separation and stabilizes a disordered phase. Such a disordered phase has been reported for ultrathin films of lamella-forming diblock copolymers.[38]

5. Asymmetric Conditions

In most supported films, the surface field, ξ, at the air and the substrate are different. Also in our experimental system (section 2), PB is preferential at the air surface, while the substrate is covered by a wetting layer of PS. This effect was considered in detail.[39] Here, we focus on the main conclusions. For thick films, the situation is similar to the previous section, although instead of only the same structure also *different* structures can form in the layer next to the surfaces (in the bulk of the film, again parallel cylinders are formed). For thin films, the asymmetry may lead to different structures or orientations at each of the surfaces, that should connect in a stable grain boundary somewhere in the bulk of the film. For thin films of lamella-forming BCP, it was already shown that asymmetric surface fields can give rise to hybrid structures[38] when the two surfaces favor different lamellar orientations. We found that cylinders can only form a connected structure that is conceptually similar to this hybrid phase in lamella-forming systems: cylinders with necks ($C_{||}C_{\perp}$). Moreover, also distinct modulations of the cylindrical *shape* were observed, indicating once again that the (bulk) cylindrical phase is easy deformable compared to the induced phases. All other observed hybrid phases were combinations of *undistorted* structures, W, $C_{\perp}$, $C_{||}$, PL or L (in the order of increasing surface field), in parallel layers separated by a B-layer, or combinations of these phases and cylinders with necks. *Confinement effects* modulate the stability regions of parallel phases as before. However, the wetting (W) layer plays a special role. When a A-wetting layer is present only at one side (always of thickness $c_0/2$), it effectively reduces the available space for the other structures in the film to a more energetically favored thickness. Wetting layers were not only found for stronger surface fields than before (larger than $\xi \approx 2$, see Fig. 3), they can be structured, in contrast to the featureless W for similar substrates. This structured wetting layer was denoted as a half structure, and could be any of the undistorted structures. The combinations with full structures in the remaining part of the film are only restricted by impossible chain conformations. Finally, the experimentally determined wetting layer at the substrate explains the good match in the previous section. When a wetting layer is present, the reduced film is effectively screened from the substrate, and experiences an effective surface field at the side where the wetting layer is located.[37] For a featureless A-wetting layer, this surface field can be expected to be (almost) equal to the $A - B$ interaction ($\epsilon_{AB} = 6.5$). If we assume that the surface field on the air side is well parametrized ($\xi = 6$), this gives rise to almost symmetric conditions. Since the effective surface field depends on the structure of the wetting layer, we concluded that the structure formation in films with dissimilar substrates is the result of a complex interplay between confinement effects, actual surface field strength and (partial) screening.

6. Dynamics of a Phase Transition

Although the static results were encouraging, the unique property of DDFT, the kinetic model, was not fully exploited. The unique experimental setup of our collaborators (*in situ* tapping mode scanning force microscopy TM-SFM) allowed for direct imaging, in real time and space, of the microstructure dynamics during annealing. The results indicated that the influence of the SFM tip on the structure dynamics was negligible. We concentrated on the dynamics of a $C_{||} \rightarrow PL$ phase transition that takes place during the formation of the lower terrace (see the images of static structures, Fig. 1(a)-(b)). To start with, a film of incommensurate thickness was spun cast on a silicon substrate. Since microphase separation takes place on a much shorter time scale than terrace formation, initially a surface structure of white (PS) dots and rather flat film is observed. Assuming symmetric conditions due to the presence of a wetting layer (screening, see the previous section), the associated 3-D film structure for this incommensurate film thickness is perpendicular cylinders ($C_{\perp}$) (see Fig. 3). At later stages, the unfavorable film thickness gives rise to the formation of terraces of commensurate thickness, by a complex interplay of microstructure transitions and mass transport at a larger scale. With decreasing film thickness, a *bright dots* $\rightarrow$ *bright stripes* $\rightarrow$ *hexagonally-ordered dark spots* transition was observed, which can be related to a $C_{\perp} \rightarrow C_{||} \rightarrow PL$ transition (see the supplementary movies in Ref. 40 for details). Apparently, the thermodynamic driving force for the $C_{||} \rightarrow PL$ transition is rather small, since both structures are experimentally found on the first terrace, and height measurements indicate that the transition is associated with only a very small change in film thickness (compared to the $C_{\perp} \rightarrow C_{||}$ transition). We concluded that we can investigate the elementary steps in the $C_{||} \rightarrow PL$ transition by means of a simulation with dissimilar substrates for constant film thickness H and large lateral dimensions ($L = 128$). We choose a commensurate $H = 10$, and cross the $C_{||}/PL$ boundary by a small change in the surface field ξ_2 at the upper surface ($\xi_1 = -1$, giving rise to a wetting layer W of thickness $c_0/2 = 3$). In particular, starting from a homogeneous mixture the simulated system is first quenched into a region where $W + C_{||}$ ($\xi_2 = 5.5$) is stable and subsequently into a region where $W + PL$ ($\xi_2 = 6.5$) is stable (see Fig. 3, we note that the effective $\xi_1 \approx 6.5$ due to screening). We found that the simulations match the experimental data in great detail. In the nucleation and growth stage, we observe the same structural defects in the $C_{||}$ phase, such as cylinder-ends and three-armed connections, in the

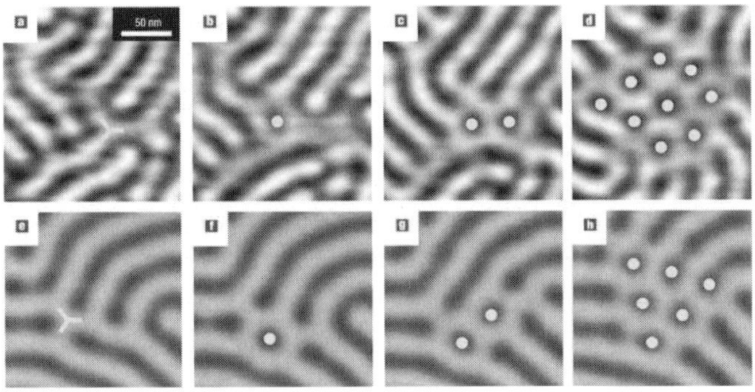

Fig. 4. Nucleation and growth. White dots indicate perforations and the star (in **a** and **e**) denotes a three-armed defect in the $C_{||}$ phase. **a-d**: snapshots of TM-SFM after 237, 267, 333 and 480 min of annealing. **e-h** corresponding snapshots from a large-scale simulation movie after 8350, 10900, 17000 and 21650 time steps. Reprinted from[40] with permission. Copyright Nature (2004).

experiments and simulations (see Fig. 4). The connections always serve as nucleation centres for the PL phase, which can be considered as a regular network of three-armed branching points. A single defect is sufficient to create a nucleus. First, three-armed connections cluster and form a single ring, which becomes the minimal nucleus of the PL phase. Near the isolated ring, additional connections between cylinders form, leading to a doubling of the perforation number. This cluster continues to grow at its border. The elementary process of this growth is the movement of kinks parallel to the $C_{||}/PL$ border. The existence of sharp borders is a clear indication that the $C_{||} \rightarrow PL$ is a first-order transition. Kink movement is similar to the zippering transition that was proposed for the $C \rightarrow L$ transition in bulk specimens.[41] The extend of the growth stage depends on the number of nucleation centres. As grains with differently oriented PL touch, only small grains of $C_{||}$ remain. The annihilation stage (Fig. 5) proceeds by movement of kinks at the phase borders, and the consecutive formation of rings. The decrease of the area A of the $C_{||}$ domain in time is shown in Fig. 6, and clearly displays the stepwise and stochastic nature of the process, with the formation of connections as the elementary step. This quantitative measure shows that the kinetics in the simulations and the experiments match very well, and that one time step in the simulations corresponds to about 6 seconds of real time. Moreover, we used the decay rate to estimate the interfacial tension between the two phases and found a very small value of $\sigma \approx 2.5 \mu N m^{-1}$,[40] showing that the thermodynamic driving force for this

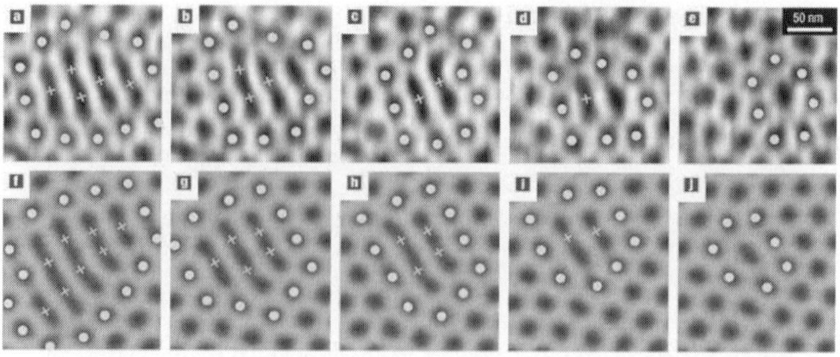

Fig. 5. Annihilation. Dots indicate perforations at the $PL/C_{||}$ boundary, crosses mark spots where connections between microdomains form with time. **a-e**: snapshots of TM-SFM after 538, 556, 607, 621 and 629 min of annealing. **e-h** corresponding snapshots from a large-scale simulation movie after 22300, 24000, 24600, 25100 and 25300 time steps. Reprinted from[40] with permission. Copyright Nature (2004).

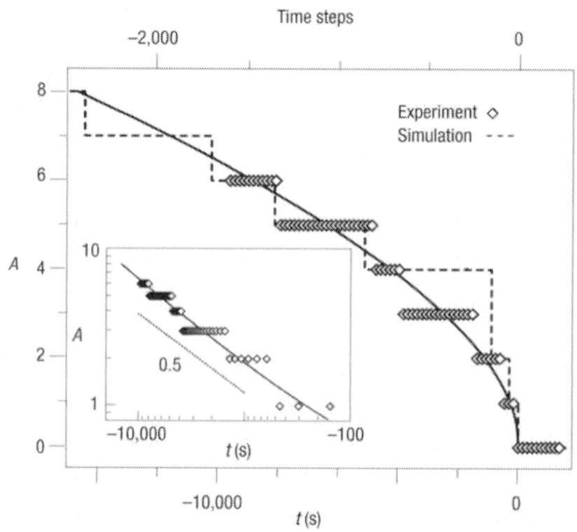

Fig. 6. Time evolution of the surface area A of the $C_{||}$ grain in Fig. 5. The area is estimated by the number of potential connections, marked by crosses. At $t = 0$, the $C_{||}$ grain disappears. The solid line is a fit to obtain the surface tension. The inset shows the experimental data and the fit on a logarithmic scale. Reprinted from[40] with permission. Copright Nature (2004).

transition is indeed small. This small surface tension σ, together with the large viscosity of the BCP explains the difficulty of annealing defects in BCP melts and solutions, and supports the characterization of order fluids as soft condensed matter.

7. Conclusion and Outlook

We have given a short introduction to the DDFT methodology, and showed how this method was used to describe and understand complex pattern formation and pattern formation kinetics in thin BCP films. Various other methods, for instance coarse-grained Monte Carlo,[42] have been used to study confined BCP systems in various conditions (including patterned surfaces), and we refer to the reviews and other chapters in this book for details. The conclusions drawn from our own studies are general, since the validated model is based on a coarse-grained chain representation and mean-field interactions. We have restricted ourselves to BCP that form hexagonally-packed cylinders in the bulk, and in particular found that surface reconstructions, PL and L, can be induced by strong surface fields. Few studies have considered the effect of confinement for other non-lamellar $A-B$ structures: gyroid (G) and spheres (S). We note that they may be anticipated from our results, since PL directly relates to a confined G structure and the $S \to C$ transition was already observed in combination with other external fields.[43] Although our kinetic model is simple and local, and does not include hydrodynamics, the results of the previous section suggest that the model is sufficient. This is a remarkable finding, as more detailed simulation methods, such as molecular dynamics, hardly cover time and length scales of 10 nm and 1 μs. We have shown here that mesoscopic models (like DDFT) can realistically model processes covering multiple length (10 nm to 1 μm) and time (1 s to 10 h) scales. We note that, because of the high viscosity in most BCP systems, one could have anticipated that hydrodynamics does not play an important role. Moreover, chain entanglements play a role in experimental systems, depending on the BCP length, but our results indicate that our particular model for chain diffusion only differs by a constant factor, which is hidden in the mobility M. In particular, we neglegt kinetic coupling between different beads by our choice of the Onsager coefficients, and consequently the bead mobility does not depend on the polymer length N. In more elaborate models for chain dynamics, for instance Rouse or reptation, the total friction experienced by a diffusing chain depends on the polymer length, and the bead mobility scales as N^{-1} and N^{-2}, respectively.[44] Surface relief (the formation of terraces) has been considered with DDFT,[45,46] but is not discussed in detail here. Besides showing a good match between experiments and simulations, including a corroboration of the time scaling found in Ref. 40 of one time step $\approx$ seconds, the studies indicate that connected structures $C_{||}C_{\perp}$ and PL play an important role in mediating phase transitions. Future research on thin films may focus on more complex morphologies in confinement (multiblock and star BCP), selective solvent, combinations with other external fields, and BCP/nanoparticle mixtures.

Acknowledgments

The collaboration between Bayreuth and Leiden University was supported for three years (2001-2004) by the NWO-DFG bilateral program. Computations were partially performed on supercomputers of SARA in Amsterdam, using time granted by Stichting Nationale Rekenfaciliteiten (NCF). GJAS acknowledges that the research reviewed in this chapter is the result of close collaboration, with K. S. Lyakhova and A.V. Zvelindovsky (DDFT, Leiden) and A. Horvat, R. Magerle, A. Knoll, G. Krausch and L. Tsarkova (application and experiments, Bayreuth).

References

1. G. Holden, N. R. Legge, R. P. Quirk, H. E. Schoeder, *Thermoplastic elastomers.* (Hanser/Gardner Publications, Cincinnati, 1996).
2. F. S. Bates, Network phases in block copolymer melts, *MRS Bull.* **30**, 525-532, (2005).
3. R. R. Netz, D. Andelman, M. Schick, Interfaces of modulated phases, *Phys. Rev. Lett.* **79**, 1058-1061, (1997).
4. H. Hasegawa, T. Hashimoto, Morphology of block copolymers near a free surface, *Macromolecules* **18**, 589-590, (1985).
5. S. H. Anastasiadis, T. P. Russell, S. K. Satija, C. F. Majkrzak, Neutron reflectivity studies of the surface-induced ordering of diblock copolymer films, *Phys. Rev. Lett.* **62**, 1852-1855, (1989).
6. G. H. Fredrickson, Surface ordering phenomena in block copolymer melts, *Macromolecules* **20**, 2535-2542, (1987).
7. M. W. Matsen, Self-assembly of block copolymers in thin films, *Curr. Opin. Colloid Interface Sci.* **3**, 40-47, (1998).
8. K. Binder, Phase transitions of polymer blends and block copolymer melts in thin films, *Adv. Polym. Sci.* **138**, 1-89, (1999).
9. M. J. Fasolka, A. M. Mayes, Block copolymer thin films: physics and applications, *Annu. Rev. Mater. Res.* **31**, 323-355, (2001).
10. A. Karim, N. Singh, M. Sikka, F. S. Bates, W. D. Dozier, G. P. Felcher, Ordering in asymmetric poly(ethylene-propylene)-poly(ethylethylene) diblock copolymer thin-films, *J. Chem. Phys.* **100**, 1620-1629, (1994).
11. L. H. Radzilowski, B. L. Carvalho, E. L. Thomas, Structure of minimum thickness and terraced free-standing films of block copolymers, *J. Polym. Sci. Part B Polym. Phys.* **34**, 3081-3093, (1996).

12. M. Konrad, A. Knoll, G. Krausch, R. Magerle, Volume imaging of an ultrathin SBS triblock copolymer film, *Macromolecules* **33**, 5518-5523, (2000).
13. Ch. Harrison, M. Park, P. Chaikin, R. A. Register, D. H. Adamson, N. Yao, Depth profiling block copolymer microdomains, *Macromolecules* **31**, 2185-2189, (1998).
14. Q. Zhang, O. K. C. Tsui, B. Y. Du, F. J. Zhang, T. Tang, T. B. He, Observation of inverted phases in poly(styrene-b-butadiene-b-styrene) triblock copolymer by solvent-induced order-disorder phase transition, *Macromolecules* **33**, 9561-9567, (2000).
15. Q. Wang, P. F. Nealy, J. J. de Pablo, Monte Carlo simulations of asymmetric diblock copolymer thin films confined between two homogeneous surfaces, *Macromolecules* **34**, 3458-3470, (2001).
16. C. J. Hawker, T. P. Russell, Block copolymer lithography: merging "bottom-up" with "top-down" processes, *MRS Bull.* **30**, 952-966, (2005).
17. A. Knoll, R. Magerle, G. Krausch, Tapping mode atomic force microscopy on polymers: where is the true sample surface, *Macromolecules* **34**, 4159-4165, (2001).
18. S. N. Magonov, J. Cleveland, V. Elings, D. Denley, M.H. Whangbo, Tapping-mode atomic force microscopy study of the near-surface composition of a styrene-butadiene-styrene triblock copolymer film, *Surf. Sci.* **389**, 201-211, (1997).
19. G. Kim, M. Libera, Morphological development in solvent-cast polystyrene-polybutadiene-polystyrene (SBS) triblock copolymer thin films, *Macromolecules* **31**, 2569-2577, (1998).
20. R. Magerle, Nanotomography, *Phys. Rev. Lett.* **85**, 2749-2752, (2000).
21. A. Knoll, R. Magerle, G. Krausch, Phase behavior in thin films of cylinder-forming ABA block copolymers: Experiments, *J. Chem. Phys.* **120**, 1105-1116, (2004).
22. B. A. C. van Vlimmeren, N. M. Maurits, A. V. Zvelindovsky, G. J. A. Sevink, J. G. E. M. Fraaije, Simulation of 3D mesoscale structure formation in concentrated aqueous solution of the triblock polymer surfactants (ethylene oxide)(13)(propylene oxide)(30)(ethylene oxide)(13) and (propylene oxide)(19)(ethylene oxide)(33)(propylene oxide)(19). Application of dynamic mean-field density functional theory, *Macromolecules* **32**, 646-656, (1999).
23. G. J. A. Sevink, A. V. Zvelindovsky, B. A. C. van Vlimmeren, N. M. Maurits, J. G. E. M. Fraaije, Dynamics of surface directed mesophase formation in block copolymer melts, *J. Chem. Phys.* **110**, 2250-2256, (1999).
24. G. H. Fredrickson, V. Ganesan, F. Drolet, Field-theoretic computer simulation methods for polymers and complex fluids, *Macromolecules* **35**, 16-39, (2002).
25. H. Huang, F. Zhang, Z. Hu. B. Du, T. He, F. K. Lee, Y. Wang, O. K. C. Tsui, Study on the origin of inverted phase in drying solution-cast block copolymer films, *Macromolecules* **36**, 4084-4092, (2003).
26. M. W. Matsen, M. Schick, Stable and unstable phases of a diblock copolymer melt, *Phys. Rev. Lett.* **72**, 2660-2663, (1994).
27. J. G. E. M. Fraaije, B. A. C. van Vlimmeren, N. M. Maurits, M. Postma,

O. A. Evers, C. Hoffmann, P. Altevogt, G. Goldbeck-Wood, The dynamic mean-field density functional method and its application to the mesoscopic dynamics of quenched block copolymer melts, *J. Chem. Phys.* **106**, 4260-4269, (1997).

28. A. Knoll, A. Horvat, K. S. Lyakhova, G. Krausch, G. J. A. Sevink, A. V. Zvelindovsky, R. Magerle, Phase behavior in thin films of cylinder-forming block copolymers, *Phys. Rev. Lett.* **89**, 035501, (2002).
29. A. J. Archer, M. Rauscher, Dynamical density functional theory for interacting Brownian particles: stochastic or deterministic?, *J. Phys. A-Math&Gen.* **37**, 9325-9333, (2004).
30. A. Onuki, *Phase transition dynamics.* (Cambridge University Press, Cambridge, 2002).
31. T. Kawakatsu, Effects of changes in the chain conformation on the kinetics of order-disorder transitions in block copolymer melts, *Phys. Rev. E* **56**, 3240-3250, (1997).
32. N. M. Maurits, J. G. E. M. Fraaije, P. Altevogt, O. A. Evers, Simple numerical quadrature rules for Gaussian chain polymer density functional calculations in 3D and implementation on parallel platforms, *Comp&Theor. Pol. Sci.* **6**, 1-8, (1996).
33. G. H. Fredrickson, L. Leibler, Theory of block copolymer solutions - nonselective good solvents, *Macromolecules* **22**, 1238-1250, (1989).
34. H. P. Huinink, J. C. M. Brokken-Zijp, M. A. van Dijk, G. J. A. Sevink, Asymmetric block copolymers confined in a thin film, *J. Chem. Phys.* **112**, 2452-2462, (2000).
35. C. D. Han, J. Kim, J. K. Kim, Determination of the order-disorder transition-temperature of block copolymers, *Macromolecules* **22**, 383-394, (1989).
36. A. Horvat, K. S. Lyakhova, G. J. A. Sevink, A. V. Zvelindovsky, R. Magerle, Phase behavior in thin films of cylinder-forming ABA block copolymers: mesoscale modeling, *J. Chem. Phys.* **120**, 1117-1126, (2004).
37. H. P. Huinink, M. A. van Dijk, J. C. M. Brokken-Zijp, G. J. A. Sevink, Surface-induced transitions in thin films of asymmetric diblock copolymers, *Macromolecules* **34**, 5325-5330, (2001).
38. M. J. Fasolka, P. Banerjee, A. M. Mayes, G. Pickett, A. C. Balazs, Morphology of ultrathin supported diblock copolymer films: theory and experiment, *Macromolecules* **33**, 5702-5712, (2000).
39. K. S. Lyakhova, G. J. A. Sevink, A. V. Zvelindovsky, A. Horvat, R. Magerle, Role of dissimilar interfaces in thin films of cylinder-forming block copolymers, *J. Chem. Phys.* **120**, 1127-1137, (2004).
40. A. Knoll, K. S. Lyakhova, A. Horvat, G. Krausch, G. J. A. Sevink, A. V. Zvelindovsky, R. Magerle, Direct imaging and mesoscale modelling of phase transitions in a nanostructured fluid, *Nature Mat.*, 886-890, (2004).
41. S. Sakurai, T. Morii, K. Taie, M. Shibayama, S. Nomura, T. Hashimoto, Morphology transition from cylindrical to lamellar microdomains of block copolymers, *Macromolecules* **26**, 485-491, (1993).
42. Q. Wang, *In: Nanostructured Soft Matter: Experiment, Theory, Simulation and Perspectives*, 498-528, (2007).

43. T. Xu, A. V. Zvelindovsky, G. J. A. Sevink, O. Gang, B. Ocko, Y. Q. Zhu, S. P. Gido, T. P. Russell, Electric field induced sphere-to-cylinder transition in diblock copolymer thin films, *Macromolecules* **37**, 6980-6984, (2004).
44. N. M. Maurits, J. G. E. M. Fraaije, Mesoscopic dynamics of copolymer melts: from density dynamics to external potential dynamics using nonlocal kinetic coupling, *J. Chem. Phys.* **107**, 5879-5889, (1997).
45. K. S. Lyakhova, A. Horvat, A. V. Zvelindovsky, G. J. A. Sevink, Dynamics of terrace formation in a nanostructured thin block copolymer film, *Langmuir* **22**, 5848-5855, (2006).
46. A. Horvat, A. Knoll, G. Krausch, L. Tsarkova, K. S. Lyakhova, G. J. A. Sevink, A. V. Zvelindovsky, R. Magerle, Time evolution of surface relief structures in thin block copolymer films, *Macromolecules* **40**, 6930-6939, (2007).

CHAPTER 4

BLOCK COPOLYMER LITHOGRAPHY FOR MAGNETIC DEVICE FABRICATION

J. Y. Cheng[*] and C. A. Ross[†]

[*]*IBM Almaden Research Center*
San Jose, CA
E-mail: chengjo@us.ibm.com

[†]*Massachusetts Institute of Technology*
Department of Materials Science and Engineering
Cambridge, MA 02139
E-mail: caross@mit.edu

Block copolymers have been used in a number of lithographic applications including the fabrication of magnetic nanostructures, which are of particular interest in data storage. This chapter reviews the materials and processing requirements for block copolymer lithography, and then describes pattern transfer from block copolymer films into magnetic materials using both additive and subtractive methods, and the magnetic properties of the resulting structures. To create structures with long-range order, templating of the block copolymer microphase separation is necessary, and the strategies for accomplishing this will be reviewed with emphasis on the use of topographically modulated templates. The chapter concludes with a summary and outlook for block copolymer patterning of nanomagnetic devices.

1. Introduction

The self-assembling nature of block copolymers makes them ideal for the generation of periodic patterns useful in nanolithography.[1-11] Conventional lithography involves exposing a resist layer with photons followed by a development step, and the feature size is therefore limited by diffraction, making the generation of sub-50 nm features difficult. To create smaller features, electron-beam lithography is commonly used,[12] but the serial nature of the process makes it expensive to pattern large areas. A number of other nanoscale patterning processes have been developed, including nanoimprint lithography,[13-15] probe-based methods such as dip-pen nanolithography,[16] interference lithography,[17,18] x-ray lithography,[19] and block copolymer lithography. The latter process has received considerable attention over the last decade because of its ability to pattern periodic features over large areas, and several reviews on this subject have been published.[1-11] In this chapter we will describe block copolymer lithography with particular reference to the formation of magnetic nanostructures and magnetic devices, in order to illustrate the processes of pattern generation and pattern transfer. We will conclude with a perspective on the application of block copolymer lithography in device fabrication.

The overwhelming majority of block copolymer (BCP) lithography research has been carried out using thin films of diblock copolymers such as poly(styrene-b-methylmethacrylate) (PS-PMMA), though triblock copolymers are occasionally used, offering the possibilities of more complex pattern geometries. When spin coated from solution to form a thin film on a substrate, diblock copolymers form the familiar spherical, cylindrical or lamellar morphologies depending on the volume fraction of the two blocks.[20] In PS-PMMA, for example, if the PS and PMMA blocks have similar volume fractions, the polymer forms lamellar domains of alternating PS and PMMA, and if the PMMA block is between 21 and 33% by volume, the structure consists of parallel cylinders of PMMA in a PS matrix. If the PMMA volume is below about 21%, the structure consists of a body centered cubic arrangement of PMMA spheres in a PS matrix. For lithography applications, thin films of BCPs can be processed similarly to a resist material. They can be spin-cast, baked, and etched or stained. Removal of one block leaves a pattern that can be used as a mask for subtractive patterning (e.g., etching) or additive patterning (e.g., liftoff or electrodeposition) of a functional material, in combination with other planar processing methods. This compatibility with semiconductor tooling and processing makes BCP lithography attractive for as an integrated part for nanofabrication.

The geometry of the patterns that can be created by BCP lithography depends on the morphology and orientation of the BCP thin films.[21] Arrays of parallel lines can be made from a lamellar BCP with the lamellae standing perpendicular to the substrate surface, or from a cylindrical morphology BCP with the cylinders lying in plane. Arrays of 'dots' or holes may be made from a cylindrical morphology BCP with the cylinders lying perpendicular to the plane, or from a spherical morphology BCP. As will be described below, significant effort has been devoted to imposing long-range order and improving uniformity in these self-assembled patterns by the use of shallow substrate topography, by chemical patterning of the substrate, or by the imposition of stimuli such as an external electric field or a shear force to guide the self-organization of the block copolymer domains. Guided self-assembly has enabled arrays of well-ordered dots and lines, as well as structures with designed aperiodicity, to be formed in BCP films and subsequently transferred into functional materials.

The patterning of magnetic nanostructures using BCP lithography has proved particularly interesting because of the potential applications in data storage.[22,23] The conventional hard disk stores data as a magnetization pattern written in a continuous thin film of a polycrystalline magnetic alloy. However, as data density increases, thermal energy can overcome the anisotropy of the material and cause erasure of the written data. It is believed that this superparamagnetic behavior will impose an ultimate limit to the data density that can be achieved in thin-film media.[24-26] A possible strategy to extend recording densities is to use patterned (discrete) magnetic media, consisting of an array of magnetic 'dots' which are physically separated from each other. Each dot is a single magnetic domain, and stores one bit of data.[27-30] For such media to be competitive, data densities on the order of 1 Tbit/in^2 and above are required,[31] corresponding to a pitch of less than about 25 nm, which imposes strict demands on the lithography method chosen to define the pattern. BCP lithography has been proposed as a possible method for creating patterned media, and a prototype with a period of 80 nm has been demonstrated by a group at Toshiba.[32]

This chapter will describe first the transfer of patterns from self-assembled block copolymers into magnetic materials to form arrays of magnetic dots, 'antidots' or wires. The desirable characteristics of BCPs for lithography applications will be described. Distinctions will be made between additive patterning, for example the electrodeposition of magnetic material into the channels of an etched BCP film, or the use of a BCP film as a liftoff mask, and subtractive patterning, where the block copolymer forms a mask through which a magnetic material is etched. We will also discuss the formation of nanoparticles of magnetic materials within the domains of a BCP, and will describe the properties of magnetic nanostructures formed by BCP lithography. Second, we will describe how long-range order and pattern registration can be imposed on self-assembled BCP domains using suitable topographical or chemical pattern substrates. Templating allows control over the placement accuracy and gives better uniformity in feature size. These are important requirements for lithography, since most devices require structures with well-defined size and location. Strategies for directing the self-assembly of BCPs will be described, including the use of topographical or chemical substrate patterns, shear flow or electric fields. Finally, we will give a perspective on the limitations and advantages of BCP lithography, and discuss possible applications in the fabrication of magnetic and other devices.

2. Pattern Transfer from Block Copolymers to Magnetic Materials

2.1. *Materials, Structural and Process Considerations for BCP Lithography*

The attractiveness of BCP films as self-assembled masks for nanopatterning became apparent in the mid-1990s,[33] and pattern transfer from block copolymer to planar substrate was first demonstrated in 1997 by Park et al.[34] Park et al. used a thin film of PS-PI as a mask and transferred the pattern from the polymer film into the underlying Si_3N_4 substrate. The chemical contrast between the domains in the BCP allowed selective removal of one specific domain. Depending on the process, either the PS block or the stained PI block can be removed to generate post or hole arrays. After pattern transfer, post or hole arrays in Si_3N_4 were made with a periodicity of 40 nm and diameter and height of

20 nm as shown in Figure 1. This demonstration of block copolymer lithography stimulated a range of experiments on pattern formation and pattern transfer of a variety of block copolymer systems. The ability to selectively remove one block is an important characteristic of a BCP used for lithography. PS-PMMA is the most thoroughly investigated systems for block copolymer lithography because of the ease of selective removal of the PMMA, material availability and well-developed full-wafer orientation control methods.[35-38] PMMA can be selectively

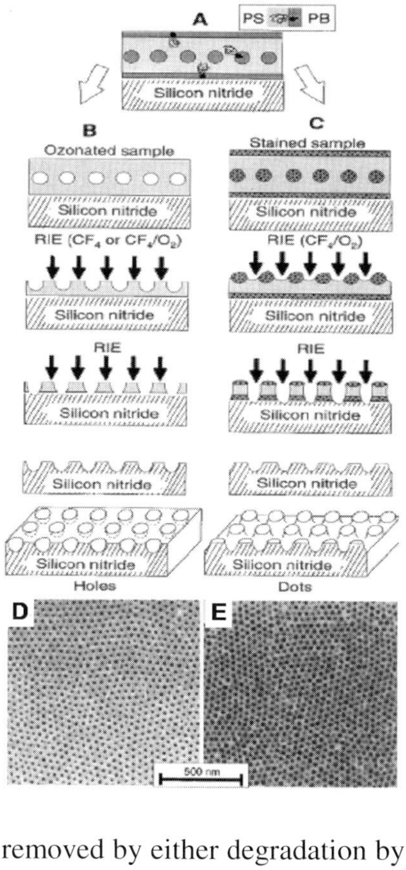

Fig. 1. (A) Schematic cross-sectional view of a nanolithography template consisting of a uniform monolayer of PB spherical microdomains on silicon nitride. PB wets the air and substrate interfaces. (B) Schematic of the processing flow when an ozonated copolymer film is used, which is used to form holes in silicon nitride. (C) Schematic of the processing flow when an osmium-stained copolymer film is used, which is used to form dots in silicon nitride. (D) An SEM image of a partially etched, ozonated monolayer film of spherical microdomains. After the continuous PS matrix at top was removed, the empty PI domains were exposed (as holes) and appear darker in the micrograph. (E) An SEM image of hexagonally ordered arrays of holes in silicon nitride on a thick silicon wafer. The pattern was transferred from a copolymer film such as that in (D). The darker regions are 20-nm-deep holes in silicon nitride, which have been etched out. (Reprinted from Ref. 34 with permission. Copyright The American Association for the Advancement of Science.)

removed by either degradation by UV or dissolution by acetic acid[37,39] or dry reactive-ion etching using oxygen, argon or CF_4.[40] The acetic acid rinse typically gives better selectivity between PS and PMMA domains than the plasma etch, but the dry etch avoids the pattern collapse common in wet etch processes. Hole arrays with high aspect ratio have been made using PS-PMMA and a wet etch.[39] A triblock copolymer PS-PMMA-polyethylene oxide (PS-PMMA-PEO), which combines the advantages of the perpendicular orientation of PS-PEO with the cleavable character of PMMA, provides another way to generate high-aspect-ratio hole arrays without a neutral underlying substrate surface.[41,42] Selective wet etching was also demonstrated in PS-polylactide (PLA) and PS-PI-PLA.[43,44] PLA can be removed by basic solvents such as water/methanol/NaOH without damaging the PS and PI domains.

A high plasma etch resistance of the unetched block can improve the profile of the resulting nanostructures and reduce process complexity. One way to increase etch resistance is to incorporate inorganic elements into one domain of the BCP. Block copolymer systems such as PS-polyferrocenyldimethylsilane (PS-PFS),[45] PI-PFS,[46,47] and PS-polydimethylsiloxane (PS-PDMS)[48] provide high etch selectivity between the organic and the silicon- or iron-containing domains. Instead of synthesizing the BCP with an inorganic component, the inorganic component can be introduced into the BCP system by blending silicon containing polymers or oligomers which are selectively miscible with one domain. For example, a hybrid block copolymer system such as PS-PEO + organosilicate (OS) has been developed in which the OS is selectively miscible with the PEO to form silicon-containing domains.[49,50] An oxygen plasma can be used to remove the organic domain and leave the inorganic-containing domains on the substrate, forming an oxidized surface layer which prevents further etching of the inorganic-containing domains.

A significant characteristic of block copolymer lithography is that the domain dimensions scale with the molecular weight of the BCP, and therefore the periodicity of the pattern can be controlled. The scalability of block copolymer lithography depends on the Flory-Huggins interaction parameter, χ, which describes the driving force for microphase separation. To obtain a microphase-separated structure, the product χN of the degree of polymerization, N, and the Flory-Huggins interaction parameter needs to be at least 10. To achieve a smaller periodicity, a low molecular weight (low N) BCP must be used, and a high χ is therefore advantageous in term of scalability because microphase separation can happen at smaller periods. Some typical room temperature χ parameter values are χ~0.04-0.06 for PS-PMMA,[51,52] χ~0.08 for PS-PEO,[53] χ~0.09 for PS-PI,[54] χ~0.18 for poly(styrene-b-2-vinylpyridine) (PS-P2VP),[54] χ~0.08 for PS-PFS[55] and χ~0.26 for PS-PDMS.[56] Based on the χ value, the minimum domain dimension in PS-PDMS can be half that achievable in PS-PMMA. In the strong segregation limit, the domain size is proportional to $N^{2/3}\chi^{1/6}$ and the interfacial width between the domains is proportional to $\chi^{-1/2}$.[20] BCPs with higher χ value therefore have smaller interfacial widths which is expected to lead to smaller high-frequency line-edge roughness.

Another important consideration for choice of a BCP is to be able to obtain appropriate orientation of the domains within the BCP thin film. Domain orientation can be controlled by a number of methods including surface functionalisation of the substrates, solvent annealing, and application of external fields such an electric field or a flow field during annealing. Interfacial interactions play a critical role in controlling domain orientation in block copolymer thin films. For example, PS and PMMA have very similar surface tension at the air/polymer interfaces, so there is no preference for one block to be present at the surface. Neutral underlayers made from random PS-PMMA copolymers or self-assembled monolayers have been used to generate perpendicularly oriented cylinders and lamellae in thin PS-PMMA films.[35-38] On the other hand, in PS-PDMS the surface energy of PDMS is much lower that of PS, and thus PDMS preferentially segregates at the air/polymer interface. The preferential segregation of PDMS at the air/polymer interface induces orientation in lamellae and cylinder-forming PS-PDMS parallel to the surface.[48] Conditions such as solvent evaporation or electric field also can be used to control the orientation of the BCP domains. Thick films consisting of perpendicular oriented cylinders have been achieved in Poly(n-butylacrylate)-b-polyacrylonitrile (PBA-PAN), PS-polybutadiene (PS-PB), PS-PFS, PS-PEO, PS-PLA, and poly(αmethylstyrene-b-hydroxystyrene (PαMS-PHOST) by controlling the solvent evaporation rate from the film.[41,58-60] Thick PS-PMMA films consisting of perpendicular cylinders were achieved by annealing the polymer film between electrodes that provide an external electric field.[39]

Finally, some consideration must be given to the processing times required to obtain well-ordered block copolymer thin films. Both thermal annealing and solvent vapor annealing have been employed to increase the polymer diffusivity and therefore help to optimize the size uniformity and domain spacing. Thermal annealing is usually carried out at a temperature higher than the glass transition temperature T_g of both blocks and lower than the order-disorder transition temperature. The higher the annealing temperature, the larger the mobility of the polymer, and therefore, a shorter time is needed to remove defects and improve long-range ordering. Well-ordered domain structures in PS-PMMA films have been achieved by annealing between 140°C to 200°C which is higher than the T_g of PS and PMMA. Typical annealing times in the literature range from a few minutes to a few days. Higher annealing temperature greatly reduces annealing time because of the increase in diffusivity. On the other hand, with a low T_g block copolymer, such as PI-PFS, well-ordered PFS spheres in a PI matrix can be achieved at room temperature within 30 min.[47]

Instead of thermal annealing, solvent annealing provides another route to optimize the domain arrangement. Solvent vapor annealing at room temperature has been demonstrated to improve structural uniformity in PS-PMMA, PS-PEO and PS-PDMS thin films.[48,61,62] Solvent annealing effects are highly dependent on the polymer, solvent, and vapor pressure, with typical annealing times from minutes to days. A neutral solvent swells both types of domain and increases the mobility of the polymer chains as well as the domain volume. Selective solvents which swell one domain more than the other can be used to adjust the volume ratio of the swollen BCP and therefore change the morphology of the film.[61] In addition, the interfacial properties of air-polymer, polymer-polymer and polymer-substrate can be changed during solvent annealing because the swollen domains have different surface energies compared to the neat polymer.[49] The annealing process can be further facilitated by introducing substrate surface modifications to reduce substrate pinning and the polymer substrate interaction. For example, in PS-PDMS, a PDMS brush layer on the silica substrate enabled more rapid diffusivity of the PS-PDMS layer compared with observations of the same BCP on a PS-coated or uncoated silica substrate.[48]

The requirement of annealing time in BCP lithography varies with the application. If BCPs are used to form a pattern on every wafer or disk, then a high throughput is critical to successful implementation of block copolymer lithography. The ideal annealing time would be less than a minute to achieve the high throughput requirement for generating magnetic patterned media. On the other hand, if block copolymers are used to make for example a nanoimprint master which is then used to make many replicas, a processing time of hours or days for annealing may be acceptable.

2.2. *Subtractive Patterning Methods*

We have discussed the desirable characteristics of block copolymers for nanolithography masks. We will now illustrate this with examples of nanostructures patterned by BCP lithography. Since the BCP patterning process has been reviewed several times in recent years,[1-11] we will focus our discussion here on the patterning of magnetic nanostructures.

Subtractive patterning refers to the removal of part of a material using an etching process, while the desired structures are protected by an etch mask. Subtractive patterning allows the deposition of the magnetic thin film to be carried out prior to the deposition and processing of the BCP film, so that the magnetic properties of the thin film and the pattern transfer process can be optimized independently. Subtractive patterning is illustrated in Figure 2 for the transfer of an array of PFS spheres into an underlying Co film to make an array of Co dots with period 55 nm and diameter 35 nm.[45] The PS-PFS thin film consists of spheres of PFS in a PS matrix. The PS is removed using an oxygen plasma (Figure 2A), then the PFS pattern is transferred into a silica layer using a CHF_3 plasma (Figure 2B). This pattern is transferred into a W layer using a CF_4/O_2 plasma (Figure 2C), and the W pattern is finally transferred into the Co layer by ion milling using Ne (Figure 2D). This process illustrates some of the issues involved in BCP pattern transfer. The silica layer ensures good wetting of the substrate by the BCP. The Co layer, in common with many transition metals, cannot be etched using a reactive ion etch process, and must be etched instead by physical sputtering (known as ion bombardment, ion beam etching or ion milling). This process is conveniently accomplished using a hard mask of W, which provides good etch contrast with Co under Ne bombardment.[63] A similar process was used to pattern NiFe films and NiFe/Cu/CoFe multilayers,[64] as well as CoCrPt/Ti perpendicular magnetic films. Magnetic measurements of

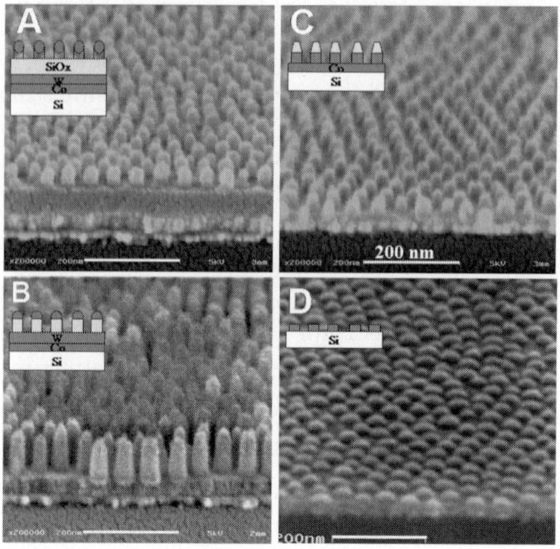

Fig. 2. Tilted SEM micrographs of the fabrication process of Co dot array using PS-PFS BCPs. (A) An O_2-RIE treated block copolymer thin film on a multilayer of silica, the metallic films and the silicon substrate. (B) Pillars of silicon oxide capped with oxidized PFS after CHF_3-RIE. (C) W (tungsten) hard mask on top of a Co layer. (D) Co dot array produced Ne ion-beam etching. (Adapted from Ref. 45 with permission. Copyright Wiley-VCH.)

the multilayer showed that the layered structure was preserved through the processing steps.

Another subtractive approach is illustrated by the fabrication of a patterned media prototype[32,65-68] using PS-PMMA as a template to pattern CoCrPt or FePt thin films. Figure 3A shows the process. The PS-PMMA was spin-cast onto the magnetic thin film, the PMMA domains were removed by an oxygen reactive ion etch, and the voids in the polymer film were then back-filled using spin-on glass. Arrays of discrete magnetic dots were then generated by argon ion beam etching the magnetic film using the spin-on glass as a hard mask. Figure 3B shows that pattern transfer can be optimised by choice of ion beam etching angle.

BCP templates have also been used as a mask to pattern antidot arrays, which are films containing an array of holes. Fe/FeF_2 antidot arrays were made by ion beam etching using a PS-PMMA cylindrical morphology BCP as the mask, after removing the PMMA cylinders with

UV and acetic acid.[69] Similarly, $Ni_{80}Fe_{20}$ antidot arrays were made by coating films of PS-PI-PLA onto $Ni_{80}Fe_{20}$–coated substrates to form perpendicular PLA cylinders in a PI/ PS matrix. After removing the PLA with a basic aqueous solution, Ar ion beam etching was used to etch the $Ni_{80}Fe_{20}$ film forming 30 nm diameter holes.[44]

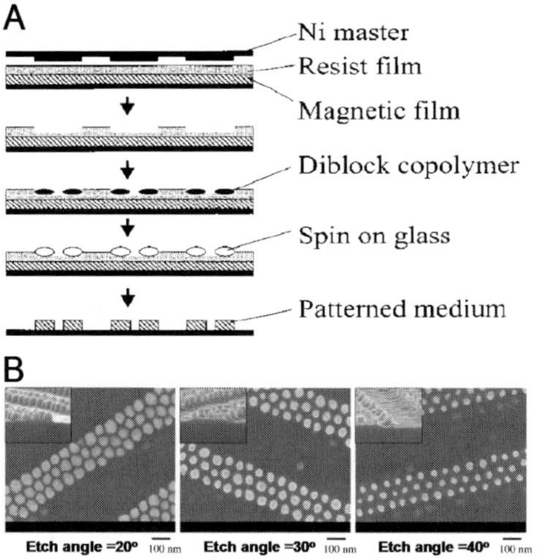

Fig. 3. (A) Scheme of the preparation method of a prototype patterned medium. A Ni master disk possessing spiral patterns was pressed, at room temperature and a pressure of 1000 bar, onto a resist film that was top of either a $Co_{74}Pt_{26}$ or $Co_{74}Cr_6Pt_{20}$ magnetic film. The magnetic film was deposited by dc magnetron sputtering onto a 40-nm Ti underlayer. A PS-PMMA thin film was cast into the obtained grooves, and then annealed to prepare ordered dot structures aligned along the grooves. The PMMA dots were selectively removed by the oxygen plasma. The resulting holes were filled by spin-on-glass (SOG). The underlying magnetic films were patterned by ion beam etching using the SOG dots as a mask. (Figure 1 from Ref. 32) (B) Plan-view and cross-sectional SEM images of a magnetic dot array fabricated with an etching angle of 20°, 30°, and 40°. (Reprinted from Ref. 40 with permission. Copyright Institute of Pure and Applied Physics.)

Additive Patterning Methods

In additive patterning, the magnetic materials are deposited within or on top of a BCP template, avoiding the need for etching the magnetic material. Perpendicularly oriented cylindrical domains can be used for generating close-packed arrays of magnetic dots or wires, by removing the cylindrical domain and then coating the pores by sputtering or filling them by electrodeposition. Both thin films and thick films of cylinder-forming PS-PMMA have been used as templates, for example to form the Co nanowires shown in Figure 4A by electrodeposition.[39] The orientation of the c-axis of the grains and the grain size can be controlled by tuning the pH of the solution and the deposition conditions.[39,70,71]

Magnetic dot arrays can be made by sputtering or evaporating magnetic materials onto etched BCP templates, then lifting off (dissolving) the BCP to leave an array of dots at the locations of the pores in the template. Xiao et al.[72] sputtered a 10 nm Co layer onto an etched PS-PMMA film. A standard lift-off process, which requires a long ultrasonic treatment to remove the continuous Co layer and the PS matrix, results in many defects because of the detachment of magnetic dots (Figure 4B), but successful lift-off was accomplished using mild ion beam etching followed by an oxygen etch (Figure 4C). In addition to PS-PMMA, cylindrical morphology PS-PLA and PS-PI-PLA BCPs have been used as templates. The etch resistivity of PS was enhanced by

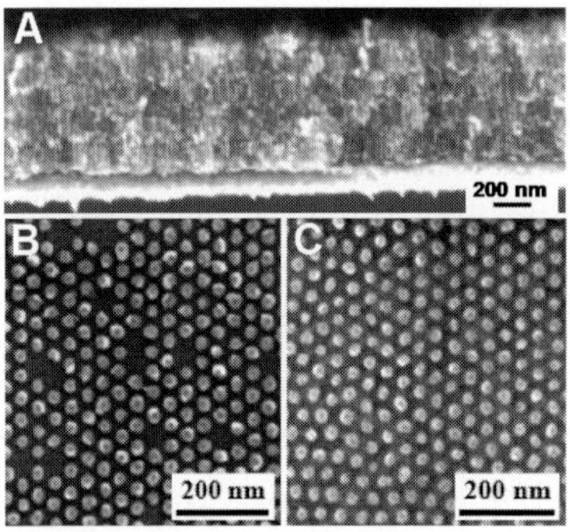

Fig. 4. (A) Cross-section SEM micrograph of an array of Co nanowires grown within an array of nanopores formed from PS-PMMA. The growth of the nanowires was terminated before the template was completely filled. (Reprinted with permission from Ref. 38. Copyright The American Association for the Advancement of Science.) (B&C) Plan-view SEM images of magnetic Co dot arrays after a sputtering and lift-off process using PS-PMMA templates. (B) 'Poor' metal dot arrays on a flat substrate, after a wet etching process. (C) 'Good' metal dot arrays on a flat substrate, after a dry etching process. (Reprinted from Ref. 72 with permission. Copyright Institute of Physics and IOP Publishing Limited.)

staining PS with OsO_4 and then the BCP film was etched in an oxygen plasma to remove PLA, followed by deposition of $Ni_{80}Fe_{20}$ and liftoff.[44] Black et al.[73] deposited 2 nm Ti/6 nm Co over a porous template made by removing PMMA cylinders from a PS-PMMA BCP with the cylinders perpendicular to the substrate. Jeong et al.[74] deposited CoCr(18at%)Pt(14at%) films over a film of PS-P2VP micelles formed by dissolving the block copolymer in toluene (a good solvent only for the PS) then spin-coating the resulting micelles on a Si surface. The micelle array had period 40 nm and a height variation of 4 nm, so the overlying magnetic film had a significant roughness.

Antidot arrays may be made by depositing magnetic material over an etched BCP film. Chuang et al.[75] prepared antidot arrays by ion-beam-sputter depositing single layer 3.5 nm Co films and Co 3.5 nm/Cu 5 nm/NiFe 3 nm pseudo-spin-valve films over hole arrays formed from a PS-PEO/OS hybrid BCP system. The PS-PEO/OS was spin-coated on top of a 40 nm polymeric underlayer. An oxygen reactive ion etch removed PS spheres and the uncovered parts of the underlayer to create channels of period 26 nm or 40 nm, and diameters 12 nm or 17 nm.

2.3. *Block Copolymer Templates for Selective Deposition*

Chemical differences between the blocks can be useful for selective sequestration or deposition of magnetic and other nanoparticles. For example, nanoparticles of Fe_2O_3, Co, FeCo, Fe and CoNi have been selectively deposited within the P2VP blocks of PS-P2VP[76-78] by a chemical reaction that occurs preferentially within the P2VP; similar strategies have been applied to a range of semiconductor and other nanoparticles. Electroless deposition may also be used for selective deposition of materials, for example Ni was deposited into the continuous channels formed by removing the PI domains of a blend of PS-PI with a PS homopolymer.[79] Such processes have so far been applied to bulk BCPs or thick films, but may be extendable to thin films of BCPs to make 2D magnetic nanostructures.

The small scale topography resulting from pattern transfer from a BCP can be used to direct the location of nanoparticles. Figure 5A-C shows a process to deposit Co nanocrystals into Si templates made by pattern transfer from a perpendicularly oriented PS-PMMA thin film. A drop of hexane containing a dilute solution of Co nanocrystals (diameter ~10 nm) was placed on the substrate and allowed to dry. Figure 5D demonstrates that the nanocrystals locate preferentially in the pits of the templates.[73] Nanoparticle location can also be controlled by the chemical differences between the blocks, which can result in selective surface adsorption. This was demonstrated using a film consisting of UV-treated PS-PMMA semicylinders.[80,81] Oleic acid-capped FePt nanocrystals adsorb into the treated PMMA regions with near-perfect selectivity to form two-dimensional nanoparticle arrays as shown in Figures 5E and

5F.[80] These techniques may be useful in making patterned magnetic media from nanocrystals, but the strong absorption between the nanocrystals and the surface can limit the applicability of annealing techniques to improve ordering.[81] An additional challenge is the control of the crystalline orientation of the magnetic particles.

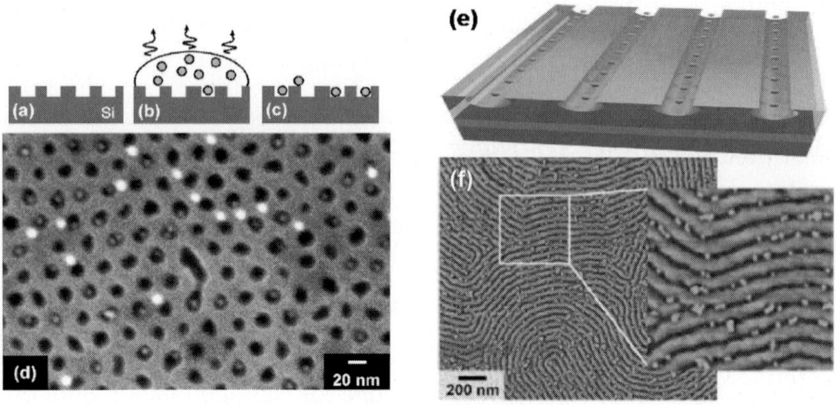

Fig. 5. Magnetic nanoparticles and block copolymer templates. (a-c) Schematic of nanocrystal separation process. (a) The starting etched silicon substrate. (b) Nanocrystals deposited on the substrate by evaporation of the carrier solvent. (c) The resulting substrate with deposited nanocrystals. (d) Top-down SEM image of nanocrystals residing in etched pits on silicon substrate. (Reprinted from Ref. 73 with permission. Copyright Materials Research Society.) (e) Schematic of nanocrystal deposition on chemically heterogenous templates made from BCP thin films. (f) AFM phase images showing 99% selective adsorption of FePt nanoparticles onto the photochemically modified PS-PMMA domains. (Reprinted from Ref. 80 with permission. Copyright Wiley-VCH.)

2.4. *Magnetic Properties of Nanostructure Arrays Made by Block Copolymer Lithography*

As mentioned previously, magnetic materials have been patterned using both subtractive and additive methods. The 2D magnetic structures described to date consist primarily of particle arrays, dot arrays, wire arrays and antidot arrays. In this section we will describe the properties of these structures, including pattern fidelity, switching field distributions and interactions.

2.4.1. *'Dot' or nanoparticle arrays*

Cheng and coworkers described the properties of dot arrays made from thin evaporated/sputtered films of Co, Ni, NiFe/Cu/CoFe[64] and Ti/CoCr(22at%)Pt(12at%)[82] which were patterned subtractively using a PS-PFS spherical morphology BCP. The dots had thicknesses of 5-20 nm, diameters of 35 nm and period of 55 nm. Changes in the magnetic properties (magnetization and coercivity) could be used to detect the endpoint of the pattern transfer. Co and NiFe films and dot arrays showed in-plane magnetization, as a result of shape anisotropy, with the out-of-plane direction a hard axis. The in-plane coercivity was higher for the dot arrays than for the unpatterned films, but was still relatively low (100 – 200 Oe). Measurements of switching volume, which is the volume of magnetic material that participates in the magnetization reversal process, showed it to be several times greater than the physical volume of a single dot (Figure 6A). This was interpreted as showing collective behavior of the dots, i.e. they reverse in small groups. This is a result of the magnetostatic interactions between adjacent dots, which are strong compared with the coercivity. A considerably different result was obtained from Ti/CoCrPt patterned into similar geometry arrays.[82] In this system, the Ti underlayer causes the Co-alloy to grow with its c-axis, and hence its magnetic easy axis, perpendicular to the plane of the substrate. Both the unpatterned film and the dot array had an out-of-plane easy axis, but the dot arrays had considerably higher coercivity, e.g. 1600 Oe for 10 nm thick dots. This is large compared with the magnetostatic interactions between adjacent dots, and as a result the switching volume of the dots was close to their physical volume. This implies that the dots are single magnetic domains which reverse independently, making them a possible candidate for patterned media.

The authors analysed the size and shape distribution of the Co and NiFe dots, showing an average ellipticity of 6% and a standard deviation in dot diameter of 9%. The slight ellipticity, as well as edge roughness and the effects of polycrystallinity in the case of high anisotropy materials such as Co, were sufficient to explain the non-zero in-plane coercivity, since a perfectly circular and homogeneous dot would have

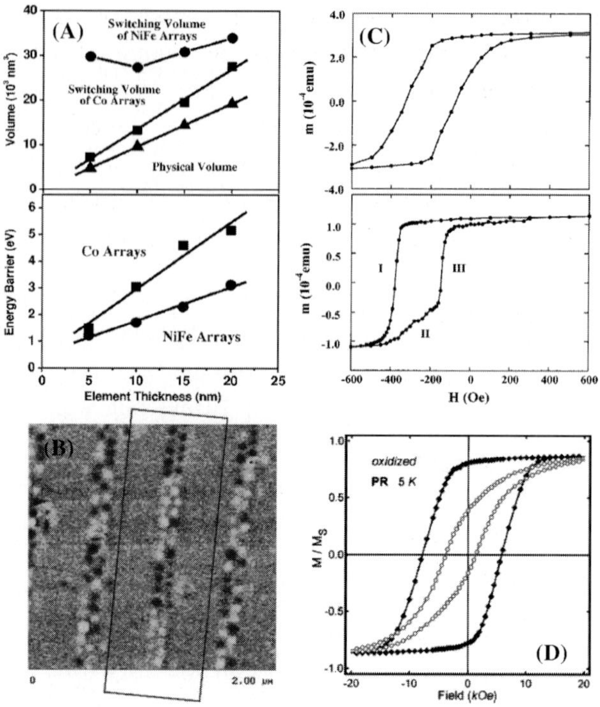

Fig. 6. A compilation of magnetic data from structures patterned using block copolymer lithography. (A) The switching volume measured for NiFe and Co dot arrays as a function of dot thickness (top panel). For the NiFe arrays the switching volume significantly exceeds the physical volume of the dots, indicating collective reversal. The lower panel shows the energy barrier for reversal. (Reprinted from Ref. 64 with permission. Copyright American Institute of Physics.) (B) A magnetic force microscope image of patterned media tracks consisting of CoCrPt dots with perpendicular magnetization. The black and white contrast represents 'up' or 'down' magnetization states of individual dots. The black rectangle indicates the trajectory of the recording head, which has been used to write a pattern on one track. (Reprinted from Ref. 68 with permission. Copyright Institute of Physics and IOP Publishing Limited.) (C) Hysteresis loops of a 15 nm Fe/20. FeF_2 continuous film (top panel) and a 15 nm Fe antidot array/20 nm continuous FeF_2 film (bottom panel) measured at 10K, showing exchange pinning from the antiferromagnetic FeF_2. (Reprinted from Ref. 69 with permission. Copyright American Institute of Physics.) (D) Hysteresis loops of a Co wire array measured at 5K after exposure to air for one month, leading to CoO formation. The solid symbols are for the field applied perpendicular to the plane, while the open symbols are for an in-plane field. The exchange pinning from the CoO is evident, as is the strong anisotropy due to the shape of the wires. (Reprinted from Ref. 70 with permission. Copyright American Institute of Physics.)

zero in-plane coercivity. The NiFe/Cu/CoFe dots showed remanent states in which the CoFe and NiFe layers are coupled antiparallel, via magnetostatic interactions. The dots were coated with gold and electrically tested, showing a giant magnetoresistance which correlated with the hysteresis loop, and which was comparable to that of the unpatterned film taking into account the effect of current shunting by the gold. Thus the patterning process did not significantly disrupt the layered structure.

Naito et al. described magnetic dots made by etching CoCr(6at%)Pt(20at%) or FePt films.[32,65-68] A Ti underlayer gave the CoCrPt film a c-axis out-of-plane orientation, so the magnetic easy axis was also out-of-plane. CoCrPt dots could be magnetized with a recording head (Figure 6B), and behaved independently; the average coercivity was 4 kOe with a distribution between 0.5 and 7.5 kOe and standard deviation 1.4 kOe. This variability was attributed to variations in the size of the PMMA domains as well as polycrystallinity in the magnetic film, since each dot contains a small number of grains from the original CoCrPt film (grain size 23 nm).[68] High anisotropy FePt L_1^0–structured films showed very high out-of-plane anisotropy with a coercivity that increased from 4.5kOe in the unpatterned film to at least 13kOe in the dot array.[67] Olayo-Valles et al.[43] reported on 5 nm thick NiFe dot arrays formed by a liftoff process using PS-PLA BCP films as the template. The dots had diameters of 49 nm, and higher coercivity (80 Oe) and lower squareness than an unpatterned film.

The work presented to date is significant in showing the fabrication and properties of both ordered and disordered dot arrays made from complex alloys and layer sequences. Although individual dots can act as single magnetic domains with coherent reversal, the dot arrays show a wide range of switching fields, in common with arrays of magnetic nanostructures made by other forms of lithography. This is most likely due to the polycrystallinity of the film, and edge roughness or size variations, which have major effects on the switching field of individual dots. Further work will need to address these issues, since a tight switching field distribution is required for patterned media.

2.4.2. *Antidot arrays and non-planar films*

An antidot array is a thin film containing an array of holes.[83-84] These structures are interesting because the holes modify the magnetic properties of the film profoundly, for example they can act as pinning sites for domain walls, and have even been proposed as data storage structures.[85] Liu et al.[69] described the properties of antidot arrays of Fe and FeF_2/Fe. The FeF_2 is an antiferromagnet below 80K, and the bilayer film at low temperature shows exchange bias of 200 Oe or more, and asymmetric loops indicating different reversal mechanisms for magnetization reversal in the ascending branch (coherent reversal) and the descending branch (domain wall nucleation and incoherent rotation) as shown in Figure 6C. Kubo et al.[44] described an antidot array made from a 5 nm thick NiFe film capped with 5 nm gold, with 31 nm diameter, 61 nm period holes. The change in coercivity, anisotropic magnetoresistance and resistivity was correlated with the etch depth in the magnetic film.

Chuang et al.[75] prepared antidot arrays of single layer 3.5 nm Co films and pseudo-spin-valve Co 3.5 nm/Cu 5 nm/NiFe 3 nm films, deposited over porous BCP films with pore period 26 nm or 40 nm, and diameter 12 nm or 17 nm. Compared with a continuous thin film, the coercivities of the Co and Co/Cu/NiFe antidot arrays were higher, and the saturation field was considerably increased. The giant magnetoresistance of the multilayer was 0.24% for the continuous film and 0.26% or 0.21% for the two antidot arrays. In the multilayer, unlike the continuous film, the NiFe reverses at positive fields due to the strong magnetostatic interactions between the Co and NiFe layers present near the holes. As the inter-hole spacing is decreased, both experiment and micromagnetic simulation results show that the coercivity and switching field distribution is reduced, unlike the behavior seen in films with micron-sized holes.[84] The simulation shows in particular the trapping of 360 degree domain walls between rows of holes, which are eliminated only at fields above 5000 Oe.

Jeong et al. reported the properties of 3 – 50 nm CoCrPt films deposited over a BCP micelle array.[74] The films appeared to be continuous but with significant roughness and local thickness variations, and had a higher coercivity and perpendicular anisotropy compared to a smooth film. The change in magnetic properties was attributed to surface roughness of the film, which reduced the demagnetizing factor and pinned domain wall motion.

2.4.3. *Wire arrays*

Arrays of parallel nanoscale magnetic wires can be prepared by deposition within a cylindrical-morphology block copolymer. Depending on the thickness of the BCP film, such wires can have extremely long aspect ratios, and their magnetic properties are very different from those of dot arrays because the shape anisotropy of the wires generally leads to a magnetic easy axis parallel to the wire length. Wire arrays have been proposed for data storage, though the high aspect ratio and close spacing makes it challenging to write data by reversing wires individually, compared to writing on a dot array.

Tuominen et al.[39,69-71,86-88] produced electrodeposited Co wire arrays in PS-PMMA BCP films with wire diameters of 15 nm and period of 24 nm. The Co crystal orientation could be controlled through the electrodeposition conditions enabling the net anisotropy to be controlled,[71] and wire arrays with high coercivity (2.7 kOe) and remanence were obtained.[39,70,86] Exchange bias appeared at low temperatures due to CoO formation[71] as demonstrated in Figure 6D. Ferromagnetic resonance was used to measure anisotropy.[88] Wire arrays made of Co/Au were deposited in selected areas of a BCP film, and magnetoresistance measurements showed evidence of anisotropic and giant magnetoresistance.[87] Magnetic wire arrays have also been produced by electrodepositing within an anodic alumina template, typically with slightly larger periodicities, and analogous behavior has been found.

3. Directed Self-Assembly of Block Copolymers

3.1. *Approaches to Directed Self-Assembly*

For magnetic devices such as patterned media or cross-point memories, the long-range ordering and placement accuracy of each active element are critical to device performance. However, thin films of block copolymers typically have many defects in their domain arrangements. Directed self-assembly of BCPs, taking advantage of the spatial precision achievable from lithographically-defined patterns to guide the self-assembly of the BCP, provides a method to generate nanostructures with high spatial resolution as well as good placement accuracy.

There are two major directed self-assembly (or graphoepitaxy) methods used in BCP thin films: topographical epitaxy and chemical epitaxy. In the topographical epitaxy method, self-organization of block copolymers is guided by prepatterned substrates with topographical features. The lateral confinement of the BCP thin films induces long-range ordering of the domains. In the chemical epitaxy method, the self-assembly of block copolymers is guided by chemical patterns on the substrate. The affinity between the chemical patterns and the BCP domains results in the placement of BCP domains on specific locations.

Both topographical epitaxy and chemical epitaxy can be applied to fabricate large-area well-ordered nanostructures. A major difference between topographical epitaxy and chemical epitaxy is the relative length scale between the prepatterned substrate and the natural periodicity of the BCP. The length scale of the topographically patterned substrate can be much larger than the periodicity of the BCP, and therefore, the BCP domains subdivide the topographical prepatterns to form sub-lithographical features. Until recently, in chemical epitaxy, the length scale of chemical patterns was close to the periodicity of the BCP, but recent work has overcome this limitation and used a chemical pattern 2 – 4 times larger than the block copolymer period. In chemical epitaxy, the 'self-healing' nature of the BCP pattern is believed to reduce the variability in sizes and the line-edge roughness compared to those of the initial chemical pattern. Directed self-assembly of BCP films has been reviewed recently.[8-11] In this section we will discuss advances, limitations and applications of these two methods.

3.2. *Epitaxy of Block Copolymers on Chemical Patterns*

Research in recent years has advanced our understanding of the interaction between block copolymers and nanometer scale chemically patterned substrates, and demonstrated that block copolymers can be precisely guided. Similar to epitaxy in a crystalline thin film, the commensurability and chemical affinity of the microdomains to the patterned substrate both play important roles. Additionally, blending low molecular weight homopolymers into block copolymers in order to alter the thickness of the respective domains gives a wider process window for precise microdomain epitaxy.

Rockford et al. investigated a symmetric PS-*b*-PMMA block copolymer dip-coated onto chemically heterogeneous substrates comprised of alternating silicon dioxide and gold stripes.[89] The SiO_2 and Au stripes are wet by PMMA and PS, respectively. Perpendicularly oriented PS-*b*-PMMA lamellae form with the highest in-plane order parameter on the patterned substrates when the lamellar period, L_0 matches the substrate stripe period L_S. The defect density increases with the mismatch between L_S and L_0. In a thin film, the interaction between the microdomains and surface stripes is sufficiently strong to orient lamellar microdomains perpendicular to the substrate even when there is as much as 40% mismatch. Further studies showed a smaller window of orientational control in a thicker film (> 500 nm), where even a 10% mismatch caused loss of orientational control.[90]

Nealey and coworkers observed similar results in symmetric PS-*b*-PMMA on flat substrates with alternating hydrophilic/hydrophobic stripes of an e-beam or extreme ultraviolet interference lithography patterned self-assembled monolayer[91-93]. Well-registered perpendicularly oriented lamellae formed over large areas when $L_S \sim L_0$. When $L_S = (0.95 - 1.15)\ L_0$, lamellae formed perpendicular to the substrate, but dropped out of registry with the substrate surface pattern as L_S and L_0 diverge. As expected, defect levels increase as the difference between L_S and L_0 increases.[92,93] By varying the affinity between the chemical patterns and the BCP domains, it was found that a well-registered surface-directed epitaxial block copolymer film can be formed if the interfacial energy gain from preferential wetting of each block is sufficient to compensate the strain energy that results from deviation between L_S and L_0.[93,94] Beyond simple periodic line/space patterns, chemical patterns with bent stripes could be replicated by block copolymers using a ternary blend of homopolymer and BCP.[94] The blend system shows advantages in enabling the block copolymer lamellae to conform to substrate stripe arrays with sharp bends. In the unblended block copolymer system, a high strain builds up at the sharp corners because the corner-to-corner distance is much larger than the natural periodicity of the block copolymer. The successful replication of arrays with 45° and 90° angles was a result of the redistribution of the homopolymer. Homopolymer is depleted in the adjacent regions and concentrated at the boundary region to reduce the strain from incommensurability.[95]

One of the major advantages of chemical epitaxy is the reduction of variations in feature size and edge roughness due to the self-healing behavior of block copolymers. If the chemical pattern has an area fraction of one block different from the volume fraction in the BCP, the lamellar boundaries tilt, but their dimensions at half-height are invariant with respect to the area fraction of the chemical pattern.[96] This suggests that irregularities in the chemical pattern do not necessarily translate into irregularities in the BCP domain arrangement. In addition, edge roughness in the BCP may be lower than that in the chemical pattern, especially for high χ. This relieves the requirements on the perfection of the chemical pattern, which can be limited by, for example, the stochastic nature of the e-beam lithography process used to make the chemical pattern. Epitaxy of BCPs on chemical patterns may also be used to improve the uniformity in feature sizes in order to make high-density imprint molds for mass production of patterned media. Recent work has shown how chemical patterns can be used to template BCPs with a period 2 – 4 times smaller than that of the chemical pattern.[97,98] A more complete discussion of chemical epitaxy may be found in Chapter 2.

3.3. *Topographical Epitaxy of Block Copolymers*

3.3.1. *Pattern formation in topographical templates*

Topographical epitaxy of block copolymers has been demonstrated in templates with various length scales and with one-dimensional, two-dimensional and three-dimensional topographical features. The width of the topographical patterns, L_S may be tens or hundreds of times greater than L_0, the period of the BCP. Segalman et al. were the first to demonstrate the formation of long range ordered sphere arrays on a grating substrate, by floating a BCP film onto a patterned substrate.[99] A monolayer of PVP domains of a spherical-morphology PS-PVP film was placed on the photolithographically-patterned substrate and then was annealed to generate ordered structures propagating several microns from the sidewalls of the grooves and the edges of the mesas (Figure 7). The effects of incommensurability are negligible in this case since $L_S >> L_0$, and the spacing of the spheres in the groove and on the mesa is indistinguishable from that on a smooth substrate. In addition to linear confinement, single crystals of PS-PVP sphere arrays have been achieved in hexagonal wells whose width was a few micrometers.[100]

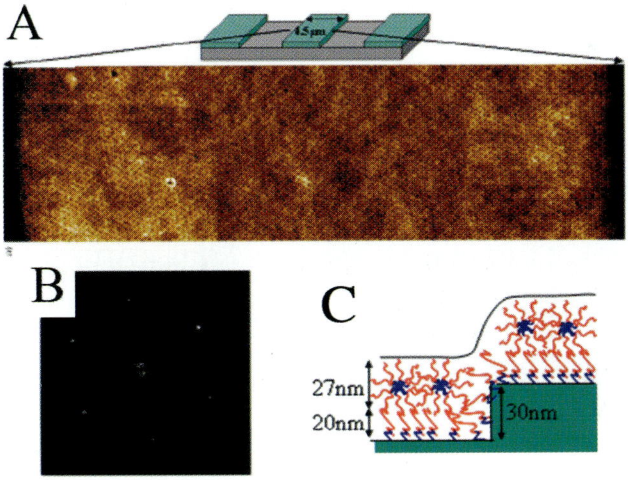

Fig. 7. (A) AFM image of a 2D ordered array of spheres in a PS-PVP film on top of a mesa. (B) The six sharp diffraction peaks in the associated Fast Fourier Transform indicate that the entire mesa region is well ordered with hexagonal symmetry. (C) The cross sectional schematic shows a polyvinylpyridine (blue) brush on the SiO_2 surfaces (green) and a monolayer of PVP spheres encased in a styrene matrix (red). (Reprinted from Ref. 99 with permission. Copyright Wiley-VCH.)

In narrow templates whose width is a few times L_0, incommensurability is neither negligible as it is in the wide template, nor as critical as it is during epitaxy in templates with width $L_S \sim L_0$. Ordered spherical, cylindrical and lamellar microdomains have been observed in narrow templates regardless of the incommensurability. To drive the in-plane long-range order of spherical and cylindrical microdomains in thin films, lithographically-defined grooved substrates,[32,65,101-115] soft PDMS stamps and hard imprint molds[116,117] have been used as templates to impose lateral confinement. In-plane cylinders typically line up along the edges of the templates, while spheres and perpendicularly-oriented cylinders align their most densely-packed rows parallel to the edges. Cheng et al. demonstrated ordered domains of sphere-forming PS-PFS in templates with various widths (see Figure 8A).[103] The number of rows n in the arrays is determined by the confinement width, L_S. Similar to lamellae confined between two rigid plates, defect-free arrays with n rows of domains are observed in narrow grooves with confinement width

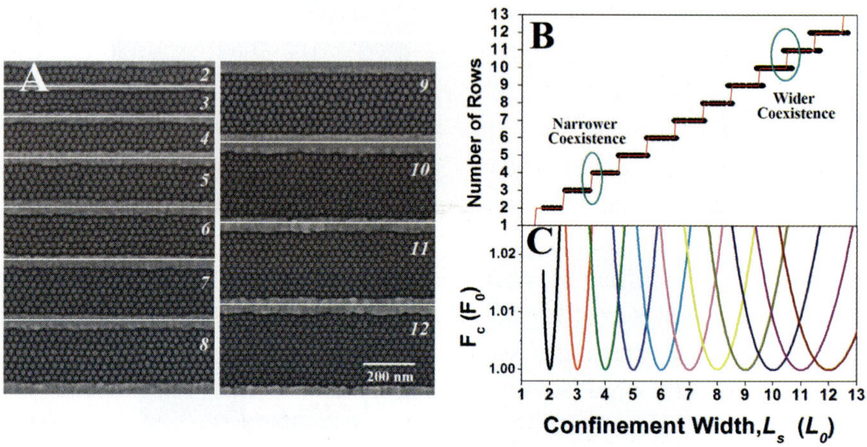

Fig. 8. Directed self-assembly of spherical PFS domains within 1D templates of varying width.(A) SEM micrographs of ordered arrays of PFS domains with N = 2 to 12 rows. (B) The number of rows in the groove, N, vs. confinement width, L_S, showing the widths at which arrays with N rows are stable. The confinement width is expressed in terms of L_0, the equilibrium row spacing, 24.8 nm in this polymer. The overlap region at $L_S \sim (N+0.5)\ L_0$ is circled to show the coexistence of two possible states, N rows and N+1 rows. (C) Energy vs. confinement width of a block copolymer system. The confined block copolymer system, of given L_S, will select the value of N with the lowest free energy. A transition in the number of rows from N to N+1 occurs when W ~ $(N+0.5)d_o$, in agreement with the experimental data of Figure 4A. The model also predicts wider coexistence regions at larger values of Ls. The free energy of the confined block copolymer (F_C) is presented relative to the free energy of the bulk block copolymer (F_0). (Reprinted from Ref. 103 with permission. Copyright Nature Publishing Group.)

L_S when $(n\text{-}0.5)L_0 < L_S < (n+0.5)\ L_0$ as shown in Figure 8B. When $L_S \sim (n\pm0.5)\ L_0$, both n row and n+1 row arrays are found at the same value of L_S, and the range of groove widths at which this coexistence occurs increases with increasing L_S. Sundrani and Xiao observed similar behavior for cylinders in PS-PEP and PS-PMMA.[106,108] Imprinting and molding techniques can also create ordered arrays of spheres and cylinders, but these approaches show more defects and poorer size uniformity, which may be a consequence of non-uniform filling of the mold, or a distribution of mechanical stresses during processing.[116,117]

The free energy of a confined sphere or cylinder microdomain system can be treated analogously to that of lamellae confined between parallel surfaces.[118,119] The ratio between the free energy per polymer chain F_c of the confined array of spheres in the template, and the free energy per polymer chain F_0 in the unconfined bulk block copolymer spherical morphology, is approximated as a function of normalized row spacing λ ($\lambda = L_S / n\ L_0$).[120,121] A plot of free energy F_c vs. confinement width L_S can be constructed for each value of n as shown in Figure 8C. F_c has its local minimum when L_S equals nL_0. At a given L_S, a confined block copolymer system will ideally select the integer value of n with the lowest free energy. A transition in the number of rows from n to n+1 is expected to occur when $L_S = (n+0.5)\ L_0$. If small energy fluctuations are available to the system, the coexistence of n rows and n+1 rows for $nL_0 < L_S < (n+1)L_0$ becomes more probable for large n, in agreement with data in Figure 8B. The similarity in energy between an n row and an n+1 row configuration at large confinement width indicates that precise control of the pattern can only occur if the confinement width is relatively small. For the polymer and annealing condition in Ref. 103, the limit is $L_S \approx 10L_0$.

An interesting situation arises when the confinement width L_S is similar to the natural periodicity of the BCP L_0. For example, a sphere-forming PS-PFS with row spacing d_0 was spin-cast into trenches of width, W, varying from W=0.5d_0 *to* W=1.6d_0. If the template width is the same as d_0, the block copolymer forms a single row of spheres inside the trench, but if the template is incommensurate with d_0, the spheres distort to fit within the template, leading to ellipsoidal shapes (Fig. 9A).[113] Figure 9B shows that the diameter of the domains, measured perpendicular to the channel, increases with the groove width, while the diameter parallel to the channel remains constant. Figure 9C shows the aspect ratio of the domains vs W/d_0 also increases with template width. By contrast (Figure 9D), the period of the 1D array parallel to the template remains constant and equal to the equilibrium periodicity of the unconfined block copolymer film. In this way a row of spheres or ellipsoids with variable aspect ratio and major axis orientation (parallel or perpendicular to the row) can be created; this shape control may be useful in patterning rows of magnetic nanostructures because it can introduce in-plane shape anisotropy. A recent extension of topographical templating uses a sparse array of pillars, made by electron-beam lithography, to template a sphere array in PS-PDMS.[122] Unlike groove templates, the pillars are incorporated seamlessly into the final BCP pattern.

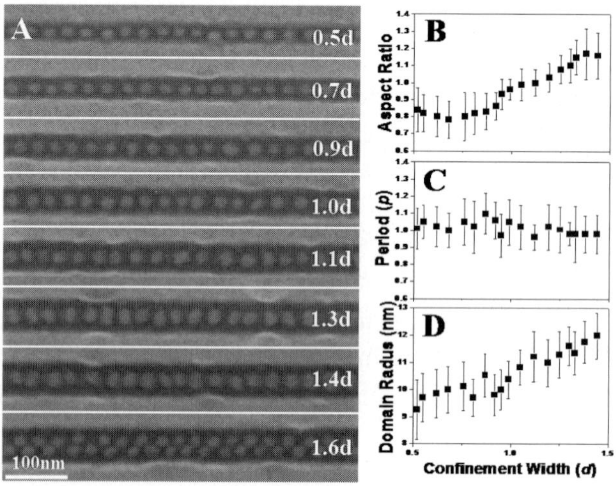

Fig. 9. (A) A composite image of PFS block copolymer domains within channels of different confinement widths W. A single row of spheres forms for $0.6 < W/d_0 < 1.5$, where d_0 is the equilibrium row spacing of this block copolymer. (B) The domain dimensions parallel and perpendicular to the channel were found by fitting ellipses to the domain images in the micrographs and measuring the projected major and minor axes. (C) The aspect ratio of the domains calculated from the data of (B), with a linear best fit to the data for W/d_0 in the range 0.65-1.5. (D) The periodicity of the 1D array is equal to p_0, the center-to-center spacing of the unconfined block copolymer, and independent of confinement width. (Reprinted from Ref. 113 with permission. Copyright American Chemical Society.)

Cylindrical domains have been oriented on a substrate with a sinusoidal height variation, although the degree of order obtained by this approach is limited. The deposited block copolymer film covers both troughs and crests of the substrate, and the parallel alignment of domains in the troughs leads to the orientation of domains on the crests.[105] A soft-molding method to orient in-plane cylinders using a PDMS mold with a sinusoidal grating has also been developed, in which the cylindrical domains align perpendicular to the grating lines of the mold, due to shear.[111] Three-dimensional arrays of BCP domains may also be templated, for example by spin-coating thick films of BCPs over topographical templates or by confining BCPs within cylindrical pores,[123-125] droplets,[126,127] sphere interstices[128] or V-shaped grooves.[112] These structures are less easily used in planar pattern transfer, so will not be discussed here in any detail.

3.3.2. *Thermal annealing of block copolymers in topographical templates*

Many factors determine the ordering kinetics of BCP domains in topographical templates, including molecular weight, the Flory-Huggins parameter χ, annealing history and domain morphology. Templating effects on the ordering behavior of 2D spherical block copolymer domains have been characterized via the translational correlation function and the orientational order correlation function.[115,129-132] For example, in the case that the domains in a thin film of a particular spherical block copolymer have liquid-like packing, imposing a topographical constraint induces ordering at the template edges, and the translational correlation length decreases algebraically with distance from an edge. The ordering propagates perpendicular to the edge, similar to the surface-induced layer ordering of thick films perpendicular to an unpatterned substrate. Both thermodynamics and kinetics play important roles in the in-plane ordering. Kramer and coworkers have demonstrated that the segregation energy associated with the minority block, χN_{min}, where N_{min} is the degree of polymerization of the minority block, can be used to describe the ordering of the system.[131] A large value of χN_{min} gives a 'polycrystalline' structure because of slow diffusion. Systems with an intermediate value of χN_{min} can form a near 'single crystalline' structure with few defects, whereas for a low χN_{min} value, fluctuations are more stable, so that dislocation pairs and disclination clusters are generated and the templated block copolymer forms hexatic and liquid phases. Therefore, given an appropriate molecular weight and interaction parameter, there is a processing window to optimize both the diffusivity and microphase segregation of the block copolymer thin film to reach the near-single crystalline state in a wide topographical template.

The results of thermal annealing can be also improved by choice of annealing history. It is suggested that annealing above the order-disorder transition temperature for a short time before annealing at a lower temperature helps to improve long-range order by 'resetting' the isotropic state.[54] Zone annealing is a very efficient method to improve ordering and to control the in-plane orientation of block copolymer domains. Zone-annealing involves processing by a moving hot-cold temperature gradient zone, hot being below the order-disorder transition temperature but above the glass transition temperature. A significant increase in the ordering kinetics of the BCP occurs, where morphologies characteristic of isothermal annealing times approaching a day are created in minutes using a moving thermal front.[133] The morphology of block copolymers also play a role in the efficiency of defect removal. Defect removal is much slower in confined lamellae than in confined cylinders of the same lateral periodicity. This might be a result of different film stress values in lamellar and cylindrical materials during defect annihilation or the different activation energies for dislocation displacement perpendicular to the layers.[115]

The ordering process for BCP microdomains in trenches using ex-situ and in-situ time-lapse atomic force microscopy was studied by Sibener and coworkers.[105] The process for domain rearrangement differs from the ordering process in an unconfined film. Figure 10 shows that the ordering of polymer domains begins with one or two cylinders aligned with the edge, then at random locations along the groove, regions of parallel cylinders start to grow. The final stage of alignment is achieved by merging of misoriented regions, and it therefore takes longer for cylinders in wider grooves to align compared with narrow grooves.

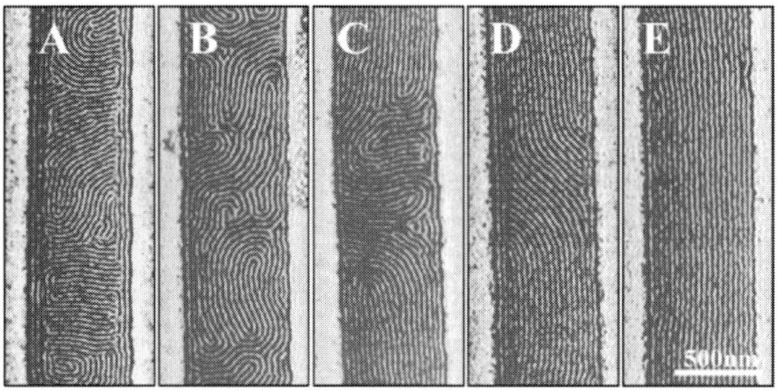

Fig. 10. Sequential phase-contest AFM images of a film of cylindrical PS-PEP BCP prepared on a 95 nm deep and 600 nm wide template with annealing times of (A) 9, (B) 14, (C) 19, (D) 24, and (E) 33 h at 130 °C. (Reprinted from Ref. 105 with permission. Copyright American Chemical Society.)

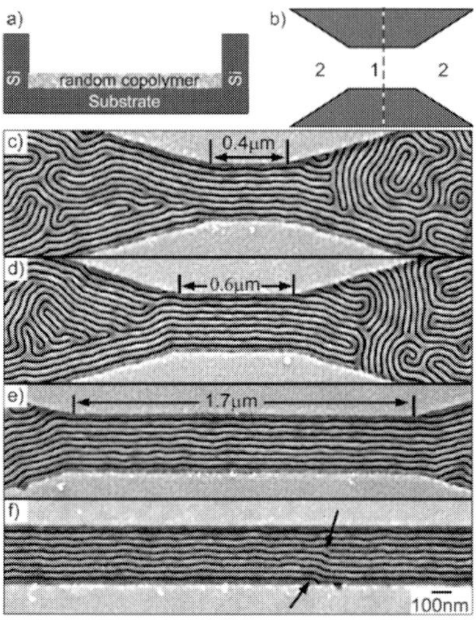

Fig. 11. (A) Schematic representation of a trench cross section prior to polymer deposition taken along the dashed line shown in (B). (B) Schematic representation of a fabricated silicon channel. White areas are recessed by ~20 nm with respect to the gray areas. (C-F) Lithographic channels with narrow regions of width $w = 9\ L_0$ and lengths of (C) 0.4 μm, (D) 0.6 μm, (E) 1.7 μm, and (F) 5 μm. (Reprinted from Ref. 134 with permission. Copyright Wiley-VCH.)

The efficiency of thermal annealing can be further improved by the design of the topographical templates. Because of the extremely limited degree of pattern coarsening in perpendicular lamellar materials, Ruiz et al. developed specially shaped templates to increase the rate of defect removal.[134] The defect formation energy is a function of groove width, such that a defect in a wider groove is less energetically costly than one in a narrower groove. Figures 11A and B show a trench with a funnel at both ends. Well-aligned and defect-free perpendicular lamellae domains from PS-PMMA can be achieved in trenches up to 5 μm long (as shown in Figures 11C-F). Defects are preferentially located in the funnel area where the defect energy is smaller; the shape of the trench also promotes shear flow which aligns the lamellae.

3.3.3. *Solvent vapor annealing of block copolymer films within topographical substrates*

As an alternative to thermal annealing, solvent vapor annealing can be used to improve long-range ordering in topographical. During solvent vapor annealing, the BCP thin film is swelled by the solvent, affecting the diffusivity and the surface energy of the blocks and enabling control of BCP morphology and domain orientation.[48,49,61,62] Solvent vapor annealing can be used to improve long-range ordering of high molecular-weight, low-diffusivity BCPs, but it can also lead to domain volume fractions and morphologies not seen in thermally annealed films.

Long-range ordered perpendicularly oriented cylinders of PS-PEO within trenches have been demonstrated by Kim et al.,[49] where the ordering improves dramatically and defect density drops significantly within first few hours of solvent annealing. At long anneal times, however, the films dewet from the substrates. Well-ordered cylinders in a

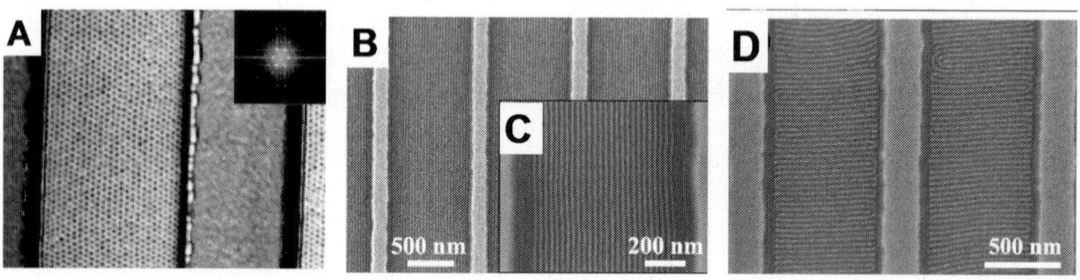

Fig. 12. (A) An AFM phase image of PS-PEO spin-casted on a topographical trench of 875 nm width. Upon solvent vapor annealing the film dewets in the region outside the channels and the dewetted PS-PEO is trapped within the channels. Solvent evaporation produces highly aligned arrays of perpendicular cylinders. (from Ref. 62) (B) and (C) SEM images of parallel cylinders on trench substrates with narrow mesas (W_{mesa} = 125 nm and W_{trench} = 875 nm) under a high vapor pressure of toluene and (D) perpendicular cylinders in a wide-mesa pattern (W_{mesa} = 270 nm and W_{trench} = 730 nm) at a lower vapor pressure. The annealing time was 15 h. (Reprinted from Ref. 48 with permission. Copyright American Chemical Society.)

875 nm wide trench produced by annealing under benzene and water vapor at room temperature are shown in Figure 12A. Jung et al. demonstrated that in-plane PDMS cylinders in confined PS-PDMS thin films can be oriented perpendicular or parallel to trench edges by solvent vapor annealing (Figure 10B-10D).[48] The direction of cylinder orientation depends on the trench geometry and annealing conditions. Trenches with wide mesas and a relatively low vapor pressure during the annealing lead to cylinders oriented across the trenches. Perpendicular alignment of cylinders in PS-PEP in local areas of trenches has been attributed to capillary flow of BCP from the mesas to trenches, perpendicular to the trench.[135]

3.3.4. *Placement accuracy and pattern uniformity of block copolymer domains*

Successful implementation of BCP lithography to make devices such as patterned media requires the ability to make uniform patterns with high placement accuracy of the features. A size distribution of the BCP domains will translate into a distribution of magnetic properties of dots including a distribution of the net magnetic moment (and therefore the readback signal), and a distribution of switching fields which introduces variability in the writing process. In addition, a size distribution of the BCP domains will degrade the spatial coherency of the BCP features leading to placement error in the magnetic dots and introducing jitter into the readback process.

There are some reports on the size uniformity of a monolayer of BCP domains in a thin film. Cheng et al. found a 9% standard deviation of diameters in magnetic dots made using a PS-PFS spherical BCP as an etch mask.[64] Guarini et al. showed that the size distribution of a perpendicularly oriented PS-PMMA cylindrical BCP increased with the molecular weight of the BCP. For 64k molecular weight PS-PMMA with 20 nm diameter perpendicular cylinders, the standard deviation of cylinder diameter was 10% after a long thermal anneal.[136] Xiao and Sun reported a standard deviation of 4.3 – 5.3% in cylindrical PS-PMMA of molecular weight 67k.[137] A monolayer of spheres or perpendicularly-oriented cylinders forms a closed packed array in which most domains have six-fold coordination, while defects have either five-fold or seven-fold coordination. Hammond et al. characterized the size of 5-fold, 6-fold and 7-fold coordinated cylinders in vertically oriented poly(cyclohexylethylene)-*b*-poly(ethylene)-*b*-poly-(cyclohexylethylene) (PCHE-PE-PCHE). The average diameter of a 5-fold coordinated PE domain was 10% smaller than that of a 6-fold coordinated cylinder while a 7-fold coordinated domain was 10% larger. Therefore, removal of 5-fold and 7-fold coordinated defects can reduce the size variation. On the other hand, within the 6-fold coordinated cylinders the standard deviation in diameter is ~5% from experiment and 3% from simulation.[138] Therefore, even if defects can be removed, there could still be a 3%-5% variation in the size of 6-fold coordinated BCP domains.

Besides pattern uniformity, placement accuracy is another critical factor for nanomagnetic devices. In a patterned recording media, the readback jitter (noise), as well as the quality of the written pattern, depend on the exact locations of the magnetic dots. In chemical epitaxy, the placement accuracy of the BCP domains depends on the placement accuracy of the guiding chemical patterns and on random alignment defects. In topographical epitaxy, the placement accuracy is determined by both template edge roughness and fluctuations in the size and spacing of the polymer domains. Cheng et al. found that while placement errors of spherical BCP domains near a template edge are proportional to the template roughness, the relatively soft domain-domain potential ensures that the portion of the array further from the edge of the template is almost unperturbed by the template roughness.[9,102,139] Away from the template edge the intrinsic fluctuations in spacing and size of BCP domains arise from the limitations on phase separation and on compositional and molecular weight polydispersity effects, which determine the ultimate placement accuracy in templated self-assembled BCPs. Pair distribution functions were used to characterize the positional distribution of the confined polymer domains. In Figure 13A, the amplitude of oscillation in the 2D pair distribution function of an array of spherical domains of PS-PFS on flat substrates decreases rapidly with distance, and the correlation length is approximately 10 L_0. In contrast, the 1D pair distribution function parallel to the groove wall of the same polymer in a groove (Figure 13B) shows that precise domain position can be predicted over distances exceeding 35 L_0. A standard deviation of ~0.1L_0 in a spherical PS-PFS block copolymer is found in this particular system, which implies that the array can be located to within ±6 nm (~0.2L_0) of a registration mark, with 95% confidence. More studies of the effects of materials and annealing conditions on the pattern uniformity and placement accuracy will help determine the ultimate precision and uniformity achievable in templated self-assembled BCPs for lithography.

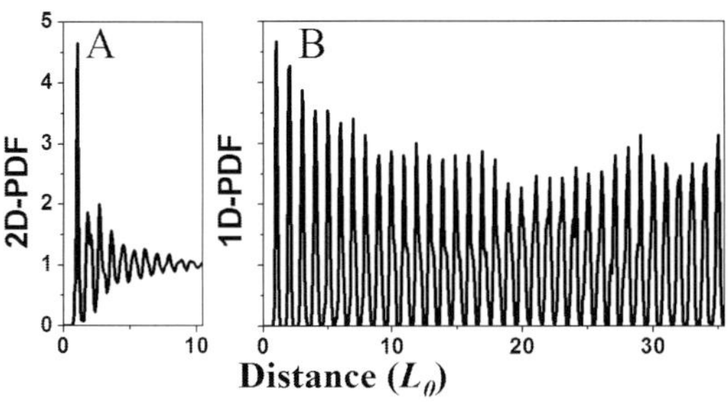

Fig 13. (A) 2D pair distribution function (PDF) of spherical domains in an as-cast film of PS-PFS on a flat substrate. (B) 1D PDF of annealed PS-PFS spheres in a 1D groove. The width of the peaks in the PDF gives the lateral placement accuracy of the spherical domains. (Reprinted from Ref. 9 with permission. Copyright Wiley-VCH.)

4. Summary and Outlook for Block Copolymer Patterning of Nanomagnetic Devices

In combination with pattern transfer processes, block copolymer patterning provides a simple and low-cost route to fabricate arrays of nanoscale structures. Discrete magnetic dots are particularly interesting because of their geometry-dependent magnetic properties as well as practical applications such as high-density patterned magnetic data storage media. We have highlighted here (i) the fabrication by BCP patterning and the magnetic properties of arrays of nanoscale magnetic elements, and (ii) the possible methods for improvement of long-range order and placement accuracy through directed self-assembly of BCPs.

In the review of nanomagnet fabrication, we saw that both subtractive and addictive methods have been used in combination with BCP patterning to generate nanoscale magnetic structures, and the magnetic properties of a limited number of systems have been explored. These fabrication methods will continue to be improved, giving structures with better pattern definition and size distribution. We also showed that long-range ordering of the self-assembled BCP arrays can be accomplished using chemically or topographically patterned substrates, enabling specific orientation, alignment, pattern registration, and even new microdomain geometries to be formed.

However, there are still many issues to address in order to demonstrate the full potential of block copolymer patterning. Most research in directed self-assembly of block copolymers has focused on the generation of ordered periodic patterns. Although periodic patterns are useful in some applications, the ability to precisely control domain positions and the purposeful placement of 'defects' (i.e. aperiodicities) will permit the encoding of additional spatial information into directed self-assembled systems. This may be valuable in, for example, the formation of servo-patterns in patterned media. More studies are needed to illuminate the role of the templates in defect formation and defect control. Further exploration of the kinetics of the ordering within the imposed boundary conditions will provide insights into the roles of diffusion, molecular weight and the χ-parameter and will benefit system design and optimization. Further experimental and theoretical studies on placement accuracy and pattern uniformity of directed block copolymers are needed: the combined effects of external fields such as electric, optical, stress, flow fields and temperature gradients combined with substrate patterns and annealing conditions will bring more control factors into play and may allow better manipulation of ordering and defects.

There is also considerable latitude in BCP design to give better control over domain-domain interactions and available pattern geometries. The use of triblock and more complex BCPs such as star copolymers can increase the geometries available, for example ring-shaped structures or square arrays of features may be generated by using triblock copolymers. In addition, the use of functional blocks, such as photosensitive materials, liquid crystals, or materials that can serve as precursors, can increase the functionality of the self-assembled pattern beyond its simple use as an etch mask. Increasing complexity allows more information to be coded into the self-assembly system. Compared to bio-macromolecules, synthetic block copolymers are comprised of simple sequences with relatively simple monomer-monomer interactions, but the possibilities inherent in more complex systems are becoming more apparent. A beautiful demonstration of high-yield, programmable complex self-assembly was shown by Rothemund,[140] who generated squares, triangles, happy faces and other shapes using DNA self-assembly. The patterned substrates developed for directed self-assembly of block copolymers could be used for the self-assembly of synthetic co-polypeptides and even bio-molecules.

Advances in BCP patterning will contribute directly to our ability to fabricate useful magnetic devices. BCP patterning has been used to make transistors, capacitors, patterned media and other simple devices, but we may anticipate the construction of more complex structures based on precisely templated nanostructures. There has as yet been little use of templated BCP arrays to form magnetic nanostructures other than early demonstrations of patterned media, and it will be interesting to see what magnetic or magnetoelectronic devices become possible as templated BCP patterning techniques become more mature.

References

1. C. Park, J. Yoon, E. L. Thomas, *Polym.,* 44, 6725 (2003).
2. I. W. Hamley, *Nanotechnology* 14, R39 (2003).
3. G. Krausch, R. Magerle, *Adv. Mater.,* 14, 1579 (2002).
4. M. Lazzari, M. A. Lopez-Quintela, *Adv. Mater.*, 15, 1583 (2003).
5. I. W. Hamley, *Angew. Chem. Int. Ed.*, 42, 1692 (2003).
6. R. A. Segalman, *Mater. Sci. Eng. Rep.*, 48, 191 (2005).
7. M. Li, C. Coenjarts, C. K. Ober, in Block Copolymers II, Advances in Polymer Science, V. Abetz, ed., 190, 183, 2005 (Springer, Heidelberg, 2005)
8. C. J. Hawker, T. P. Russell, MRS Bulletin, 30, 952 (2005).
9. J. Y. Cheng, C. A. Ross, E. L. Thomas, H. I. Smith, *Advanced Mater.*, 18, 2505 (2006).
10. M. P. Stoykovich, P. F. Nealey, *Materials Today*, 9, 20 (2006).
11. S. B. Darling, *Prog. Polymer Sci.*, in press (2007).
12. M. A. McCord, M. J. Rooks, Electron-beam lithography, in *Handbook of Microlithography, Micromachining and Microfabrication, Volume 1, Microlithography*, 140 Ed. By Choudhury PR, SPIE Optical Engineering Press (1997).
13. S. Y. Chou, P. R. Krauss, P. J. Renstrom, *Science*, 272, 85 (1996).

14. M. D. Stewart, C. G. Willson, *MRS Bulletin*, 30, 947 (2005).
15. Z. Z. Bandić, E. A. Dobisz, T. Wu, T. R. Albrecht, *Solid State Tech.* (2006).
16. R. D. Piner, J. Zhu, F. Xu, S. H. Hong, C. A. Mirkin, *Science*, 283, 661 (1999).
17. T. A. Savas, M. L. Schattenburg, J. M. Carter, H. I. Smith, *J. Vac. Sci. Tech. B*, 14, 4167 (1996).
18. H. H. Solak, C. David, J. Gobrecht, V. Golovkina, F. Cerrina, S. O. Kim, P. F. Nealey PF, *Microelec. Eng.*, 67, 56 (2003).
19. H. I. Smith, D. C. Flanders, *J. Vac. Sci. Tech.*, 17, 533 (1980).
20. F. S. Bates, G. H. Fredrickson, *Annual Review of Physical Chemistry*, 41, 525 (1990).
21. M. J. Fasolka, A. M. Mayes, *Ann. Rev. Mater. Res.*, 31, 323 (2001).
22. B. D. Terris, T. Thomson, *J. Phys. D*, 38, R199 (2005).
23. S. B. Darling and S. D. Bader, *J. Mater. Chem.*, 15 4189 (2005).
24. S. H. Charap, P. L. Lu, Y. He, *IEEE Trans. Magn.*, 33 978 (1997).
25. D. Weller and A. Moser, *IEEE Trans. Magn.*, 35, 4423 (1999).
26. M.H. Kryder, R. W. Gustafson, *J. Magn. Magn. Mater.*, 287 449 (2005).
27. R. L. White, R. M. H. New, R. F. W. Pease, *IEEE Trans. Magn.*, 33 990 (1997).
28. C.A. Ross, *Ann. Rev. Mater. Sci.*, 31, 203 (2001).
29. G. Hughes, in *The Physics of Ultra-High-Density Magnetic Recording*, ed. M.L. Plumer, J. van Ek, D. Weller (Springer, Heidelberg, 2001).
30. H. J. Richter, A. Y. Dobin, R. T. Lynch, D. Weller, R. M. Brockie, O. Heinonen, K. Z. Gao, J. Xue, R .J. M. de Veerdonk, P. Asselin, M.F. Erden, *Appl. Phys. Letts.*, 88 222512 (2006).
31. H. J. Richter, *J. Phys. D*, 40, R149 (2007)
32. K. Naito, H. Hieda, M. Sakurai, Y. Kamata, K. Asakawa, *IEEE Trans. Magn.*, 38, 1949 (2002).
33. P. Mansky, P. Chaikin, E. L. Thomas, *Journal of Materials Science*, 1987 (1995).
34. M. Park, C. Harrison, P. M. Chaikin, R. A. Register, D. H. Adamson, *Science*, 276, 1401 (1997).
35. P. Mansky, Y. Liu, E. Huang, T. P. Russell, C. J. Hawker, *Science* 1997, 275, 1458
36. R. D. Peters, X. M. Yang, T. K. Kim, B. H. Sohn, P. F. Nealey, *Langmuir*, 16, 4625 (2000).
37. T. Thurn-Albrecht, R. Steiner, J. DeRouchey, C. M. Stafford, E. Huang, M. Bal, M. Tuominen, C. J. Hawker, T. P. Russell, *Adv. Mater.*, 12, 787 (2000).
38. D. Y. Ryu, K. Shin, E. Drockenmuller, C. J. Hawker, T. P. Russell, *Science*, 308, 236 (2005).
39. T. Thurn-Albrecht, J. Schotter, C. A. Kastle, N. Emley, T. Shibauchi, L. Krusin-Elbaum, K. Guarni, C. T. Black, M. T. Tuominen, T. P. Russell, *Science*, 290, 2126 (2000).
40. K. Asakawa, T. Hiraoka, *Jpn. J. Appl. Phys.*, 41, 6112 (2002).
41. Z. Lin, D. H. Kim, X. Wu, L. Boosahda, D. Stone, L. LaRose, T. P. Russell, *Adv. Mater.*, 14, 1373 (2002).

42. J. Bang, S. H. Kim, E. Drockenmuller, M. J. Misner,T. P. Russell, C. J. Hawker, *J. Am. Chem. Soc.*, 128, 7622 (2006).
43. R. Olayo-Valles, M. S. Lund, C. Leighton, M. A. Hillmyer, *J. Mater. Chem*, 14, 2729 (2004).
44. T. Kubo, J. S. Parker, M. A. Hillmyer, C. Leighton, *Appl. Phys. Lett.*, 90, 233113 (2007).
45. J. Y. Cheng, C. A. Ross, V. Z. H. Chan, E. L. Thomas, R. G. H. Lammertink, G. J. Vancso, *Advanced Materials*, 13, 1174 (2001).
46. R. G. H. Lammertink, M. A. Hempenius, J. E. van der Enk, V. Z-H. Chan, E. L. Thomas, G. J. Vancso, *Advanced Materials*, 12, 98 (2000).
47. M. Roerdink, M. A. Hempenius, G. J. Vancso, *Chem. Mater.*, 17, 1275 (2005).
48. Y. S. Jung, C. A. Ross, *Nano Lett.*, 7, 2046 (2007).
49. E. M. Freer, L. E. Krupp, W. D. Hinsberg, P. M. Rice, J. L. Hedrick, J. N. Cha, R. D. Miller, H. C. Kim, *Nano Lett.*, 5, 2014 (2005).
50. L. Sundstrom, L. Krupp, E. Delenia, C. Rettner, M. Sanchez, M. W. Hart, H. C. Kim, Y. Zhang, *Appl. Phys. Lett.*, 88, 243107 (2006).
51. T. L. Bucholz, Y. L. Loo, *Macromolecules*, 39, 6075 (2006).
52. T. P. Russell, R. P. Hjelm, P. A. Seeger, *Macromol.*, 23, 890 (1990).
53. H. Frielinghaus, N. Hermsdorf, K. Almdal, K. Mortensen, L. Messe, Corvazier, L.; Fairclough, J. P. A.; Ryan, A. J.; Olmsted, P. D.; Hamley, I. W. *Europhys. Lett.*, 53, 680 (2001).
54. M. R. Hammond, E. Cochran, G. H. Fredrickson, E. J. Kramer, *Macromolecules*, 38, 6575 (2005).
55. H. B. Eitouni, N. P. Balsara, H. Hahn, J. A. Pople, M. A. Hempenius, *Macromolecules*, 35, 7765 (2002).
56. Nose, T. *Polymer*, 36, 2243 (1995).
57. C. Tang, A. Tracz, M. Kruk, R. Zhang, D.-M. Smilgies, K. Matyjaszewski, T. Kowalewski, *J. Am. Chem. Soc.*, 127, 6918 (2005).
58. G. Kim, M. Libera, *Macromolecules*, 31, 2569 (1998).
59. K. Temple, K. Kulbaba, K. N. Power-Billard, I. Manners, A. Leach, T. Xu, T. P. Russell and C. J. Hawker, *Adv. Mater.*, 15, 297 (2003).
60. P. Du, X. Li, K. Doki, M. Li, S. Gruner, D. Smilgies, U. Wiesner, C. Ober, *Adv. Mater.*, 16, 953 (2004).
61. Yu Xuan, Juan Peng, Liang Cui, Hanfu Wang, Binyao Li, and Yanchun Han, *Macromolecules*, 37, 7301 (2004).
62. S. H. Kim, M. J. Misner, M. Kimura, T. Xu and T. P. Russell, *Adv. Mater.*, 2004, 16, 226 (2004).
63. M. E. Walsh, Y. Hao, C. A. Ross, H. I. Smith, *J. Vac. Sci. Technol. B*, 18, 3539 (2000).
64. J. Y. Cheng, W. Jung, C.A. Ross, *Phys. Rev. B*, 70, 064417 (2004).
65. K. Asakawa, T. Hiraoka, H. Hieda, M. Sakurai, Y. Kamata, *J. Photopolym. Sci. Tech.*, 15, 465 (2002).

66. Y. Kamata, A. Kikitsu, H. Hieda, M. Sakurai, K. Naito, *J. Appl. Phys.* 2004, 95, 6705 (2004).
67. H. Hieda, Y. Yanagita, A. Kikitsu, T. Maeda, K. Naito, *J. Photopolym. Sci. Tech.*, 19, 425 (2006).
68. Y. Kamata, A. Kikitsu, H. Hieda, M. Sakurai, K. Naito, J. M. Bai, S. Ishio, *Jpn. J. Appl. Phys.*, 46, 999 (2007).
69. K. Liu, S. M. Baker, M. Tuominen, T. P. Russell, I. K. Schuller, *Phys. Rev. B*, 63, 060403 (2001).
70. A. Ursache, J.T. Goldbach, T.P. Russell, M.T. Tuominen, *J. Appl. Phys.*, 97, 10J322 (2005).
71. A. Ursache, M. Bal, J. T. Goldbach, R. L. Sandstrom, C. T. Black, T. P. Russell, M. T. Tuominen, *Proc. Mater. Res. Soc.*, 421 E3.3 (2002).
72. S. Xiao, X. Yang, E. W. Edward, Y. La, P. F. Nealey, *Nanotech.*, 16, S324 (2005).
73. C. T. Black, K. W. Guarini, R. L. Sandstrom, S. Yeung and Y. Zhang, *Proc. Mater. Res. Soc.*, 728, S4.9 (2002).
74. J. R. Jeong, M. C. Choi, M. W. Kim, S. C. Shin, *Phys. Stat. Sol. B*, 241 1609 (2004).
75. V. P. Chuang, W. Jung, J. Y. Cheng, Oun-ho Park, Ho-Cheol Kim, C. A. Ross, submitted.
76. B. H. Sohn, R. E. Cohen, G. C. *Journal of Magnetism and Magnetic Materials*, 182, 216 (1998).
77. J. I. Abes, R. E. Cohen, C. A. Ross, *Chem. Mater.*, 15 1125 (2003).
78. J. I. Abes, R. E. Cohen, C. A. Ross, *Mat. Sci. Eng. C*, 23, 641 (2003).
79. T. Hashimoto, K. Tsutsumi, Y. Funaki, *Langmuir*, 13, 6869 (1997).
80. S. B. Darling, N. A. Yufa, A. L. Cisse, S. D. Bader, and S.J. Sibener, *Adv. Mater.*, 17, 2446 (2005).
81. S. B. Darling, *Surf. Sci.,* 601, 2555 (2007).
82. F. Ilievski, C. A. Ross, GJ Vancso, *J. Appl. Phys.* 103, 07C520 (2008)
83. F. J. Castano, K. Nielsch, C. A. Ross, J. W. A. Robinson, R. Krishnan, *Appl. Phys. Lett.*, 85 2872 (2004).
84. L. J. Heyderman, F. Nolting, D. Backes, S. Czekaj, L. Lopez-Diaz, M. Klaui, U. Rudiger, C. A. F. Vaz, J. A. C. Bland, R. J. Matelon, U. G. Volkmann, P. Fischer, *Phys. Rev. B*, 73, 214429 (2006).
85. R. P. Cowburn, A. O. Adeyeye, J. A. C. Bland, *Appl. Phys. Lett.*, 70, 2309 (1997).
86. T. Shibauchi, L. Krusin-Elbaum, L. Gignac, C. T. Black, T. Thun-Albrecht, T. P. Russell, J. Schotter, G. A. Kastle, N. Emley, M. T. Tuominen, *J. Magn. Magn. Mater.*, 226, 1554 (2001).
87. M. Bal, A. Ursache, M. T. Tuominen, J. T. Goldbach, T. P. Russell, *Appl. Phys. Letts.*, 81, 3479 (2002).
88. O. Yalcon, F. Yildiz, M. Ozdemir, B. Aktas, Y. Koseoglu, M. Bal, M. T. Tuominen, , *J. Magn. Magn. Mater.*, 272, 1684 (2004).

89. L. Rockford, Y. Liu, P. Mansky, T. P. Russell, M. Yoon, S. G. J. Mochrie, *Phys. Rev. Lett.*, 82, 2602 (1999).
90. L. Rockford, S. G. J. Mochrie, T. P. Russell, *Macromol.*, 34, 1487 (2001).
91. X. M. Yang, R. D. Peters, T. K. Kim, P. F. Nealey, *J. Vac. Sci. Technol. B*, 17, 1203 (1999).
92. X. M. Yang, R. D. Peters, P. F. Nealey, H. H. Solak, F. Cerrina, *Macromol.*, 33, 9575 (2000).
93. S. O. Kim, H. H. Solak, M. P. Stoykovich, N. J. Ferrier, J. J. de Pablo, P. F. Nealey, *Nature*, 424, 411 (2003).
94. E. W. Edwards, M. F. Montague, H. H. Solak, C. J. Hawker, P. F. Nealey, *Adv. Mater.*, 16, 1315 (2004).
95. M. P. Stoykovich, M. Muller, S. O. Kim, H. H. Solak, E. W. Edwards, J. J. de Pablo, P. F. Nealey, *Science*, 308, 1442 (2005).
96. E. W. Edwards, M. Muller, M. P. Stoykovich, H. H. Solak, J. J. de Pablo, P. F. Nealey, *Macromol*, 40, 90 (2007).
97. J. Y. Cheng, C. T. Rettner, D. P. Sanders, H. C. Kim, W. D. Hinsberg, *in press, Advanced Materials* (2008).
98. R. Ruiz, H. Kang, F. A. Detcheverry, E. Dobisz, D. S. Kercher, T. R. Albrecht, J. J. de Pablo, P. F. Nealey, in press, *Science*, (2008).
99. R. A. Segalman, H. Yokoyama, E. J. Kramer, *Adv. Mater.*, 13, 1152 (2001).
100. G. E. Stein, E. J. Kramer, X. Li, J. Wang, *Phys. Rev. Lett.*, 98, 086101 (2007).
101. J. Y. Cheng, C. A. Ross, E. L. Thomas, H. I. Smith, G. J. Vancso, *Appl. Phys. Lett.*, 81, 3657 (2002).
102. J. Y. Cheng, C. A. Ross, E. L. Thomas, H. I. Smith, G. J. Vancso, *Adv. Mater.*, 15, 1599 (2003).
103. J. Y. Cheng, A. M. Mayes, C. A. Ross, *Nature Mater.*, 3, 823 (2004).
104. C. T. Black, O. Bezencenet, *IEEE Trans. Nanotech.*, 3, 412 (2004).
105. D. Sundrani, S. B. Darling, S. J. Sibener, *Nano Lett.*, 4, 273 (2004).
106. D. Sundrani, S. B. Darling, S. J. Sibener, *Langmuir*, 20, 5091 (2004).
107. X. Yang, S. Xiao, C. Liu, K. Pelhos, K. Minor, *J. Vac. Sci. Tech. B.*, 22, 3331 (2004).
108. S. Xiao, X. Yang, E. W. Edward, Y. La, P. F. Nealey, *Nanotech.*, 16, S324 (2005).
109. Q. Zhang Q, D. C. Lee, L. Yu, S. J. Sibener, *Materials Research Society Proceedings*, 854E (2005).
110. T. Yamaguchi, H. Namatsu, *J. Photopolym. Sci. Tech.*, 18, 421 (2005).
111. L. Li, H. Yokoyama, *Adv. Mat.*, 17, 1432 (2005).
112. V. P. Chung, J. Y. Cheng, T. A. Savas, C. A. Ross, *Nano Lett.*, 6, 2332 (2006).
113. J. Y. Cheng, F. Zhang, A. M. Mayes, C. A. Ross, *Nano Lett.*, 6, 2099 (2006).
114. S. M. Park, R. Ruiz, Y. Zhang, C. T. Black, P. F. Nealey, *Adv. Mater.*, 19, 607 (2007).
115. R. Ruiz, R. L. Sandstrom, C. T. Black, *Adv. Mater.*, 19, 587 (2007).

116. T. Deng, Y. Ha, J. Y. Cheng, C. A. Ross, E. L. Thomas EL, *Langmuir*, 18, 6719 (2002).
117. H. Li, W. T. S. Huck, *Nano Lett.*, 4, 1633 (2004).
118. P. Lambooy, T. P. Russell, G. J. Kellogg, A. M. Mayes, P. D. Gallagher, S. K. Satija, *Phys. Rev. Lett.*, 72, 2899 (1994).
119. G. J. Kellogg, D. G. Walton, A. M. Mayes, P. Lambooy, P. D. Gallagher, S. K. Satija, *Phys. Rev. Lett.*, 76, 2503 (1996).
120. V. Abetz, R. Stadler, L. Leibler, *Polym. Bull.*, 37, 135. (1996).
121. R. Stadler, C. Auschra, J. Beckmann, U. Krappe, I. Voigt-Martin, L. Leibler, *Macromolecules*, 28, 3080 (1995).
122. I. Bita, J. K. W. Yang, Y. S. Jung, C. A. Ross, E. L. Thomas, K. K. Berggren, *Science,* in press (2008).
123. Y. Wu, G. Cheng, K. Katsov, S. W. Sides, J. Wang, J. Tang, G. H. Fredrickson, G. D. Stucky, *Nature Mater.*, 3, 816 (2004).
124. Q. Y. Liu, F. Gao, S. Komarneni, T. E. Mallouk, *J. Am. Chem. Soc.*, 126, 8650 (2004).
125. H. Xiang, K. Shin, T. Kim, S. I. Moon, T. J. Macarthy, T. P. Russell, *Macromolecules*, 38, 1055 (2005).
126. J. R. Reffner, *The influence of surfaces on structure formation: phase separation of block copolymers and polymer blends under nonplanar surface constraints.* Ph.D. Thesis, University of Massachusetts at Amherst (1992).
127. E. L. Thomas, J. R. Reffner, J. Bellare, *J. de Phys.*, 51, C7363 (1990).
128. A. C. Aresenault, D. A. Rider, J. I. Chen, N. Coombs, G. A. Ozin, I. Manners, *J. Am. Chem. Soc.*, 127, 9954 (2005).
129. R. A. Segalman, A. Hexemer, E. J. Kramer, *Macromolecules*, 36, 6831 (2003).
130. R. A. Segalman, A. Hexemer, E. J. Kramer, *Phys. Rev. Lett.*, 91, 196101 (2003).
131. R. A. Segalman, A. Hexemer, R. Hayward, E. J. Kramer, *Macromolecules*, 36, 3272 (2003).
132. A. Hexemer, G. E. Stein, E. J. Kramer, S. Magonov, *Macromolecules*, 38, 7083 (2005).
133. B. C. Berry, A.W. Bosse, J. F. Douglas, R. L. Jones, A. Karim, *Nano Lett.*, 7, 2789 (2007).
134. R. Ruiz, N. Ruiz, Y. Zhang, R. L. Sandstrom, C. T. Black, *Adv. Mater.*, 19, 2157 (2007).
135. D. Sundrani, S. J. Sibener, *Macromolecules*, 35, 8531 (2002).
136. K. W. Guarini, C. T. Black, S. H. I. Yeuing, *Adv. Mater.*, 14, 1290 (2002).
137. S. Xiao and X. Yang, *J. Vac. Sci. Technol. B*, in press (2007).
138. M. R. Hammond, S. W. Sides, G. H. Fredrickson, E. J. Kramer, J. Ruokolainen, S. F. Hahn, *Macromolecules*, 36, 8712 (2003).
139. J. Y. Cheng, F. Zhang, H. I. Smith, G. J. Vancso, C. A. Ross, *Adv. Mater.*, 18, 597 (2006).
140. P. W. K. Rothemund, *Nature*, 440, 297 (2006).

CHAPTER 5

HIERARCHICAL STRUCTURING OF POLYMER NANOPARTICLES BY SELF-ORGANIZATION

Masatsugu Shimomura*[abcde], Hiroshi Yabu[abcd], Takeshi Higuchi[bc], Atsunori Tajima[c] and Tetsuro Sawadaishi[b]

[a]*Institute of Multidisciplinary Research for Advanced Materials (IMRAM) Tohoku, University, 2-1-1, Katahira, Aoba-ku, Sendai 980-8577, Japan*
[b]*Frontier Research System, The Institute of Physical and Chemical Research (RIKEN), 2-1, Hirosawa, Wako, Saitama 351-0198, Japan*
[c]*Graduate School of Science, Hokkaido University, N10W8 Kita-ku, Sapporo 060-0810, Japan*
[d]*Core Research for Evolutional Science and Technology (CREST), Japan Science and Technology Agency (JST), 4-1-8, Honcho, Kawaguchi, Saitama 332-0012, Japan*
[e]*WPI Research Center: Advanced Institute for Materials Research, Tohoku University, 2-1-1, Katahira, Aoba-ku, Sendai 980-8577, Japan*
E-mail: shimo@tagen.tohoku.ac.jp

Spontaneous formation of polymer nanoparticles and their mesoscopic assemblies are achieved by using self-organization processes. Spherical and hemispherical polymer nanoparticles are formed from a clear solution, containing a non-volatile poor solvent, by slow evaporation of a volatile good solvent. Homogeneous nucleation and successive growth of polymer particles is emerged during dynamic non-equilibrium process of the solvent evaporation at the solution/air interface. Micro phase separation is observed not only in the block copolymer particles but also in the particles of polymer blend. Hierarchical structuring is successfully achieved by combining nanoparticles and self-organized honeycomb-patterned polymer matrix, and also by using "dissipative structures" formed in the evaporating front of aqueous polymer particles dispersions.

1. Introduction

Hierarchical self-organization of molecules from nanoscopic to macroscopic scale in biological systems is structural bottom-up strategy of spatial functions for many biological processes associated with their temporal developments. Biological membranes are fundamental functional "molecular devices" in biological systems and multi-component complex systems mainly consisting of lipid bilayers and biological macromolecules. Furthermore, biomembranes are elemental structural components of mesoscopic submicrometer-sized subcellular apparatuses, i.e. organelles. Biological cells are micrometer-size assemblies of the organelles, and basal constructing units of biological tissues and living organs in macroscopic scale. Hierarchical structuring of biomolecules from nanoscopic to macroscopic scale in biological systems is a vital and constitutional base of living organs. Then the hierarchical structuring of functional molecules is an indispensable strategy of the biomimetic approach in the advanced materials fabrication.

In the last century, by using the self-assembling nature of artificially designed molecules, chemists have succeeded to construct many sorts of nanometer-size molecular assemblies, e.g. surfactant bilayer assemblies,[1] Langmuir-Blodgett films,[2] self-assembled monolayers,[3] polymer assemblies,[4,5] and so on. "Supramolecular Chemistry"[6] is nanoscopic chemistry beyond molecules and has suggested fundamental aspects of designing novel materials based on self-assembly. Driving forces of molecular assembling employed in the nanoscopic world are many weak physico-chemical intermolecular interactions, e.g. van der Waals force, hydrogen bonds, hydrophobic interaction, electrostatic force, and so on. Assembling the nanostructured molecular architectures up to mesoscopic-scaled organizates is the next challenging step of the hierarchical structuring of molecules for novel materials fabrication.

Efforts have been made to fabricate mesoscopic molecular organizations. It is well known that some mesoscopic patterns of polymer assemblies have been found as micro- and macrophase separated structures of block copolymers,[7,8] and polymer blends, respectively. The main driving force for the mesoscopic self-structuring in the polymer films is the compatibility of each block of the copolymers or blended polymers. However, the essential driving forces of the mesoscopic self-structuring in the artificial systems are distinctively different from the biological systems. In the biological systems, the hierarchical structuring is employed by dynamic "self-organization" processes with energy consumption in an open system. As a general physical phenomenon, we know self-structuring in the open system based on the energy-consuming dynamic "self-organization" process, which is the "dissipative structure" coined by Ilya Prigogine.[9] Belousov-Zabotinsky reaction and Rayleigh-Bénard convection are well known examples of the dissipative structures that spontaneously produce "spatio-temporal patterns" in various spatial scales from sub-micrometer to kilometer size.

Nowadays many attentions have been attracted to the terms of "self-organization" and "self-assembly" especially in nanotechnology. These are the strategic basis of the bottom-up methods for the nanomaterials fabrication. "Self-assembly" is the fundamental principle of integration where the components spontaneously assemble on all scale from atoms, molecules to galaxies, including living organs. A crystal is a typical example of the ordered integrated states spontaneously formed by "self-assembly" as the thermodynamic energy minimum states. "Self-organization" is another "self-assembly" which is dynamically emerged in an open system far from thermodynamic equilibrium. The dissipative structure, a typical example of "self-organization", is dynamic "self-assembly" characterized by the spontaneous appearance of order structure with energy and/or matters dissipation from chaos.

We have recently demonstrated that using dissipative structures generated in casting polymer solutions formed varieties of mesoscopic polymer assemblies[10–14] such as regularly arranged dots,[15] stripes,[16] grids,[17] and honeycomb-like structures,[18–21] and so on. Because of the physical generality of dissipative structures and a diversity of materials selection, our new preparation methodology based on "self-organization" is widely applicable to nano and micro fabrication of polymer materials without conventional lithographic procedures. In this chapter we have expanded our strategy based on dynamic "self-assembly" into the fabrication of polymer nanoparticles and their self-structured organization. We have focused here to the solvent evaporation process of polymer solutions because it occurs in an open system with the solvent dissipation and succeeding physico-chemical events emerged far from thermal equilibrium.

2. Preparation of Polymer Nanoparticles by Self-Organization

Polymer nanoparticles are widely used for versatile applications in the fields of photonics,[22] electronics,[23] and biotechnology.[24] There are many preparation methods[25] of polymer nanoparticles including milling bulk materials, emulsion polymerization, and reprecipitation.[26] Reprecipitation is commonly used for collection and purification of polymer substances from their solutions. In the reprecipitation process, polymer molecules dissolved in a good solvent are immediately precipitated as particles of polymer aggregates when a small amount of polymer solution is dropped into a large amount of poor solvent. Rapid exchange of solvent molecules surrounding polymer solutes decreases polymer solubility toward precipitation of polymer aggregates. Size and shape of polymer precipitates are largely dependent on physical conditions of mixing, such as stirring speed of mixing, dropping rate of solution, etc.

Recently, we have found that nanoparticles with narrow size distribution can be formed by evaporation of a good solvent from a polymer solution containing good and poor solvents.[27,28] In our method, precipitation of polymer particles starts from a clear solution of mixed solvent. Initially, the solution is homogeneously transparent because the amount of the poor solvent contained in the polymer solution is not enough large for polymer precipitation. With gradual evaporation of the good solvent at room temperature the solubility of the polymer solution reduces to form precipitation of the polymer molecules as fine particles.

2.1. *Self-Organized Particle Emergence by Good Solvent Evacuation*

Contrary to the reprecipitation method, in our procedure, a poor solvent is first added to a polymer solution of a good solvent and then the good solvent is slowly evacuated by evaporation. Finally polymer particles are dispersed in the poor solvent. The combination of the solvents used for this method requires that two solvents are miscible and boiling point of the good solvent is lower than that of the poor solvent. Figure 1 shows typical experimental procedures of polymer nanoparticle formation. Polystyrene (1.0 mg) was first dissolved in 2 ml of tetrahydrofuran (THF) as a good solvent, and 2 ml of water, a poor solvent, was slowly added into the THF solution through a syringe pump with dropping speed of 1 ml/min under stirring. The polymer solution was still optically transparent after mixing. If the receptacle was closed tightly to prevent the good solvent evaporation, the solution kept clear at room temperature ever after. When the mixed solution was left open for gradual evaporation of the good solvent, the solution kept optical transparency for a long while and then turned to turbid[28] (see a time course of transmittance in Figure 1). The long induction for turbidity generation in our method strongly suggests that particle formation mechanism is different from the reprecipitation method. Instantaneous clouding just

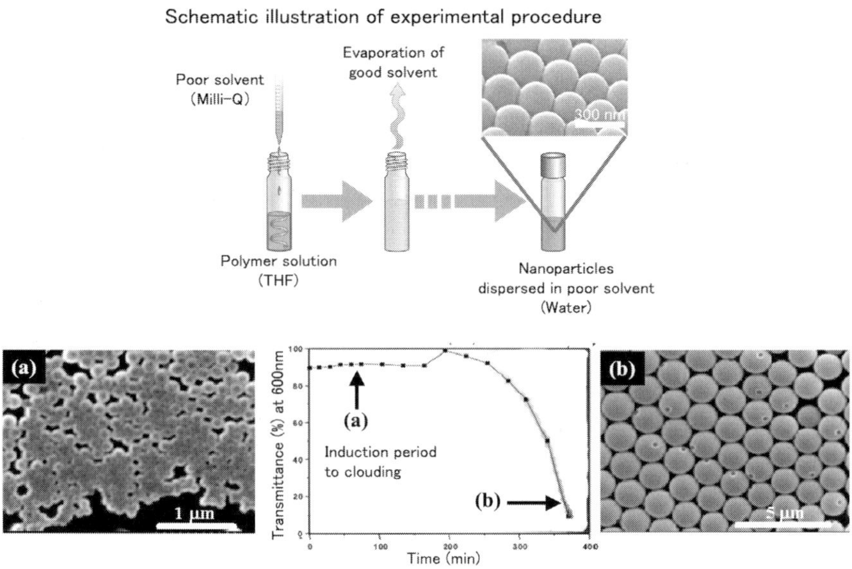

Fig. 1. Schematic illustration of experimental procedure of particle preparation and time course of the turbidity change after solvent evaporation. SEM images of nanoparticles indicating particle growth.

after mixing with the poor solvent observed in the reprecipitation process indicates that formation of polymer particles immediately occurs in the solution. Two images of a scanning electron microscope (SEM) of polystyrene particles are shown in Figure 1, too. Uniform sized (ca. 200 nm of diameter) spherical particles were observed when a small amount of polymer solution was picked up from the transparent solution (a). Since the particle size was in the sub-wavelength region the solution was optically transparent. The polymer solution abruptly became opaque ca.250 minutes after starting of evaporation. The SEM image (b) clearly shows that the particle size exceeds one micrometer.

From these experimental results, a plausible schematic model of the nanoparticle formation mechanism during the good solvent evaporation is proposed in Figure 2. Due to solvation by the good solvent, polymer molecules are dissolved as random coil form. Increasing the poor solvent, desolvation of polymer molecules starts to form the compact folding of the polymer chain and polymer molecules grow to small nuclei of particles. The equilibrium between random coils and nuclei dynamically shifts toward nuclei with decreasing the good solvent during its evaporation. The equilibrium changed from moment to moment during the evaporation may be a driving force of the homogeneous nuclei formation and successive growth of polymer nanoparticles. Particles grow with decreasing polymer solubility of the mixed solution, and the

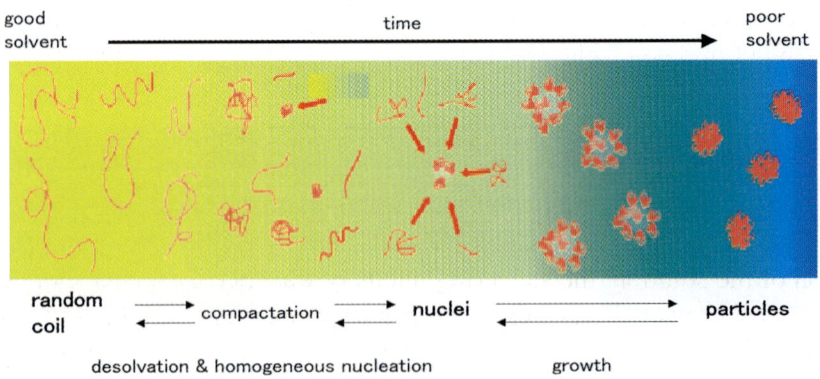

Fig. 2. Schematic representation of formation mechanism of uniform sized nanoparticles during dynamic process of solvent evaporation.

solution turns turbid when the particle size exceeds the wavelength region. Finally after complete evaporation of the good solvent polymer nanoparticles are dispersed in the poor solvent.

The final size of polymer nanoparticles is affected by the polymer concentration and the mixing ratio of the poor solvent. The particle size decreases with decreasing the polymer concentration as well as with increasing the mixing ratio of the poor solvent. Since the particle formation is based on general physical process, a large variety of polymers, e.g., engineering plastics, electro-conductive polymers, biodegradable polymers and biopolymers, etc. can be used for nanoparticle formation.

2.2. *Preparation of Hemispherical Polymer Nanoparticles*

As shown in Figure 1, a long induction period is required for the turbidity formation even though the solution is kept standing in open for the good solvent evaporation without any stirring. With careful observation of the external appearance of the polymer solution during the good solvent evaporation, we have found noticeable effect of the solution concentration on polymer particle formation.[29] The concentration of the polymer solution affects not only particle size but also its shape. In the case of polystyrene, spherical particles were formed when the concentration of the polystyrene solution was higher than 0.4 g/L. Their particle size were controlled from several tens nm to several tens μm. We have found that hemispherical particles were formed if the solution was diluted lower than ca. 0.2 g/L (Figure 3(a)).

After the long induction period, the turbidity starts from the surface of the solution whose concentration is lower than 0.2 g/L (Figure 3(b) left). We observed the time courses of the scattering intensity at upper and bottom regions of the low concentration solution. At the upper region of the solution, the scattering intensity was increased after mixing of the poor solvent (water). On the other hand, the scattering intensity at the bottom region was gradually increased to reach the same scattering intensity of the upper region for 10 hours after mixing the poor solvent. Since the good solvent, THF, evaporates from the surface of the solution,

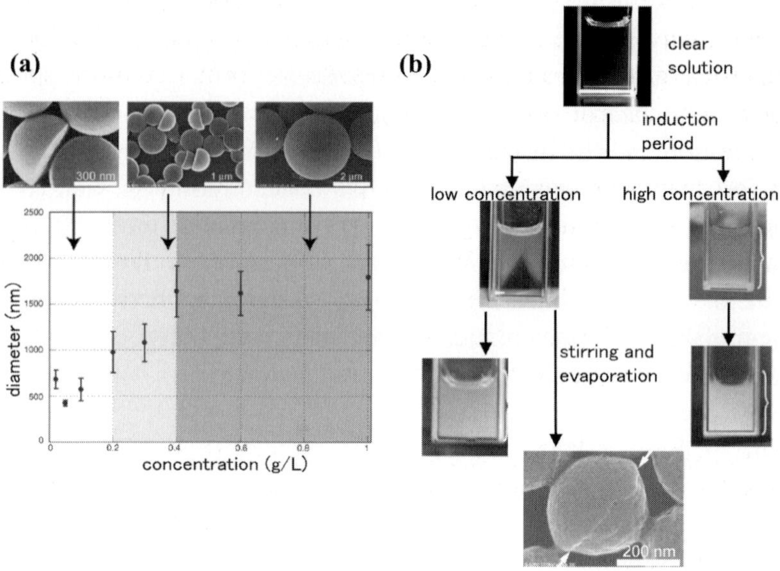

Fig. 3. Effect of polymer concentration on particle size and shape (a) and turbidity change of the polymer solution during THF evaporation (b).

the polystyrene was aggregated at the air/solution interface. Thus, the upper region of the solution firstly turned turbid. If the nucleation of the polymer particle starts at the air/solution interface, it is suggested that the particles grew up asymmetrically as the hemispherical particles.

In the bottom region of the solution, there was enough THF to dissolve polystyrene. The convectional flow, which was driven by the temperature and the concentration gradient due to the THF evaporation, transported the hemispherical particles or nuclei down to the bottom region of the solution. The hemispherical particles partially dissolved again in THF rich solution but were supposed to act as asymmetrical shaped nuclei of growing polymer particles. Finally, the hemispherical shaped particles grew and dispersed in water rich medium with keeping of their unique shape.

In the case of higher concentration than 0.4 g/L, the turbidity spread uniformly in the solution after the induction period for clouding (Figure 3(b) right). Nucleation and particle formation might start homogeneously in the solution after decreasing the good solvent to the critical concentration for the nucleation. Homogeneous nucleation and symmetrical growth from the nuclei results the formation of the spherical shaped polymer particle. Switching the asymmetrical growth at the interface to the symmetrical growth in the solution induces the spherical particle shape. In our particle preparation procedure, the polymer solution is kept standing without any mechanical stirring during the good solvent evaporation. When the evaporating solution of lower concentration was stirred after surface clouding, asymmetrical growth was switched to the symmetrical growth. A SEM image of a particle in the bottom of Figure 3(b) clearly shows boundary in the particle (white arrows), which means spherical growth starts from the hemispherical particle. It is remarkably mentioned that if the evaporation of the good solvent is proceeded under mechanical stirring from the beginning, polymers aggregate as bulk amorphous precipitates but not regular shaped particles.

Nucleation and asymmetrical growth of polymer particles at the evaporating solution surface provide unique inner nanostructures in the particles of block copolymers and polymer blends described in session 3.

3. Unique Nanostructures of Polymer Particles Prepared by Self-Organization

3.1. *Inner Nanostructures in Block-Copolymer Nanoparticles*

Block-copolymers are very important materials for nanotechnology. Block-copolymer has two or more different polymer segments covalently attached linearly with end-to-end, or comb-shaped with end-to-side connections. Since these polymer segments are immiscible each other, unique nanoscale structures based on their micro phase separation, depending on their molecular shape and segment size, are spontaneously emerged. Many types of block-copolymers have been synthesized and their characteristic micro-phase separation structures have been reported.[30,31] Nanoparticles of block-copolymers are also interesting building components of nanomaterials because of their unique microstructures. It is intensely interesting to know what kinds of micro-phase separation structures are formed in nanoparticles of block-copolymers.

Emulsion polymerization is most conventional method to prepare polymer nanoparticles. Some sorts of amphiphilic block-copolymers bearing water-soluble polymer blocks have been prepared by the emulsion polymerization.[32] These block-copolymers can be dissolved in water as nanoparticles with core-shell type structure like surfactant micelles.[33] The emulsion polymerization, which is carried in an aqueous solution, is not applicable for preparation of hydrophobic block-copolymers nanoparticles, because most hydrophobic block-copolymers are prepared by anionic polymerization, which requires highly controlled non-aqueous polymerization conditions.

To our knowledge, only the reprecipitation method was a way to prepare nanoparticles of hydrophobic block-copolymers. We have applied our new method for nanoparticle formation of block-copolymers. Poly(styrene-block-isoprene) (PSt-b-PI) is used because there are many data about phase diagram in bulk film of this copolymer. Two types of PSt-b-PI, one has same segment ratio in molecular weight to form lamella structure as in its bulk film (Mn: PSt(17800)-b-PI(12000), Mw/Mn =1.02), and another forms cylindrical phase in the bulk film (Mn: PSt(40800)- b-PI(10400), Mw/Mn: 1.06), were used.

As shown in Figure 4, electron micrographs clearly show formation of regular-sized nanoparticles from block-copolymers by our solvent evaporation method. Figure 4(a) shows surface image of the lamella forming PSt(17800)-b-PI(12000) particles observed by STEM (scanning transmission electron microscope). Bright parts show the presence of the isoprene block, because the electrons are scattered by heavy metal ion, osmium, used for selective staining of the carbon-carbon double bond in the isoprene unit. The repeating period of the lamella is corresponds to the twice of the calculated polymer length. Periodic distribution of osmium located in the bright parts of the particle surface was found in the EDX (energy dispersive x-ray) spectrum, too. Figure 4(b) shows a TEM (transmission electron microscope) image of particle's cross section prepared by an ultra-microtome.

A STEM image in Figure 4(d) suggests that hexagonal packing of cylindrical phase is formed in the nanoparticle of PSt(40800)-b-PI(10400). A cross sectional SEM image shown in Figure 4(e) clearly indicates the bundle structure of block copolymer cylinders in the particle. Figure 4(f) simultaneously shows the parallel alignment of the cylinder long axis and hexagonal packing of the cylinder edges. Size of the hexagonally packed lattice of the isoprene cores in each cylinders is almost same as the calculated value from the molecular weight of PSt(40800)-b-PI(10400). These results strongly indicate that micro-phase separation is formed in the block copolymer nanoparticles as well as in its bulk film. It is worthy of note that polymer particles having well developed lamella structures with narrow size distribution are not formed by the conventional reprecipitation method.

Recently, we have succeeded to prepare nanoparticles from the amphiphilic block copolymers, poly(styrene-block-sodium acrylate)[34] and poly(styrene-block-4-vinylpyridine).[35] As shown in Figure 5, SEM reveals spherical and uniform-sized particle formation. There are some particles having a single hole (white arrow in Figure 5(a) and (c)). Cross

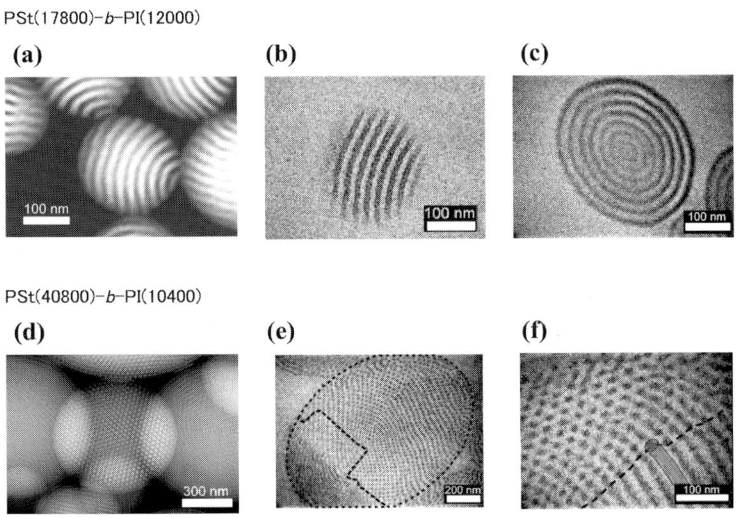

Fig. 4. STEM and cross sectional TEM images of lamella (a–c) and cylinder (d–f) phased PSt-b-PI block copolymer nanoparticles.

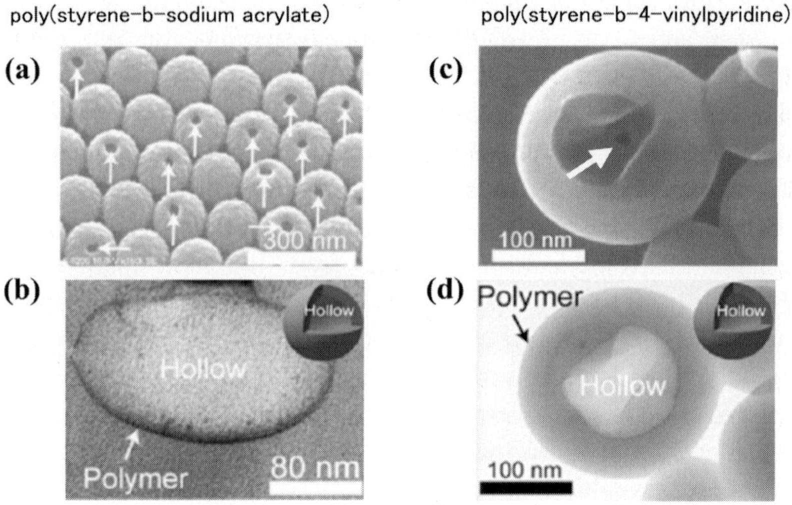

Fig. 5. Surface and cross sectional STEM images of amphiphilic block copolymers particles.

sectional STEM images of the thin film samples scratching from PVA embedded nanoparticles by FIB (focused ion beam) clearly show a large hollow in individual particles. By using our method, these amphiphilic block copolymers form hollow particles but not simple micelle-like core-shell particles.

3.2. *Phase Transition of Block Copolymer Nanoparticle*

In order to prepare the highly ordered micro-phase separation structures in bulk films of block-copolymers, films are annealed over glass transition temperature (T_g) for a long time and then gradually cooled down below T_g[36] or exposed with vapor of a good solvent.[37] We have investigated the effect of annealing on the phase separation structures formed in the block-copolymer nanoparticles.[38] Water suspension of lamellar-structured nanoparticles prepared by our method at room temperature (~20°C) was annealed at 30°C and 50°C for 10 hours, respectively. After annealing, these suspensions were cooled down at room temperature for 30 min. and then were stained with osmium tetraoxide. Electron microscopy revealed that the lamellar structure was kept after annealing at 30°C and transformed to random web-like structure after annealing at 50°C.

Hashimoto et al.[39] and Floudas et al.[40] reported that the lamellar phase in the bulk film turns to disorder phase by annealing over its T_g described as

$$T_{ODT} = 213 + 1.00742Nn \quad (1)$$

where T_{ODT} and Nn are the order-disorder transition temperature and the number average degree of polymerization of the block copolymers, respectively. The annealing experiment shows that T_{ODT} in the particles is much lower than that in the film whose T_{ODT} is estimated to 290°C from equation (1). These results show the lamellar structure formed in the particles is not stable.

This instability of the lamellar structures formed in the particles relates to the particle formation mechanism of our method. It is known that the phase separation structure of block-copolymers is affected by the interface. For example, a hydrophilic polymer segment tends to face the hydrophilic substrate in a phase separation structures. By using this property, spontaneous patterning of block-copolymers can be realized.[41] Russell et al. reported that when the block-copolymer is filled in the anodized alumina pores, the concentric lamellar structures along the walls of pores were formed.[42] Considering the phase separation structures in the block-copolymer particles based on this concept, the onion-like lamellar structure is the most suitable to reduce the surface free energy.

It is, however, shown in Figure 4(a) and (b) that the nanoparticles prepared by our method consist of piled-up lamellar layers one-directionally oriented in each particle. As described in section 2.2, we have found that the polymers were gradually aggregated as particles at the air/solution interface to form the hemispherical shaped particles. When formation of lamellar structure and particle growth are simultaneously occurred at the air/solution interface, anisotropic lamellar structures can be formed in the particles. We are now finding preparation conditions of the onion-like concentric lamella particles. Concentration and temperature are important experimental factors determining nanostructures in the block copolymer particles. Figure 4(c) is a typical cross sectional TEM image of the PSt(17800)-b-PI(12000) particle prepared from the high concentration solution. Details about the determinant factors are now under investigation. The anisotropic lamellar structure is assumed to thermodynamically be less stable than the isotropic onion-like concentric sphere structures.

3.3. *Structural Approach of Anisotropic Lamellar Formation in Particles*

Broken symmetry reflected not only in the one-directionally piled-up lamellar structure in the block copolymer particles but also in the hemispherical shaped polymer particles strongly suggests the particle formation mechanism at interface. This section describes structural approach of the particle formation mechanism. Osmium tetraoxide used for selective staining of the isoprene moiety selectively reacts with the carbon-carbon double bond to bridge two polymer chains (see reaction scheme in Figure 6). Due to the cross linking by osmium the polyisoprene moiety can not dissolve in THF again. Only the polystyrene

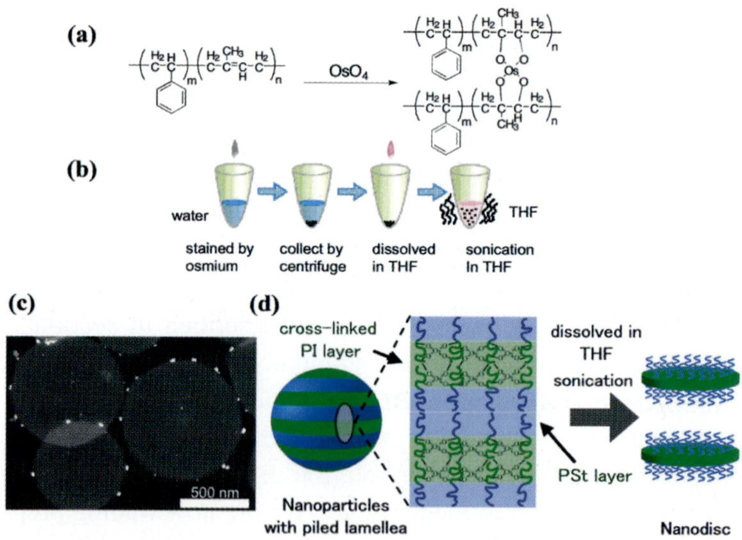

Fig. 6. Cross-link reaction of PI by osmium tetraoxide (a), schematic representation of SISE method (b) and nano disk formation from lamella phase nanoparticles (c,d).

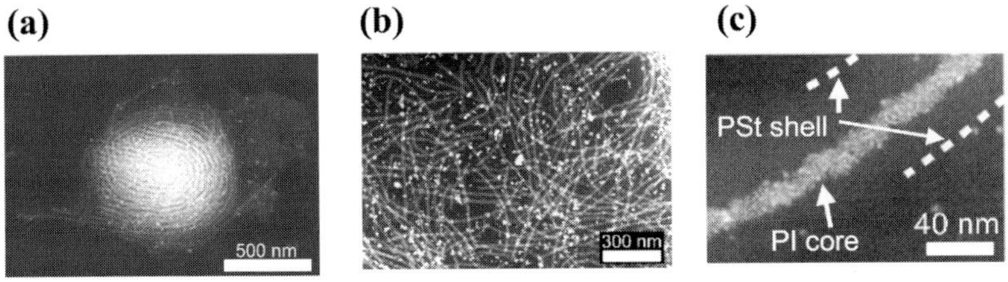

Fig. 7. Isolation of the cylinder phases from the PSt-b-PI block copolymer nanoparticle by SISE method.

moiety can dissolve in THF after staining. According to the procedure schematically shown in Figure 6(b), the stained nanoparticles are redispersed in THF under ultrasonication. Figure 6(c) clearly shows SEM image of discotic fragmental plates after sonication of the stained particles in THF. The disc consists of PSt-PI-PSt layers because it is pealed from the spherical lamellar particles (Figure 6(d)).

We are now developing the procedure of metal ion staining and dissolution as a novel preparation method of nanostructured materials. During the process represented in Figure 6(b), selective immobilization of the isoprene moiety is achieved by the metal complex formation and in the following process only the untreated polystyrene moiety selectively dissolves in organic solvent. Combination of the selective immobilization and the selective elution (SISE) of each polymer segments in the block copolymer polymer particle provides novel polymer structure as the nanodisc. Figure 7 shows the results of SISE experiment of nanoparticle having cylinder phase. Entangled polymer cylinders in the particle are effectively isolated by ultrasonication. Careful observation of single thread of polymer reveals that the polystyrene layer envelops the immobilized isoprene core (Figure 7(c)).

3.4. *Nanoparticles of Polymer Blends*

Polymer blends are another candidate of nano- and micro-structured polymer assemblies. What kind of phase separation is emerged in polymer nanoparticles when two different polymers are employed for our solvent evaporation method? We prepared two polymer mixtures, poly (methyl methacrylate) (PMMA) and polyisoprene (PI), and polystyrene (PSt) and PI. A STEM image clearly shows that core–shell type particles, where the PI core (white region) is surrounded by the PMMA shell (gray region), are formed from the mixed solution of PI and PMMA in THF after addition of water as poor solvent (Figure 8(a)). Since the solubility of PI to the poor solvent is lower than PMMA, the PI core is firstly formed during the good solvent (THF) evaporation and then PMMA shell deposited to form shell on the PI core. Particle shape depends on the polymer concentration as well as the polystyrene experiment shown in Figure 3, the hemispherical core-shell particles are mainly formed in the dilute solution.

PSt has similar THF solubility of PI but is a little more hydrophobic. Figure 8(b) clearly shows that "Janus" particles are formed in this combination. Each particle is composed by two polymers with clear phase separation. Interesting finding is that the volume of each polymer

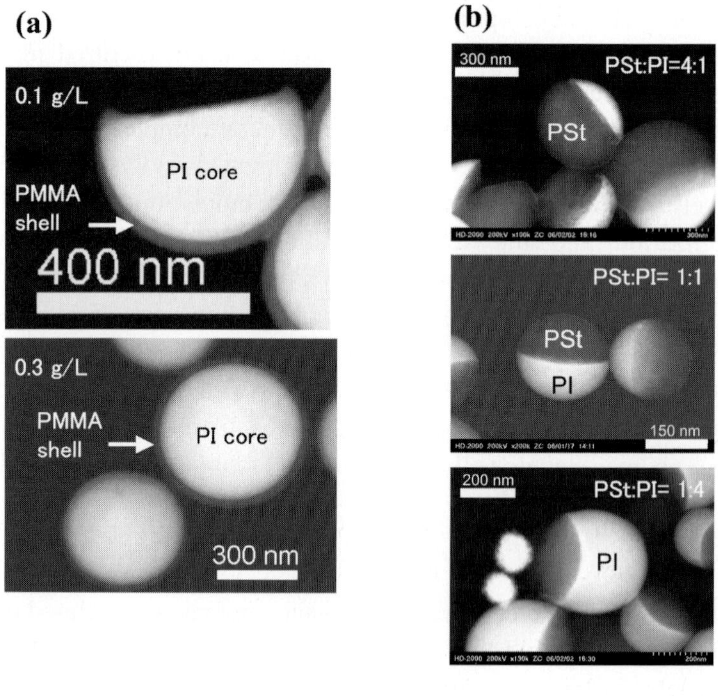

Fig. 8. STEM images of the polymer blend nanoparticles of PI/PMMA (a) and PSt/PI (b).

is almost parallel to the mixing weight ratio in the THF solution when the total polymer concentration is 0.1 g/L.

4. Hierarchical Assembling of Polymer Nanoparticles in Self-Organized Honeycomb Mesh Template and by Dissipative Processes

Assembling the nanostructured polymer particles up to mesoscopic-scaled organizates is the next step of the hierarchical structuring of novel materials based on nanoparticles. Here we show the preparation of honeycomb-nanoparticles hybrid structures by simple casting of nanoparticle dispersion onto the self-organized honeycomb-patterned films. Dissipative structure formation in the casting nanoparticle solution is another driving force of the hierarchic assembling of particles. Two experimental results of regular pattern formation of nanoparticle assemblies are described in this section.

The polystyrene honeycomb-patterned films were prepared by the "Breath Figure" method.[13,43–45] Hexagonally packed water micro-droplets are formed by evaporative cooling on the surface of the casting solution, and these water droplets are transferred to the solution front by convectional flow or capillary force. When the water droplets condense on the solution surface, the clear solution surface turns opaque. After solvent evaporation, a honeycomb-patterned polymer film is formed with the water droplet array acting as a template; the water droplets themselves evaporate soon after the solvent. Here, we show the fabrication of periodic micro-structured films having multiple periodicities by hybridization of the honeycomb-patterned films and polymer nanoparticles.

To embed polymer nanoparticles in the regular pores in the honeycomb matrix, a dispersion of polystyrene particles was simply cast onto the honeycomb-patterned films. However, the surface of the honeycomb-patterned polystyrene films is too hydrophobic[20,46] to incorporate aqueous dispersion of polymer particles into its regular pore. To improve the surface wettability of the honeycomb matrix, UV-ozone treatment was performed. After surface treatment, the submicron sized polystyrene particles are embedded only in the honeycomb pores. Figure 9 clearly shows the number of particles embedded in each pore can be controlled by changing the size of particles.[47]

By using "dissipative structure" formed in evaporating particle solution, mesoscopic regular patterns of nano-particles are formed on solid substrate by simply casting or dipping of their aqueous dispersions. Regular pattern formation based on the "dissipative structure" is usually observed in our daily life. Periodic liquid stripes formation in a wine glass, called "wine legs" or "tear of wines" is a typical "dissipative structure". Due to faster evaporation of ethanol in the thin liquid film of water-ethanol mixture climbing the wine glass surface, convectional flow, so-called Maringoni convection, is induced by changing local surface tension in the front of thin liquid film. With increasing surface tension, the water-rich liquid film turns small droplets and eventually the droplet crawl down as periodic stripes after breaking the balance of surface tension and gravity. The "wine legs" phenomenon is called a fingering instability in physics.[48]

We have found that the fingering instability and regular stripe formation were formed in the casting dilute polymer solution on solid surfaces.[49,50] Dynamic regular structures, periodic "wine legs" formed in casting polymer solutions are fixed as regularly arranged striped polymer patterns after solvent evaporation.[51] Periodic stripes are formed from the casting aqueous dispersions of polymer nanoparticles. Interesting finding

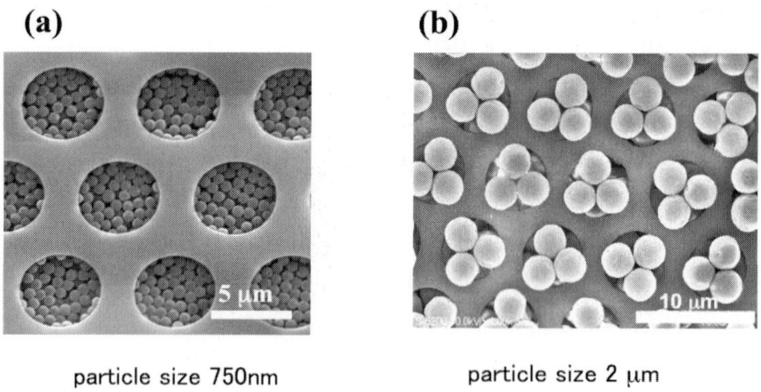

Fig. 9. SEM images of nanoparticles embedded honeycomb-patterned polymer matrix (pore size 5µm).

in spontaneous phase separation is occurred when an aqueous mixture of different sized particles is cast on mica surface. Figure 10(a) and (b) shows two types of phase separation in the striped patterns. When an aqueous mixture of different sized polystyrene particles (32 nm and 100 nm) is cast on mica substrate, periodically arranged stripes of particle assemblies are generated from the receding front of the casting solution. Figure 10(a) clearly shows that these two sized particles (32 nm and 100 nm) spontaneously separate in a single stripe. Magnified SEM image shows the large particles are surrounded by the small sized particle. Complete phase separation is observed when small (25 nm) and large (100 nm) silica particles are used. As shown in Figure 10(b), these particles form alternatively arranged individual stripes. Due to dewetting, stripes of the small particles are shrunk to regularly align dot-shaped particle assemblies. Mesoscopic patterns of particles assemblies generated from receding front of casting or dipping solution are largely independent on experimental conditions,[52–54] e.g. concentration, temperature, surface wettability, etc. Recently we have found that high

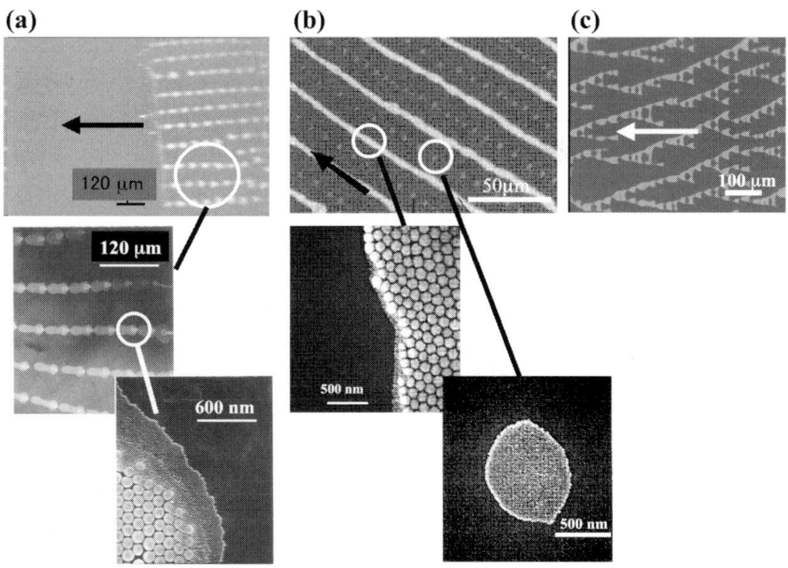

Fig. 10. Mesoscopic particles patterns formed from dissipative structures in the casting particle solutions. Arrows mean the receding direction of the casting solution front.

humidity of the dipping experiment provided "Sierpinski Gasket" typed fractal patterns of nanoparticle assembly of polystyrene (Figure 10(c)). Details about mesoscopic patterning of particle assemblies by using "dissipative structure" are now under investigation.

5. Conclusion

We have described a novel and simple method of spontaneous formation of polymer nanoparticles based on non-equilibrium processes emerging in slowly evaporating polymer solutions. The method consists of two procedures, (1) gentle mixing of the poor solvent to the polymer solution of good solvent, and (2) gradual evaporation of the good solvent. Miscibility of two solvents is essentially required to this method. In order to evaporate the good solvent preferentially, boiling point of the good solvent must be lower than that of the poor solvent. Particle size is affected by the mixing ratio of two solvents as well as the polymer concentration. Due to the physical generality of the formation mechanism, this method can be applicable for a wide variety of polymers. Unique micro-phase separation structure, regularly piled-up lamellae, was formed in the nanoparticles of the block copolymer.

Formation mechanism of nanoparticles resembles that of crystallization. Dynamic equilibrium between dissolved polymer molecules and nuclei of polymer particles, which shifts every moment with the temporal change of the solvent composition, may causatively induce the uniform nucleation of particles. Formation of hemispherical particles and anisotropic phase separation of lamella typed block copolymer nanoparticles strongly suggest that the dynamic events of nuclei formation emerge at air/solution interface.

Hierarchic structuring of nanoparticles is achieved when the honeycomb-patterned self-organized polymer matrix and "dissipative structure" are combined with nanoparticles. The hybrid and hierarchic structures of nanoparticles can be utilized as novel materials not only for photonic band gap materials but also for optical display devices, micro-electrodes, novel cell-culture devices, and other practical applications. A novel preparation method of supra-polymer assemblies from phase separated polymer nanoparticles, the selective immobilization and selective elution method (SISE), is proposed in this article. Isolated polymer nano disks and cylinders are novel candidates of building polymer blocks of the hierarchical mesoscopic polymer organizations. Broken symmetry and spontaneous phase separation in the nano scale space as well as in the mesoscopic world are challenging mysteries to clarify the intrinsic and essential nature of "self-organization".

References

1. T. Kunitake, *Angew. Chem. Int. Ed. Engle.*, 31, 709-726 (1992).
2. G. Roberts, "Langmuir-Blodgett Films", Plenum Press, New York (1990).
3. A. Ulman, "An Introduction to Ultrathin Organic Films from Langmuir-Blodgett to Self-Assembly", Academic Press, New York (1991).
4. H. Ringsdorf, B. Schlarb and J. Venzmer, Angew. *Chem. Int. Ed. Engl.*, 27, 113-158 (1988).
5. M. Shimomura, *Prog. Polym. Sci.*, 18, 295-339 (1993).
6. J. M. Lehn, *Proc. Natl. Acad. Sci.* USA, 99, 4763 (2002).
7. I. W. Hamley, "Developments in Block Copolymer Science and Technology", John Wiley & Sons Inc, New York (2004).
8. "*Amphiphilic Block Copolymers: Self-Assembly and Applications*", Ed. by P. Alexandridis and B. Lindman, Elsevier, Amsterdam, (2000).
9. G. Nicolis and I. Prigogine, "*Self-Organization in Non-Equilibrium Systems*", Wiley, (1977).
10. M. Shimomura, "Architechturing and Applications of Films Based on Surfactants and Polymers" in "*Supramolecular Polymers*", Ed. by A. Ciferri, Marcel Dekker, 471-504 (2000).
11. M. Shimomura and T. Sawadaishi, *Current Opinion in Colloid & Interface Science*, 6, 11-16 (2001).
12. M. Shimomura, "Dissipative Structures and Dynamic Processes for Mesoscopic Polymer Patterning" in "*Nanocrystals Forming Mesoscopic Structures*", Ed. by M. P. Pileni, Wiley-VCH, pp. 157-171 (2006).
13. N. Maruyama, T. Koito, T. Sawadaishi, O. Karthaus, K. Ijiro, N. Nishi, S. Tokura, S. Nishimura and M. Shimomura, *Supramol. Sci.*, 5, 331-336 (1998).
14. M. Shimomura, T. Koito, N. Maruyama, K. Arai, J. Nishida, L. Grasjo and O. Karthaus, *Mol.Cryst.Liq.Cryst.*, 322, 305-312 (1998).
15. O. Karthaus, K. Ijiro and M. Shimomura, *Chem.Lett.*, 821-822 (1996).
16. O. Karthaus, L. Grasjo, N. Maruyama and M. Shimomura, *Chaos*, 9, 308-314 (1999).
17. H. Yabu and M. Shimomura, *Adv. Func. Mater.*, 15, 575-581 (2005).

18. O. Karthaus, N. Maruyama, X. Cieren, M. Shimomura, H. Hasegawa and T. Hashimoto, *Langmuir*, 16, 6071-6076 (2000).
19. H. Yabu, M. Tanaka, K. Ijiro and M. Shimomura, *Langmuir*, 19, 6297-6300 (2003).
20. H. Yabu, M. Takebayashi, M. Tanaka and M. Shimomura, *Langmuir*, 21, 3235-3237 (2005).
21. H. Yabu and M. Shimomura, *Langmuir*, 21, 1709 -1711 (2005).
22. Y. Yin and Y. Xia, *Adv. Mater.*, 14, 605 (2001).
23. X. Wang, C. J. Summers, Z. L. Wang, *Nano Lett.*, 4, 423 (2004).
24. Y. Bae, S. Fukushima, A. Harada and K. Kataoka, *Angew. Chem. Int. Ed.*, 42, 4640 (2003).
25. F. Caruso, "Colloids and Colloid Assemblies", WILEY-VCH Verlag GmbH & Co. KGaA, Weinheim, (2004).
26. H. Kasai, H. S. Nalwa, H. Oikawa, S. Okada, H. Matsuda, N. Minami, A. Kakuta, K. Ono, A. Mukoh and H. Nakanishi, *Jpn. J. Apply. Phys.*, 31, L1132 (1992).
27. H. Yabu, T. Higuchi and M. Shimomura, *Adv. Mater.*, 17, 2062 (2005).
28. H. Yabu, T. Higuchi, K. Ijiro and M. Shimomura, *Chaos*, 15, 047505-1-7, (2005).
29. T. Higuchi, H. Yabu and M. Shimomura, *Colloids and Surfaces A*, 284-285, 250-253 (2006).
30. I. W. Hamley, "Developments in Block Copolymer Science and Technology", John Wiley & Sons Inc (2004).
31. "*Amphiphilic Block Copolymers: Self-Assembly and Applications*", Ed. by P. Alexandridis and B. Lindman, Elsevier, Amsterdam (2000).
32. M. Q. Chen, A. Kishida, T. Serizawa and M. Akashi, *J. Polym. Sci. Part A: Polym. Chem.*, 38, 1811-1817 (2000).
33. G. Battaglia, A. J. Ryan, *J. Am. Chem. Soc.* 127, 8757 (2005).
34. H. Yabu, T. Higuchi and M. Shimomura, *Int. J. Nanosci.*, 5, 195-198 (2006).
35. T. Higuchi, H. Yabu and M. Shimomura, *J. Nanosci. Nanotech.*, 7, 856-858 (2007).
36. U. Jeong, D. Y. Ryu, J. K. Kim, D. H. Kim, X. Wu and T. P. Russell, *Macromolecules* 36, 10126 (2003).
37. S. H. Kim, M. J. Misner and T. P. Russell, *Adv. Mater.* 16, 2119 (2004).
38. T. Higuchi, H. Yabu, S. Onoue, T. Kunitake and M. Shimomura, *Colloids and Surfaces A,* 313-314, 87-90 (2008)
39. H. Tanaka and T. Hashimoto, *Macromolecules*, 24, 5713 (1991).
40. T. J. Millar, P. R. Farquhar and K. Willacy, *Astron. Astrophys. Suppl.* , 139 (1997).
41. G. M. Wilmes, D. A. Durkee, N. P. Balsara and J. A. Liddle, *Macromolecules*, 39, 2435 (2006).
42. H. Xiang, K. Shin, T. Kim, S. I. Moon, T. J. McCathy and T. P. Russell, *Macromolecules*, 37, 5660 (2004).
43. G. Widawski, M. Rawiso and B. Francois, *Nature*, 369, 387 (1994).
44. U. H. F. Bunz, *Adv. Mater.*, 18, 973–989 (2006).
45. M.Shimomura, T.Koito, N.Maruyama, K.Arai, J.Nishida, L.Grasjo and O.Karthaus, *Mol.Cryst.Liq.Cryst.*, 322, 305-312 (1998).

46. H. Yabu and M. Shimomura, *Chem. Mater.* 17, 5231 (2005).
47. T. Higuchi, H. Yabu and M. Shimomura, *Colloids and Surfaces*, A, 284-285, 250-253 (2006).
48. A. Cazabat, F. Heslot, S. Troian and P. Carles, *Nature*, 346, 824 (1990).
49. O. Karthaus, L. Grasjo, N. Maruyama and M. Shimomura, *Chaos*, 9, 308 (1999).
50. M. Nonomura, R. Kobayashi, Y. Nishiura and M. Shimomura, *J. Phys. Soc., Jpn.*, 72, 2468 (2003).
51. H. Yabu and M. Shimomura, *Adv. Func. Mater.*, 15(4), 575-581 (2005).
52. T. Sawadaishi and M. Shimomura, *Mol.Cryst.Liq.Cryst.*, 406, 159-162 (2003).
53. T. Sawadaishi and M. Shimomura, *Colloids and Surfaces A*, 257-258, 71-74 (2005).
54. T. Sawadaishi and M. Shimomura, *Mol.Cryst.Liq.Cryst.*, 464, 227-231 (2007).

CHAPTER 6

WRINKLING POLYMERS FOR SURFACE STRUCTURE CONTROL AND FUNCTIONALITY

Edwin P. Chan* and Alfred J. Crosby**

*Harvard-MIT Division of Health Sciences and Technology
Massachusetts Institute of Technology
77 Massachusetts Avenue, E25-342 Cambridge, MA 02139
E-mail: epc@mit.edu

**Polymer Science & Engineering Department, University of Massachusetts
120 Governors Drive, Amherst, MA, USA
E-mail: crosby@mail.pse.umass.edu

Surface wrinkling offers a unique, novel, and robust approach to define the topography of polymer surfaces. Not only can structure be defined with discrete, predictable length scales, but the direct control of surface properties can be achieved. Significant efforts have been demonstrated, but this field is only emerging in the broad context of a new design paradigm for polymer surfaces and many challenges remain. In this chapter, we present a brief overview of the general mechanics that govern surface wrinkling and an experimental discussion on strategies that use swelling of a constrained elastomer to generate wrinkled surfaces. We discuss factors that dictate the length scale, alignment, and general morphology of wrinkled surfaces, and we end with an overview of potential applications for surfaces defined by this strategy.

1. Introduction

The local structure of a material's surface defines its interactions with the environment and determines the physical properties of a material, ranging from optical reflectivity to cell adhesion. Thus, the ability to define surface structures across multiple length scales is a powerful strategy for property control and serves as one of the primary initiatives for nano- and bio-technology. In nature, elastic instabilities are ubiquitous in determining shape, structure, and function. Examples include ridges on plant leaves, wrinkles on human skin, and blebs in cell membranes.[1-5] In synthetic materials, elastic instabilities have classically been studied in the context of materials failure and deformation, mostly focused on designing materials to avoid initiation of instabilities.[6-8]

At an increasing rate, over the past decade researchers have exploited the elastic instability of surface wrinkling as a mechanism for controlling polymer surface morphology, enabling technical advances in areas from enhanced metrology for materials properties to flexible electronics.[9-25] However, unleashing the full potential of surface wrinkling requires a fundamental understanding of the material properties that define kinetic and equilibrium structures.

This understanding will define a new materials design paradigm to control the morphology of polymer surfaces, to arrange these structures in complex, long range hierarchies, and to capitalize on their unique ability to define surface properties that can adapt to their environment. This paradigm lies at the intersection of chemistry, physics, biology, and engineering and provides scalable strategies for tuning physical properties.

Several review articles have been written to summarize the current understanding of surface wrinkling and the numerous strategies that have been developed to induce and control them.[3,26] Here it is not our intention to repeat these reviews, rather we discuss the physics and design of surface wrinkles by the development of an osmotic stress in a polymer film. This process offers significant advantages in the definition of polymer film morphologies with advanced property control and function. We begin the discussion with a brief background on the mechanics of surface instabilities and an overview of available strategies with the aim of highlighting the advantages of osmotic pressure-driven surface wrinkling. Next, we discuss the mechanics of surface wrinkles induced by osmotic pressure and the experimental details for two demonstrated strategies. We then discuss the ability to define hierarchical surface structures and conclude by describing technological applications of surface wrinkles and the challenges that remain unsolved.

2. Background

2.1. *Surface Wrinkling Instability Mechanics*

A film or sheet will deform out-of-plane at the imposition of a critical in-plane compressive stress (Fig. 1(a)). This out-of-plane deformation is associated with the onset of an elastic instability, where the total energy is best minimized by the film bending rather than straining in-plane. This elastic instability has been studied classically,[8] especially in the context of failure modes for structural panels, and its onset can be determined in a straightforward manner by considering the work done by external loads and the strain energy due to bending. At the point of instability, the increase in strain energy due to bending will be less than the work done by external loads for in-plane compression. For a square film loaded equibiaxially, this condition is marked by a critical in-plane stress related to the reduced modulus ($E^*=E/(1-\nu^2)$), the thickness (h), and the lateral dimension (b) of the film:[8]

$$\sigma_{cr} = \frac{\pi E^*}{6}\left(\frac{h}{b}\right)^2 \tag{1}$$

Intuitively, thin, long films buckle at lower stresses. This critical condition is valid for a free sheet or film, but many wrinkling examples involve a thin film that is attached to a compliant substrate (Fig. 1(b)). In this geometry, the compliant substrate exerts a stress on the film, and its strain energy must be accounted in the total energy of the system. Accordingly, the critical stress for out-of-plane buckling is:[27,28]

$$\sigma_{cr} = \frac{E_f^*}{4}\left(3\frac{E_s^*}{E_f^*}\right)^{2/3} \tag{2}$$

where subscripts f and s denote film and substrate, respectively. This critical condition is valid for any loading mode (e.g. uniaxial, biaxial) that produces deformations with a critical wavenumber:[27,28]

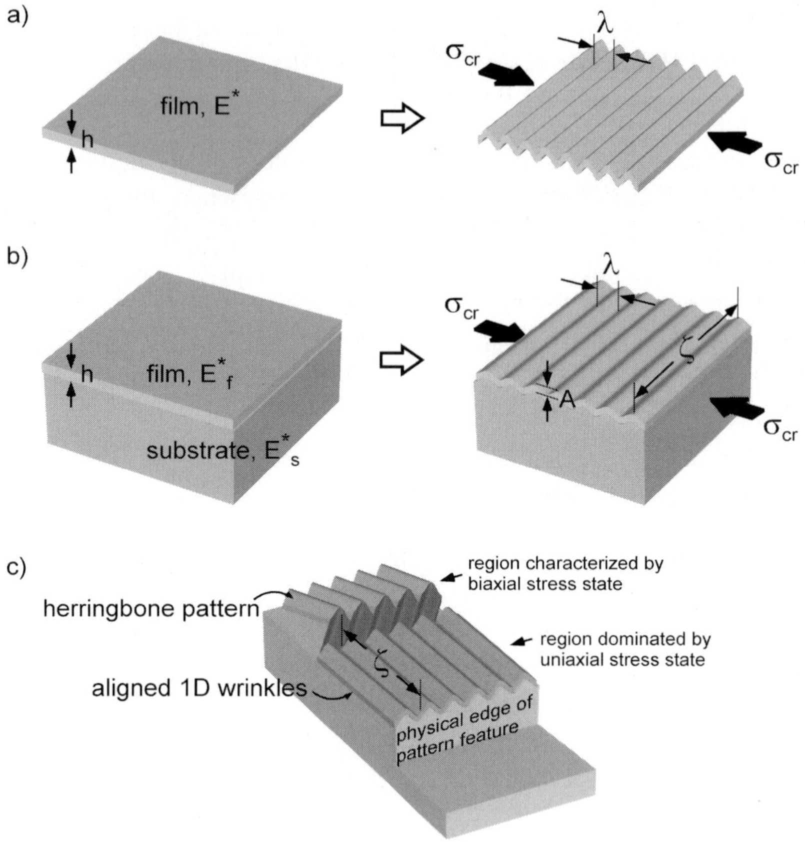

Fig. 1. Examples of general approaches used to develop surface wrinkles. a) Uniaxial mechanical compression of a film with thickness (h) and elastic modulus (E^*). At a critical compressive stress (σ_{cr}), the film deforms by wrinkling with a periodic profile as described by a characteristic wavelength (λ). b) The uniaxial compression of a film with elastic modulus (E^*_f) adhered to an elastomeric substrate with elastic modulus (E^*_s). This example is the typical approach used to develop polymeric surface wrinkles since the elastomer can facilitate the generation of the required compressive stress. Again, above σ_{cr}, wrinkles develop on the film surface with a characteristic wavelength, amplitude (A) and persistence length (ζ). c) Development of surface wrinkles by application of a biaxial stress state on a patterned elastomer. At the pattern edge, the compressive stress is predominantly uniaxial parallel to the boundary. This altered stress state leads to the development of aligned one-dimensional wrinkles that persist for a finite length as described by ζ. Beyond ζ, the stress state reverts back to biaxial, which favors the formation of herringbone wrinkles when the stress state far exceeds the critical value.[27-29]

$$k_c = \frac{1}{h}\left(3\frac{E_s^*}{E_f^*}\right)^{1/3} = \sqrt{k_1^2 + k_2^2}\,, \tag{3}$$

where h is the thickness of the top film and k_1 and k_2 are the wavenumbers in the orthogonal, in-plane axes directions. The wavenumber k is inversely related to the wavelength, λ, of the wrinkled film: $k = 2\pi/\lambda$. Comparing equations (1) and (2), it is evident that the underlying substrate imposes additional constraint that delays buckling. Additionally, the wavelength is defined by the thickness and modulus mismatch, not the lateral dimensions of the sample.

2.2. *Morphology of Surface Wrinkles*

The morphology of surface wrinkles can be characterized by the dominant length scales of the imposed pattern: wavelength (λ) and amplitude (A) (Fig. 1(b)). These length scales have been presented by Cerda and Mahadevan in a general format:[26]

$$\lambda \propto \left(\frac{D}{K}\right)^{1/4} \qquad A \propto \lambda \cdot \varepsilon^{1/2} \tag{4}$$

where D is the bending stiffness of the top film, K is the effective stiffness of the substrate, and ε is the in-plane compressive strain applied to the sample. For an isotropic, linear elastic thin film with thickness h on a semi-infinite, isotropically linear elastic substrate,[26] $\lambda \approx h(E^*_f/E^*_s)^{1/3}$, which is consistent with the uniaxial deformation mode ($k_c=k_1$) in equation (3). Equation (4) clearly demonstrates that the wavelength of the wrinkled surface is defined by the balance of geometry and materials properties, while the amplitude of the surface wrinkles is linked additionally to the extent of compressive strain.

Besides wavelength and amplitude, the arrangement of surface wrinkles and their persistence length (ζ) perpendicular to λ play a critical role in defining surface topography (Fig. 1(b)). For uniaxial loading conditions, a one-dimensional periodic wave will be observed where ζ is defined by the lateral extent, or width, of the compressed sample. For biaxial stress states, the arranged pattern of surface wrinkles is dependent upon the level of applied stress relative to the critical wrinkling stress, as Hutchinson has described in recent publications.[27,28] For stresses far exceeding the critical stress, a herringbone pattern is predicted as the lowest energy configuration and has been observed experimentally by many research groups.[14,25,27,28,30] In this configuration, the length of the herringbone "jogs" is not predicted in a straightforward manner from equilibrium mechanics and is suggested to be dependent upon the kinetics of wrinkle formation. Near an edge of a biaxially stressed material, the compressive stress parallel to the edge is predominantly uniaxial, which leads to the formation of wrinkles with a defined persistence length perpendicular to the edge (Fig. 1(c)). Groenewold has described this distance as[3]:

$$\zeta \approx \lambda \cdot \left(\frac{1}{2\varepsilon^{1/2}} + 2 \cdot \left(\frac{1}{4\left(1-\nu_f^{\,2}\right)} \frac{E_f}{E_s} \right)^{1/3} \right). \tag{5}$$

3. Using Osmotic Pressure to Drive Surface Wrinkling

The general requirement for the onset of surface wrinkling is the development of a critical compressive stress; therefore, a variety of strategies can lead to this critical stress development. Examples range from direct application of compressive stresses by external loads[21] to compressive stresses associated with differences in the coefficient of thermal expansion between composite layers that are exposed to thermal processing conditions.[25] Although all of these approaches are unique and offer advantages for specific applications, one attractive route is the use of swelling to generate surface wrinkles since most polymers have the propensity to swell. This approach is particularly attractive since the wrinkling polymer can be designed to swell in the presence of a particular agent or chemical environment; furthermore, the swelling agent can be a monomer or formulation that can be polymerized upon wrinkling. This approach not only provides stabilization of the wrinkled topography but also allows for the definition of the surface and near surface chemistry. Furthermore, osmotically-driven surface wrinkles can achieve wavelengths that range from ~100 nm to ~1 mm, and their lateral extent has been demonstrated, without effort, over several square centimeters, suggesting the scalability of the process.[14]

Similar to wrinkling in laminates with different thermal expansion coefficients, wrinkling with osmotic stress requires a multi-component "composite" with different swelling responses in each component. This "composite" can be as simple as a homogeneous polymer film fixed to a rigid substrate or as complex as a multilayer composite. In all cases, one of the components or layers restricts the swelling processes in another component. This constraint can impose compressive stresses near a free surface. Values that exceed a materials-defined critical stress will lead to the onset of the elastic instability as characterized by wrinkling of the free surface. To understand this process in more detail, we consider the details of two model systems.

Bottom-constrained materials

Consider a bilayer that consists of a thin elastomer (swellable) adhered to a rigid substrate (not swellable), as illustrated in Fig. 2(a). Upon exposure to a swelling agent, the elastomer layer attempts to expand, or swell, in all directions; however, for thin films this expansion is constrained in-plane by adhesion to the rigid substrate. Consequently, expansion is limited to the out-of-plane direction, and in-plane compressive stresses develop in the elastomer layer. If the equilibrium osmotic stress of the swollen elastomer leads to the development of an in-plane compression that exceeds the instability conditions, then surface wrinkles on the free surface will develop. One of the first observations of this process was reported by Southern and Thomas,[31] but more recently several research groups have studied this process in more detail. In particular, Hayward *et al.* have focused on the development of a related surface instability,

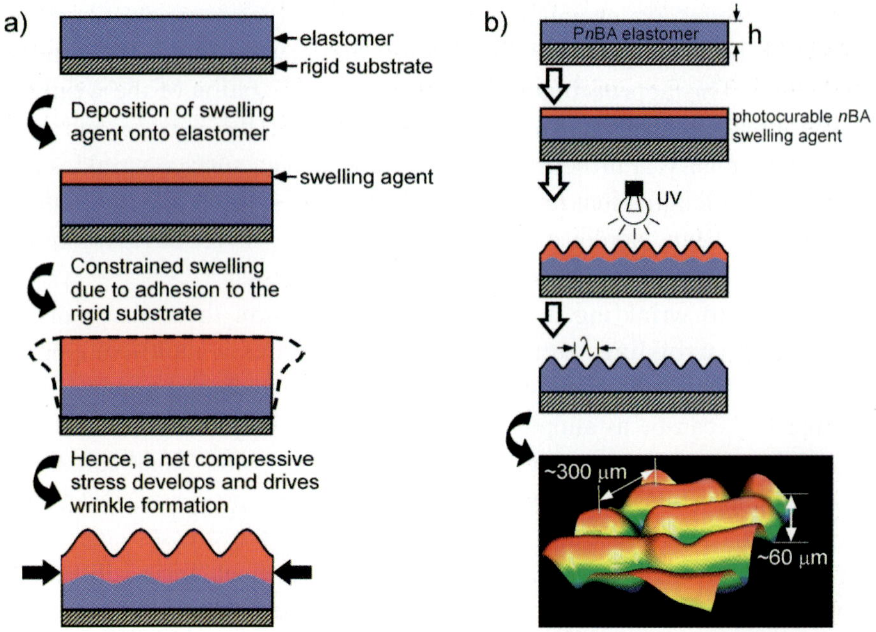

Fig. 2. a) Example of a bottom-constrained material wrinkled with an osmotic stress. b) An approach developed by Chan *et al.* for the fabrication of stable surface wrinkles on a poly(*n*-butyl acrylate) (P*n*BA) elastomer surface.[11] Their approach utilizes photochemistry in the swelling agent to stabilize the wrinkles. The micrograph is an optical profilometry image of the resultant wrinkled surface.

termed creasing, where very large local strains are associated with the instability morphology.[9]

In the research by Southern and Thomas,[31] the swelling of a confined elastomer with a *volatile* swelling agent clearly demonstrated the dynamic response of surface wrinkles. As the swelling agent evaporated, the wrinkled surface returned to its native planar state. This dynamic or transient attribute of surface wrinkle topographies is especially attractive for the design of responsive surfaces that can reversibly change properties such as adhesion, friction, or reflectivity in response to environmental stimuli. If stable wrinkles are desired, then the swelling agent can incorporate chemistry that facilitates stabilization. One approach developed by Chan *et al.* is presented in Fig. 2(b).[11] In their approach, a thin P*n*BA elastomeric film is adhered to a rigid substrate.

Next, a photocurable *n*-butyl acrylate-based formulation is used to swell the elastomer. Again, due to the development of a net compressive stress associated with confined swelling, wrinkles develop on the free surface of the elastomer. Following wrinkling, the entire assembly is irradiated with ultra-violet light that photopolymerizes the swelling agent and stabilizes the wrinkles. Other approaches such as thermal polymerization of the swelling agent can also be used; however, the use of photochemistry facilitates a more rapid and controlled stabilization route.

Top-constrained materials

Similarly, surface wrinkling can be induced in an elastomer that is not bound to a rigid substrate when the confinement layer is introduced from the top surface, i.e. a rigid, skin layer. Again, the differences in the swelling response between the skin (resists swelling) and elastomer (prefers swelling) layers lead to the development of a net compressive stress and subsequent wrinkling. Chan and Crosby have demonstrated this route with a very simple process that begins with the modification of a polydimethylsiloxane (PDMS) elastomer in the presence of UV/Ozone, as illustrated in Fig. 3.[14,15] In this process, the UV/Ozone converts the surface region of the PDMS into a gradient SiOx layer, which is significantly more rigid compared to the non-modified elastomer. This elastomer with a rigid, strongly adhered capping layer is then exposed to a swelling agent. Upon initial swelling, the capping layer restricts the swelling of the elastomer, and a net compressive stress develops near the capping layer/elastomer interface. At a critical compressive stress, the capping layer/elastomer interface buckles, or wrinkles. In contrast to surface creasing as observed by Hayward[9] and others[31,32] for bottom-confined elastomers, the high bending energy of the rigid capping layer in this process limits the morphology to more wave-like patterns. If desired, the swelling agent can be a formulation consisting of monomer, crosslinker, and initiators that can be polymerized and crosslinked to stabilize the wrinkled topography. If the swelling agent is removed, the wrinkled topography will disappear and the surface returns to initial planar state. If the swelling agent's equilibrium osmotic stress within the elastomer is significantly large, the rigid capping layer may not be

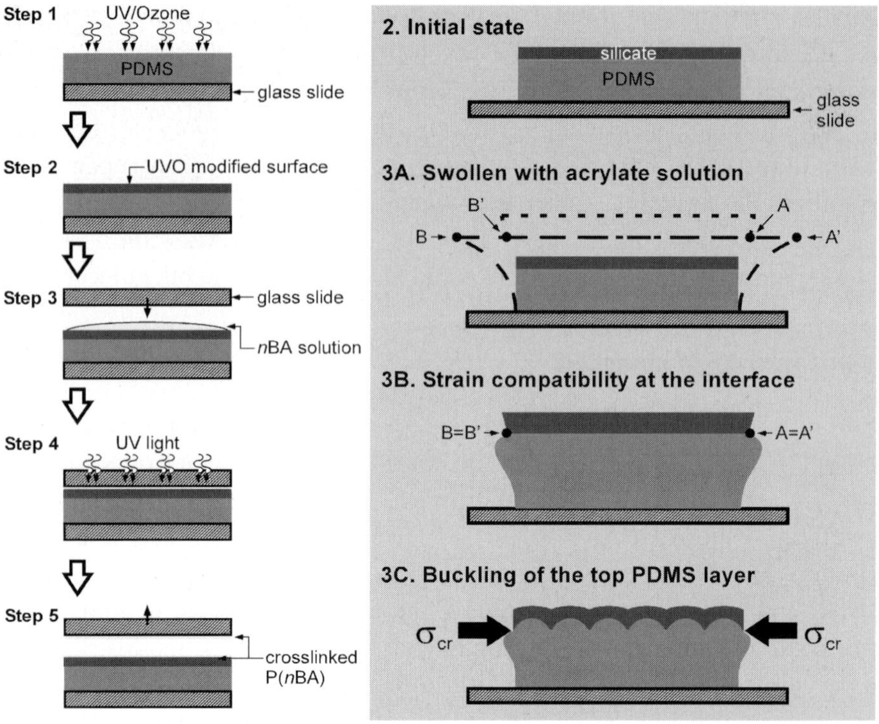

Fig. 3. Example of a top-constrained material wrinkled with an osmotic stress. An approach demonstrated by Chan and Crosby uses a polydimethylsiloxane (PDMS) elastomer as the substrate and develops the top constraint layer through a UV/Ozone (UVO) oxidation process to form an SiOx skin.[14]

capable of restricting the swelling of the underlying elastomer. Under these conditions, the rigid capping layer may fracture to relieve the stress development thereby relieving the stresses that caused the surface wrinkling.[33]

4. Controlling Wrinkle Morphology

The orientation of the surface wrinkles is associated with the direction of the applied compressive stress. This relationship between wrinkle orientation and compression direction can be easily defined in the wrinkling of skin and the wrinkling by mechanical compression, where

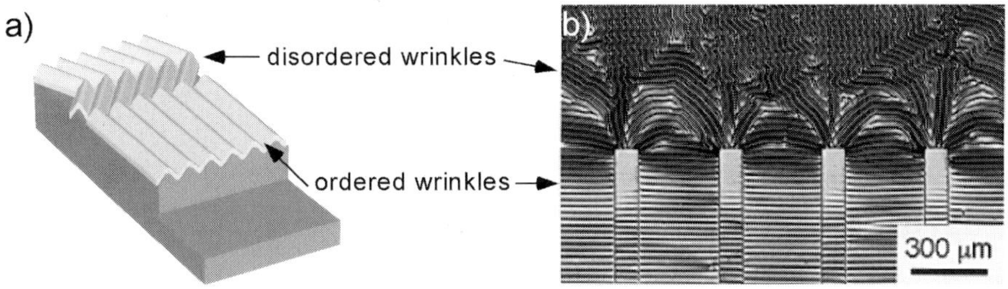

Fig. 4. Alignment of surface wrinkles with topography. a) Illustration of aligning surface wrinkles with topography. Order is observed at the physical edge of a feature. b) Example of topographical alignment of surface wrinkles. However, away from the topographic feature of the PDMS elastomer, the wrinkles became disordered. Figure reproduced by Bowden *et al.*[30]

wrinkles align perpendicular to the direction of the compression.[3,19,21] However, stimuli such as swelling and thermal mismatch are inherently triaxial processes. In the context of film geometry, these processes can be approximated as biaxial processes, which as predicted, lead to equilibrium herringbone wrinkle patterns at stresses above the critical buckling stress.[14,25,27] In order to align wrinkles developed by biaxial stress states, a local principle compressive stress must be developed. For wrinkling by thermal mismatch, one common approach involves the use of topography to develop a primary compressive stress.[25] The use of a patterned elastomer facilitates alignment since the boundary conditions defined by the physical edge of a topographic feature establish a primary compressive stress (Fig. 4(a)). Far from a pattern feature in the absence of physical boundaries, the compressive stress is biaxial. Hence, this region forms the 2-D isotropic wrinkles. Near the physical edge of a pattern feature, the discontinuity of the elastomer surface leads to the development of a compressive stress perpendicular to the feature edge. Since the elastomer is a continuous medium parallel to the feature edge, the stress is still continuous parallel to the edge. This control has been demonstrated by Bowden *et al.* on the wrinkling of gold or silicate thin films supported by a patterned PDMS elastomer. (Fig. 4(b)). Similar results have been obtained by Yoo *et al.* by using a patterned elastomer stamp to confine a thin metal film upon cooling, thus altering the local stress state to cause wrinkle alignment.[22]

An alternative means to align wrinkles without a pre-existing topographic pattern has been demonstrated by Chan and Crosby[14] using PDMS elastomer as the elastic support (Fig. 5(a)). In this approach, moduli-mismatch in the near-surface region was created through selective UV/Ozone oxidation of the PDMS through a stencil mask. This oxidation process leads to the selective conversion of regions of the PDMS surface into SiOx films, thus creating the desired moduli-mismatch between the SiOx and neighboring regions of non-modified PDMS. Upon swelling, a net compressive stress develops due to the differences in the swelling response between the PDMS and silicate layer, as described above for the top-constraint process. Due to the presence of the moduli-mismatch between the finite silicate regions and the PDMS, a primary compressive stress develops parallel to the PDMS-silicate boundary and leads to the formation of aligned surface wrinkles (Fig. 5(b)).

Regardless of the specific approach to generating aligned surface wrinkles, the alignment process is only effective over a finite distance from the stress-altering boundary. Beyond this finite length, the wrinkles become disordered and defects develop as equibiaxial morphologies dominate. This distance is described as ζ in equation (5). An experimental confirmation of this distance is shown in Fig. 5(b), where the distance between stress-altering boundaries is varied continuously in one feature.[14] At a critical separation distance, defects in the aligned wrinkles are observed and ζ can be measured directly.

In addition to 1-D alignment of surface wrinkles, both spatial and orientational control can be achieved by confining their development in two-dimensions (2-D). Extending the concept of moduli-mismatch to 2-D, Chan and Crosby confined the formation of surface wrinkles within hexagonally-shaped silicate patches (Fig. 5(c)).[15] Consistent with mechanism of 1-D alignment, oriented wrinkles were observed at the boundaries where a primary compressive stress develops. The ordering continues until interaction with neighboring wrinkles occurs, beyond which the random wrinkles dominate due to the evolution to an equibiaxial stress state. By systemically reducing the average diameter of the silicate plate (D), two new wrinkling morphologies were observed.

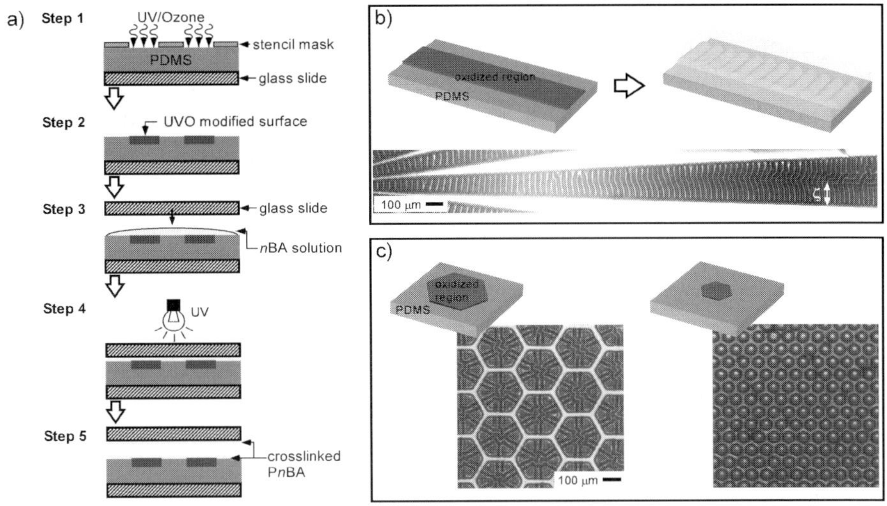

Fig. 5. Methodology to produce alignment of surface wrinkles without pre-existing topography. a) An approach developed by Chan and Crosby that utilizes selective UVO oxidation of the PDMS surface to develop a mismatch in elastic modulus in the near-surface region of the elastomer.[14] b) Demonstration of one-dimensional alignment of surface wrinkles through the formation of a SiOx strip region. c) Formation of spatially and orientationally defined surface wrinkles through the formation of SiOx patches on the PDMS surface.

The first is a dimpled surface and the second is the microlens morphology (Fig. 6). The formation of these new morphologies is related to the commensurability of the dimension of the wrinkles in relation to the plate diameter. To estimate this commensurability, the authors defined a confinement parameter that is described as the ratio of the silicate plate diameter versus the persistence length (D/ζ). For large confinement parameter values ($D/\zeta > 1$), 2-D isotropic wrinkles were formed since the dimension of a wrinkle is significantly less than the silicate plate diameter. This result is attributed to the fact that the boundary defined by the plate diameter does not significantly influence the wrinkle formation. Away from the plate edges, specifically, at the center of the plate, (v in Fig. 6), 2-D isotropic wrinkles develop. Slight reduction of the plate diameter leads to further alignment of the wrinkles

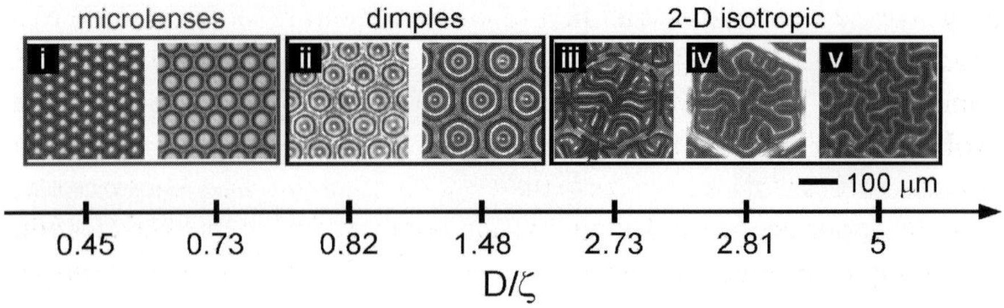

Fig. 6. Control of the surface wrinkling morphologies by defining the silicate plate diameter (D). The morphological transitions can be predicted using a experimentally-established and materials-defined confinement parameter. This confinement parameter is defined by the ratio of D versus ζ.

(iii & iv in Fig. 6). Further increases in plate confinement (i.e. when $D/\zeta \leq 1$) lead to the formation of the two new axisymmetric patterns that includes the dimple (ii) and microlens (i) morphologies. These two new morphologies can be explained by the commensurability of the materials-defined persistence length compared with the experimental defined plate diameter. Additionally, the axisymmetric nature of the morphologies results from the axisymmetic compressive stress state defined by the silicate plate; the plate is significantly small such that the net compressive stress is a hoop stress. The dimple morphology occurs when only a single wrinkle can be accommodated ($D/\zeta \sim 1$). The microlens morphology occurs at the highest confinement that we develop for our materials ($D/\zeta < 1$) as only a fraction of the wrinkle can be accommodated.

5. Hierarchical Patterning

In addition to the benefits of scalable patterning, another attribute of surface wrinkling is the potential for generating hierarchical structures, where the structure consists of multiple discrete length-scales. For osmotically-driven surface wrinkling, the formation of wrinkling patterns can occur on both planar and non-planar surfaces. As a simple demonstration of dual length-scale patterning, Chan and Crosby patterned a crosslinked PDMS hemispherical surface with the microlens array formed by surface wrinkling to create a biomimetic compound eye (Fig. 7(a)).[15] Their approach involves selectively oxidizing the PDMS hemispherical surface to generate silicate patches. This surface is then swollen with the photocurable acrylate swelling agent and then subsequently crosslinked to form the final compound lens. Alternatively, on a planar surface, wrinkle patterns can be formed with one wavelength, but multiple, discrete persistence lengths.[14] The presence of multiple persistence lengths on one surface pattern can be classified as a hierarchical pattern. An example is presented in Fig. 7(b).

Similarly, Efimenko and coworkers used a similar material system to form a nested "self-similar" wrinkled surface (Fig. 7(c)), where one persistence length develops but multiple wrinkle wavelengths are

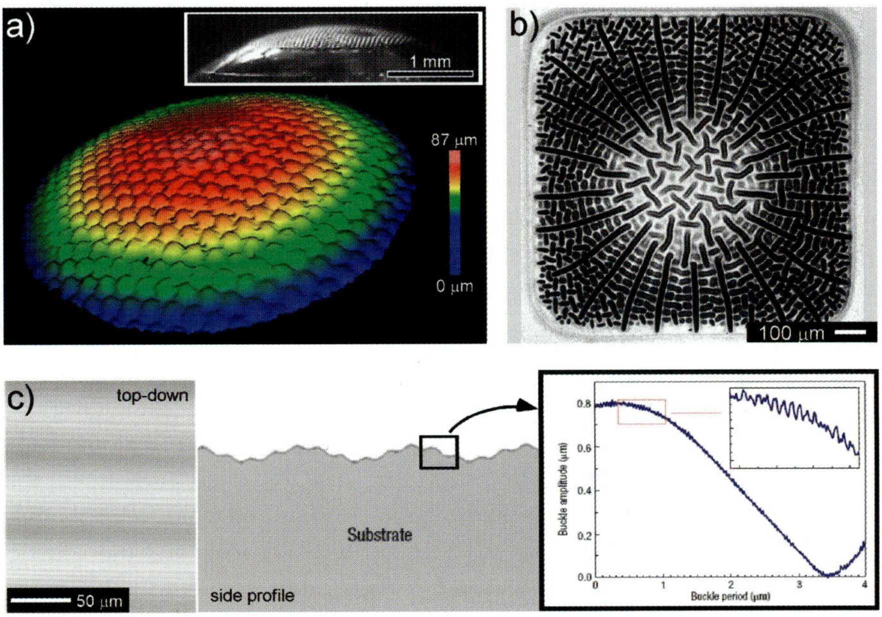

Fig. 7. a) A biomimetic compound eye fabricated by patterning a PDMS hemisphere with surface wrinkles. This structure consists of two pattern principal length-scales as defined by the radii of curvatures of the microlenses and the macroscopic hemisphere. Figure reproduced from Chan and Crosby.[15] b) Hierarchical length scale pattern consisting of multiple persistence lengths. Figure reproduced from Chan and Crosby.[14] c) A nest self-similar wrinkled surface consisting of multiple wavelengths. Figure reproduced from Efimenko *et al.*[19]

present.[19] Unlike the swelling approach used by Chan and Crosby, Efimenko *et al.* used mechanical compression to induce wrinkling. Specifically, they uniaxially stretched a PDMS elastomer and then oxidized the surface by UV/Ozone to form the silicate skin layer. Upon releasing the strain, the compressive force generated due to the moduli-mismatched materials leads to wrinkling. The authors report a hierarchy of five generations of wrinkle wavelengths for their materials. They attribute this high degree of structure hierarchy to the processing conditions: slow release of the applied strain leads to formation of the smallest wrinkles, and as the strain is further released, an "effective" skin that is thicker than the silicate develops and leads to the formation of the second generation wrinkles. Further release of the strain leads to the formation of a thicker "effective" skin which develops the higher order wrinkles.

6. Applications

Recently, there have been tremendous interests in designing surface wrinkles as functional materials. Several attributes make surface wrinkling attractive as technological applications. These attributes include ease in pattern formation, formation of well-defined and controllable pattern length-scales and amenability to a variety of materials. Hence, surface wrinkling can potentially have broad impacts in a variety of technologies. Currently, there are several applications of these instabilities including metrology of thin films, optical devices, flexible electronics and patterned adhesives (Fig. 8).

Stafford and coworkers have developed a metrology tool that utilizes surface wrinkling as an indirect way to measure the elastic modulus of polymers (Fig. 8(a)).[20] This approach, termed Strain-Induced Elastic Buckling Instability for Mechanical Measurement (SIEBIMM), measures the mechanical properties of thin films by measuring the wavelength of mechanically-induced wrinkles, which is clearly defined by equation (4).[20] Therefore, prior knowledge of the elastic modulus of the foundation and the thickness of the thin film, along with measurement of the wrinkle wavelength, allows for the determination of the elastic modulus of the thin film. Compared with

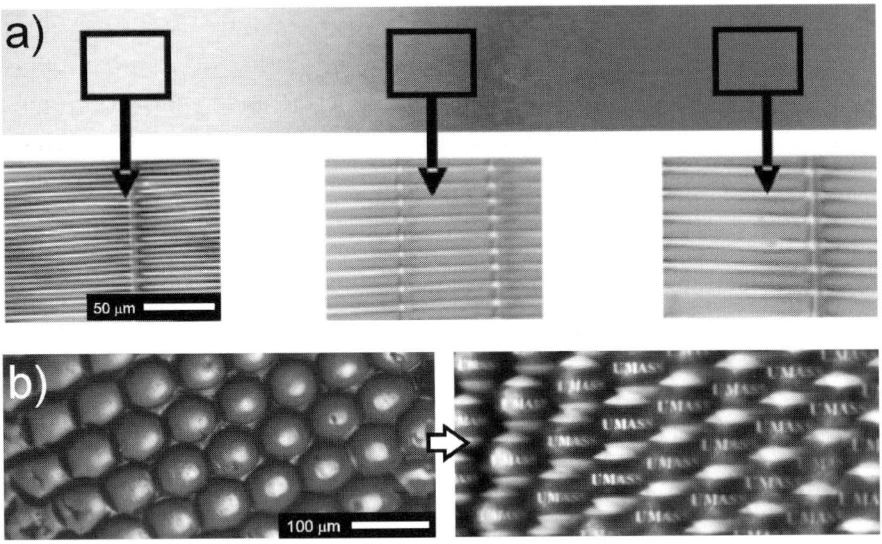

Fig. 8. a) Using surface wrinkling to measure the elastic modulus of thin films. Figure reproduced from Stafford *et al.*[20] b) Application of the wrinkled microlenses as an optical device for information capture.

other mechanical measurement of thin films such as nanoindentation, this measurement technique is very simple since the wavelength of the wrinkles can be observed optically. Conversely, if the elastic modulus and film thickness of the thin film is known, then the modulus of the elastic support can be determined with the SIEBIMM approach.[34]

Menon and Russell have developed an alternative wrinkling strategy that is especially attractive for measuring the elastic and viscoelastic properties of thin films without the aid of an elastic foundation.[35] This strategy wrinkles a thin film that is supported on a liquid foundation. The wrinkles are generated by surface tension of a liquid droplet placed on the floating thin film. The ease, simplicity, and potential for wide materials application make this approach very robust since the properties of the liquid foundation do not play a role in the wrinkle wavelength. Only the thickness and mechanical properties of the floating thin film determine the wrinkle dimensions.

For optical applications, Chan and Crosby have demonstrated that an array of microlenses created by surface wrinkling can be used as an optical device.[15] The authors demonstrated that the wrinkled microlens array are optically responsive and can be used to display information as illustrated by the simple light projection experiment shown in Fig. 8(b). The depth of focus of the lenses can be controlled by tuning the radius of curvature of the microlenses, which the authors demonstrated by their processing conditions. Additionally, the field of view of the microlens array can be enhanced by mimicking the compound eye of dragonfly, where the peripheral vision is enhanced by designing a microlens array on a spherical surface (Fig. 7(a)).

With the new drive for designing electronic circuits to adapt to non-planar platforms, flexible electronic devices are being developed to meet this demand. The Rogers research group has utilized wrinkling as a strategy to generate flexible electronic devices.[32,36] This concept is analogous to the use of pre-stressed concrete in structural engineering applications such as bridge design. In their approach, single-crystal silicon stripes are adhered onto a uniaxially stretched PDMS elastomer. Upon the release of the strain, a compressive stress develops that leads to buckling or compression of the semi-conducting stripes. In this buckled configuration, the semi-conductor is more resistant to permanent deformation.

Surface wrinkled polymers have also been used as gecko-inspired patterned adhesives.[11] The underlying mechanism of adhesion for these patterned surfaces is associated with the enhanced control of the contact line per area during separation. In other words, patterning an adhesive does not increase interfacial contact area compared with the smooth analog, rather, the patterned interface lengthens the total contact line per area during separation. Previous work has demonstrated that well-defined microfabricated pillar arrays can enhance the properties of polymer adhesives.[37,38] However, as the patterning process is lithography-based, commercial application of these materials is rather limited due to scalability of the fabrication process. Surface wrinkling provides an efficient and scalable approach to fabricating a patterned adhesive. Using the osmotic stress wrinkling approach to generate wrinkles on an acrylate elastomer surface, Chan *et al.* demonstrated that the an initially-smooth, nonpatterned adhesive can be patterned in-situ and used directly as a patterned adhesive.[11] More importantly, they were able to show that the wavelength of the wrinkles controls the properties of the adhesive that is based on the mechanism of contact line enhancement.

7. Summary

Elastic instabilities, especially in the form of surface wrinkling, offer a unique, novel, and robust approach to defining the topography of polymer surfaces. Not only can structure be defined with discrete, predictable length scales, but function in the direct control of surface properties can be achieved. Furthermore, environmental stimuli can be used to drive the onset of surface wrinkling, thus presenting an attractive strategy for the definition of responsive surfaces for dynamic surface property control. Although significant efforts have been demonstrated, this field is only emerging in the broad context of a new design paradigm for polymer surfaces and many challenges remain. Among these challenges are describing the kinetic parameters that dictate wrinkle morphology, controlling the development of defects in equilibrium morphologies, describing the mechanics of wrinkled surfaces, and identifying the dimensional limits of surface wrinkling. As these challenges are met, there is little doubt that these strategies will find application in a vast range of advanced materials.

Acknowledgments

The authors gratefully acknowledge the financial support of an NSF CAREER DMR-0349078 award and the Army Research Office Young Investigator Program.

References

1. Charras, G. T.; Yarrow, J. C.; Horton, M. A.; Mahadevan, L.; Mitchison, T. J. *Nature* **2005,** 435, (7040), 365-369.
2. Forterre, Y.; Skotheim, J. M.; Dumais, J.; Mahadevan, L. *Nature* **2005,** 433, (7024), 421-425.
3. Genzer, J.; Groenewold, J. *Soft Matter* **2006,** 2, (4), 310-323.

4. Marder, M.; Sharon, E.; Smith, S.; Roman, B. *Europhysics Letters* **2003,** 62, (4), 498-504.
5. Sharon, E.; Roman, B.; Swinney, H. L. *Physical Review E* **2007,** 75, (4).
6. Biot, M. A. *Applied Scientific Research A* **1963,** 12, 168-182.
7. Biot, M. A. *Proceedings of the Royal Society of London Series a-Mathematical and Physical Sciences* **1957,** 242, (1231), 444-454.
8. Timoshenko, S., *Theory of Elastic Stability*.
9. Trujillo, V.; Kim, J.; Hayward, R. C. *Soft Mattter* **2008,** 4, 564-569.
10. Holmes, D. P.; Ursiny, M.; Crosby, A. J. *Soft Matter* **2008,** 4, 82-85.
11. Chan, E. P.; Smith, E.; Hayward, R. C.; Crosby, A. J. *Advanced Materials* **2008,** 20, 711-716.
12. Qian, W. X.; Xing, R. B.; Yu, X. H.; Quan, X. J.; Han, Y. C. *Journal of Chemical Physics* **2007,** 126, (6).
13. Holmes, D. P.; Crosby, A. J. *Advanced Materials* **2007,** 19, 3589-3593.
14. Chan, E. P.; Crosby, A. J. *Soft Matter* **2006,** 2, (4), 324-328.
15. Chan, E. P.; Crosby, A. J. *Advanced Materials* **2006,** 18, (24), 3238-3242.
16. Mahadevan, L.; Rica, S. *Science* **2005,** 307, (5716), 1740-1740.
17. Huck, W. T. S. *Nature Materials* **2005,** 4, (4), 271-272.
18. Hayward, R. C.; Chmelka, B. F.; Kramer, E. J. *Macromolecules* **2005,** 38, (18), 7768-7783.
19. Efimenko, K.; Rackaitis, M.; Manias, E.; Vaziri, A.; Mahadevan, L.; Genzer, J. *Nature Materials* **2005,** 4, (4), 293-297.
20. Stafford, C. M.; Harrison, C.; Beers, K. L.; Karim, A.; Amis, E. J.; Vanlandingham, M. R.; Kim, H. C.; Volksen, W.; Miller, R. D.; Simonyi, E. E. *Nature Materials* **2004,** 3, (8), 545-550.
21. Harrison, C.; Stafford, C. M.; Zhang, W. H.; Karim, A. *Applied Physics Letters* **2004,** 85, (18), 4016-4018.
22. Yoo, P. J.; Suh, K. Y.; Park, S. Y.; Lee, H. H. *Advanced Materials* **2002,** 14, (19), 1383-1387.
23. Huang, R.; Suo, Z. *Journal of Applied Physics* **2002,** 91, (3), 1135-1142.
24. Huck, W. T. S.; Bowden, N.; Onck, P.; Pardoen, T.; Hutchinson, J. W.; Whitesides, G. M. *Langmuir* **2000,** 16, (7), 3497-3501.
25. Bowden, N.; Brittain, S.; Evans, A. G.; Hutchinson, J. W.; Whitesides, G. M. *Nature* **1998,** 393, (6681), 146-149.
26. Cerda, E.; Mahadevan, L. *Physical Review Letters* **2003,** 90, (7), 074302-1-4.
27. Chen, X.; Hutchinson, J. W. *Journal of Applied Mechanics* **2004,** 71, 597-603.
28. Chen, X.; Hutchinson, J. W. *Scripta Materialia* **2004,** 50, 797-801.
29. Breid, D.; Crosby, A. J. *in preparation* **2008**.
30. Bowden, N.; Huck, W. T. S.; Paul, K. E.; Whitesides, G. M. *Applied Physics Letters* **1999,** 75, (17), 2557-2559.
31. Southern, E.; Thomas, A. G. *Journal of Polymer Science Part a-General Papers* **1965,** 3, (2PA), 641-&.

32. Tanaka, T.; Sun, S. T.; Hirokawa, Y.; Katayama, S.; Kucera, J.; Hirose, Y.; Amiya, T. *Nature* **1987,** 325, (6107), 796-798.
33. Kundu, S.; Sharma, R.; Crosby, A. J. *in preparation* **2008**.
34. Wilder, E. A.; Guo, S.; Lin-Gibson, S.; Fasolka, M. J.; Stafford, C. M. *Macromolecules* **2006,** 39, 4138-4143.
35. Huang, J.; Juszkiewicz, M.; de Jeu, W. H.; Cerda, E.; Emrick, T.; Menon, N.; Russell, T. P. *Science* **2007,** 317, (5838), 650-653.
36. Khang, D.-Y.; Jiang, H.; Huang, Y.; Rogers, J. A. *Science* **2006,** 311, 208-212.
37. Arzt, E.; Gorb, S.; Spolenak, R. *PNAS* **2003,** 100, (19), 10603-10606.
38. Crosby, A. J.; Hageman, M.; Duncan, A. *Langmuir* **2005,** 21, (25), 11738-11743.

CHAPTER 7

CRYSTALLIZATION IN POLYMER THIN FILMS: MORPHOLOGY AND GROWTH

Ryan M. Van Horn and Stephen Z. D. Cheng

Department of Polymer Science, University of Akron
170 University Circle, Akron, OH 44325
E-mail: scheng@uakron.edu

Confinement of semi-crystalline polymers into thin films can affect both the morphology and the growth of the resultant crystals. Several morphologies, including polycrystalline and single crystalline, can be obtained in thin films under certain conditions. In addition, the molecular orientation and kinetics can be affected by the temperature, film thickness, and interfacial energy. This chapter is a review of various reports in thin film crystallization.

1. Introduction

Phase behavior of polymers is an important aspect in understanding their resulting macroscopic properties. These phases range from amorphous liquids and glasses to mesophases to three-dimensionally (3D) ordered crystals. The relationship between the molecular arrangement, molecular dynamics, structures in different length scales and resultant bulk properties has been studied for decades. Mechanical, optical, and electrical properties are important in the fields of microelectronics, biomedical technology, and nanotechnology. Recently, due to new science in low dimensional materials and the miniaturization of products, there has been increasing interest in these relationships in thin films.[1]

Specifically, crystallization of polymers plays a vital role in determining the final properties of the material. Crystallization can enhance a material's strength and toughness, disrupt optical clarity, or improve conductivity. Thus, it is important to understand the physics behind polymer crystallization. This topic has been studied for years in order to predict crystallization behavior.[2-16] It is well known that polymers crystallize into lamellar habits in either a folded-chain conformation (metastable state) or an extended chain conformation (equilibrium state). It is also known from the theory developed by Hoffman and Lauritzen[17] that the growth rate of these crystals, G, is ex-ponentially proportional to the inverse of the supercooling, ΔT, $\exp\{-1/\Delta T\}$, which is the distance from the equilibrium melting temperature, T_m^o. The crystal lamellar thickness and morphology are also determined by ΔT. However, there are many features of polymer crystallization that are unexplained by this theory and are currently under discussion.[15,16]

Because of the growing interest in nanotechnology, it becomes important to understand crystallization in low dimensional space under confinement. As the desired size of the product material gets increasingly smaller, it is important to determine the variations in crystal morphology, growth, or kinetics as compared to the bulk. There are two common ways to study crystallization in confinement. One is in block copolymers with at least one crystalline component; the other is in thin films. In both of the systems, the size of the domain in block copolymers or the thickness of the film approaches the length scale of the crystal size or even the chain dimensions. Thin film confinement has been shown to affect the chain dynamics and thus, the glass transition temperature, T_g, of several polymers, though the origin of this change is still being studied;[18-24] therefore, we would expect crystallization to undergo alterations. In addition to the application needs, crystallization under confinement provides a new way to study the fundamental physics of crystallization in real time.

In this chapter, we will focus on the crystallization behavior of pseudo-2D confined polymers in a thin film with a thickness between several nanometers and 1 μm. Similar to those in three dimensions, the morphology of these crystals varies from quasi-spherulites to single crystals; however, by confining these structures to two dimensions, the growth and kinetic study of general crystallization becomes more accessible. These subjects will all be covered in more detail throughout the chapter.

The processing of these films will be relegated to a few specific cases. We will not be discussing films made from conventional industrial methods, such as extrusion, rolling, and biaxial stretching. All films were made either by spin-casting or solution-casting. In both cases, a dilute polymer solution, typically between 0.1% and 1% is cast onto a substrate of mica, silicon, or glass (with some exceptions). In spin-casting, the substrate is placed on a rotating stage. The stage is spun between 2000 and 3000 rpm depending on the desired thickness. For solution-casting, a drop or two of solution is placed on the substrate. The solvent evaporates leaving a uniform film behind. It is important that the solvent has a low vapor pressure and that it wets the substrate surface. The thickness of these films is also determined by the concentration. For more uniform films, optimized spin-coating conditions typically provide the best results.

As for instrumentation and experimental techniques, there are several methods to study crystallization in thin films. Any technique that can separate the responses of crystalline and amorphous segments can be utilized here, given the amount of material is sufficient. Typically there are several main techniques to be used. They are optical microscopy (OM), atomic force microscopy (AFM) and transmission electron microscopy (TEM) in real space. Optical microscopy provides visualization of the crystal morphology as well as a method of tracking crystal growth. In reciprocal space, the grazing incidence x-ray diffraction (GIXRD) technique is commonly utilized. X-ray diffraction provides structural information caused by a difference of electron densities of the materials in different phases. In the grazing incidence geometry, specifically, it is possible to follow the crystallization at the surface, interface, and inside of the film. Additional methods include Fourier transform infrared spectroscopy (FT-IR) and ultra-violet spectroscopy (UV). Differences in crystalline and amorphous absorption bands in the IR and UV spectra provide information on the crystallinity of the sample and its dependence on crystallization time. Because of their wider use, the following will be an in-depth look at the uses of AFM and TEM in studying the crystallization in thin films.

The use of AFM provides many advantages over other experimental techniques. AFM is equipped to study crystallization *in situ* over a large temperature range.[25] Using the tapping mode, one can visualize the crystal morphology as well as determine the crystal thickness, which are both a function of ΔT. The hot stage attachment allows one to change the temperature of the system. This allows the user to determine other factors such as growth rate, kinetics, and melting and thickening mechanisms. It is important to use a minimum tapping force as to not fracture the surface or induce external energy to assist in crystallization.

TEM can be used to study the morphology and crystal structure, post-crystallization. The bright field image provides a real space representation of the crystal morphology; while, electron diffraction (ED) provides a reciprocal space representation giving the crystal structure and unit cell dimensions. Using other techniques such as dark field imaging or polyethylene decoration, special features of the crystal can be determined, such as orientation of specific crystallographic planes or folds, respectively. Transfer of the thin film on to a copper grid and shadowing of the sample with Pt/C are critical steps in preparing the film for TEM imaging.

The study of polymer crystallization and crystallization in thin films is quite extensive. Our goal is to provide an overview of the literature involving crystal morphology, growth, and kinetics in thin films. The results here will be discussed in a way that non-crystallographers studying thin films can obtain the appropriate knowledge; however, in depth analysis of general polymer crystallization, crystal structure, and crystallization theory will be minimal for simplicity. In addition, the melt state properties of the thin film will only be discussed where appropriate. These properties have a general affect on crystallization – these are unchanged by the confined dimensions – that can be found in the literature. For more general information on crystallization, see references 2, 3 and 15.

2. Spherulitic and Dendritic Morphology

Polycrystalline morphologies consist of those that have growth from a central nucleus with lots of branching. The two main polycrystalline morphologies are spherulites and dendrites.[2] Spherulites have lamellar crystals that grow from the center with sufficient branching to create a 3D sphere. The sphere has crystalline branches with amorphous sections in between. Under polarized optical microscopy, spherulites exhibit a Maltese cross pattern due to the orientation of the crystalline regions with respect to the polarizer. These crystals are typically formed at high ΔT due to the higher growth rate.

Dendrites are single-crystalline-like structures that grow due to a concentration gradient at the growth front of the crystal. The ratio of the diffusion constant, D, to growth rate, G, is critical for dendrite formation. This limitation causes branching that leads to needle-like growth. When finished, the growing crystal looks like a tree. Due to spiral dislocations, these dendrites can have multiple layers of crystalline material. Dendritic growth becomes more prominent at moderate-to-low supercoolings.

In thin films, we see formations of both of these structures. For spherulites, because of the pseudo-2D space, the branching occurs in-plane. The growing lamellae continue to branch as the material is consumed in the crystallization process. Typically, the crystals grow edge-on with respect to the substrate, meaning the fold surface is perpendicular to the substrate. Dendrites, or diffusion-limited aggregates (DLA),[26] are formed as flat-on "single crystals" with a large degree of branching. There are a few different kinds of DLA's as described by Brener and coworkers.[27] They are compact seaweed (CS), fractal seaweed (FS), compact dendrite (CD), and fractal dendrite (FD) where seaweeds are isotropic and dendrites are anisotropic. All of these morphologies are highly dependent on supercooling as well as substrate surface energy.

2.1. *Spherulites*

Several research groups have studied the spherulite morphology and their growth using *in situ* AFM and TEM for poly(ε-caprolactone) (PCL),[25,26] poly(triflouroethylene) (PF_3E),[27] poly(bisphenol A-*co*-octane) (PBAC8),[29] poly(ethylene terephthalate) (PET),[30,31] poly(ethylene oxide) (PEO),[32-34] isotactic polystyrene (i-PS),[35] and polyethylene (PE).[36] All of these studies show that the polymer crystals grow from a central nucleus. From this nucleus, a lamellar sheaf is created. As the chains attach to the crystal, they form long edge-on lamellae. It is determined that the crystals are edge-on because their width is approximately the same as the lamellar thickness in the bulk at the given isothermal crystallization temperature.[29] These lamellae continue to grow; however, due to the high growth rate, they begin to branch off. As branching continues, the crystal begins to resemble the 3D spherulite. Near the center, or eye, the lamellae appear to grow flat-on.[32] As the crystallization time increases, these lamellae begin to thicken.

Generally speaking, the cause of branching is due to screw dislocations within the primary lamellae; however, in some cases, especially in thin films, it is caused by secondary nucleation along the dominant lamellae.[29,37] As the original growth front propagates, small secondary nuclei are formed near the growing crystal. The nuclei are formed by loose loops or protruding cilia from the surface of the mother crystal. As these daughter lamellae propagate, they intersect with the mother lamella. The material between the branches acts as a reservoir for the growth front; however, the limited space between branches also facilitates the joining of two lamellae. This joining, along with the high branching density, gives the crystal the characteristic morphology.

As the spherulite grows in the thin layer of liquid, the non-crystalline material is rejected from the growth front, and transported away from this front. The distance that they can be transported depends upon the diffusion rate of the molecules and crystal growth rate. On the other hand, the crystalline chains find channels in the thin layer to diffuse to the growth front. This mechanism is vital to all crystallization in thin films and will be discussed throughout the chapter. The overall growth rate of the spherulite is constant in isothermal conditions; however, the growth of individual lamella varies slightly.

Figure 1 shows the development of a spherulite of poly(bisphenol A-*co*-octane) as observed by *in situ* AFM. The spherulite was grown at room temperature (ΔT = 59°C) from a film with a thickness of 300 nm. In Figure 1a, the lamellar sheaf is beginning to develop with flat-on crystals near the center and fibrous lamellae extending outward. Figures 1b-1f show the development of the branches from the primary lamellae.

Over time, the central area of the spherulite becomes quite dense. This shows the high branching density and the joining of the lamellae. Near the center, there remains some uncrystallized area. The lamellae at the periphery continue to splay outward, thus leaving this region uncrystallized. Spherulites of this morphology are called "hedgehog" spherulites, or hedrites. Spherulites with a more regular radial growth can be seen in references 33 and 35.

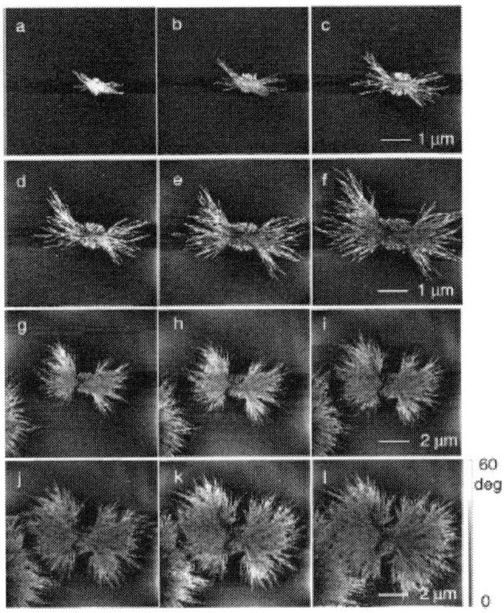

Fig. 1. Growth of "hedgehog" spherulite, or hedrite, from AFM phase image. *Source*: Li (2001)

2.2. *Diffusion-Limited Aggregates (DLA)*

In addition to spherulitic growth, dendritic growth is a polycrystalline morphology that occurs at large ΔT. Similar techniques have been used to identify the dendrite morphology along with their growth.[26,27,34,35,38,39] Dendrites are single-crystal like and grow flat-on as compared to edge-on as confirmed by the ED technique. The branching occurs along the growth faces and is caused by a difference in the diffusion rate of the chain and the growth rate. Unlike spherulites, this is the only mechanism for branching and does not include secondary nucleation from the fold surface.

There are four different DLA morphologies, compact dendrite (CD), compact seaweed (CS), fractal dendrite (FD), and fractal seaweed (FS).[28] The fractal pattern has self-similarity; whereas, the compact does not. These assessments have been made according to theoretical work. They have been observed experimentally as well. The dendrites exhibit patterns from a nearly spherulitic morphology (those with high branching density) to the single-crystal morphology (those with fewer branches). The seaweed morphology is not as common but can be found with low anisotropy or high ΔT. Figure 2 shows a range of dendrite morphologies for i-PS.[35] Figure 2(a) is similar to the spherulite morphology. Figure 2(b) is the CS morphology. Figure 2(c) is the CD morphology. The final three morphologies (2(d)-2(f)) are characteristic of the fractal patterns (FD and FS). These morphologies occur with a difference in film thickness indicating that the surface energy (anisotropy) is the key parameter. The ΔT dependence will be discussed in the single crystal section.

Fig. 2. Dendritic morphologies as a function of crystal thickness: a) 17 nm, b) 14 nm, c) 11 nm, d) 9.7 nm, e) 8.7 nm, and f) 6.1 nm. Scale bars represent 5 μm. *Source*: Taguchi (2002)

Diffusion-controlled growth can be explained by three different mechanisms:[30,31] 1) thermal diffusion, 2) impurity diffusion, and 3) low-dimensionality effect. The thermal diffusion mechanism occurs when the ratio of the thermal-diffusion constant and growth rate is small; however, in polymer crystallization, the growth rate is slow with a minimal temperature gradient. This explanation rules out thermal diffusion. For impurity diffusion, the molecular weight (MW) distribution causes a fractionation to occur during crystallization. This affects the growth of bulk spherulites through a high concentration of low-MW components at the interface. Low-dimensionality appears to be the mechanism from which these dendrites and seaweeds grow. This is due to the thickness of the film. As the film gets thinner, the quasi-2D nature induces the low-dimensionality affect, resulting in dendritic morphologies.

The growth of the branches, and thus the resultant morphology, is dependent on film thickness[35] and crystallization temperature.[39] In both of these studies, the branch width, w, was compared to the given parameter. For the film thickness (d), Taguchi *et al.*[35] showed that the branch width increases with one over the thickness ($1/d$). Above 10 nm, w is constant and then is linearly proportional with $1/d$. Zhang and coworkers[39] show that the width of CS fingers has an exponential relationship with crystallization temperature at a film thickness of 50 nm.

2.3. *Origin of Crystal Orientation*

As was discussed in the previous sections, spherulites contain an edge-on crystal orientation; whereas, the dendrites have a flat-on orientation. This orientation is created in the nucleation step and determines the morphology of the final crystal. Wang *et al.*[36] present the energetic origin for the crystal orientation.

Starting from the derivation of the crystal free energy,[2] Wang and coworkers add the substrate free energy into each term to show its effect on nucleation:

$$\Delta G = -a^2 l \Delta G_v + 2a^2 \gamma_e + 3al\gamma_l + al(\gamma_{cs} - \gamma_{ms}) \qquad (1)$$

$$\Delta G = -a^2 l \Delta G_v + 4al\gamma_l + a^2(\gamma_e + \gamma_{cs} - \gamma_{ms}) \qquad (2)$$

where ΔG is the overall change in free energy, a is the lateral size, l is the crystal thickness, ΔG_v is the free energy change on crystallization, γ_e is the interfacial energy between the fold surface and the melt, γ_l is that between the lateral surface and the melt, γ_{cs} is between the crystal and substrate, and γ_{ms} is between the melt and substrate. Equation 1 is for the case where the chains lie parallel to the substrate (edge-on), and Equation 2 is for the case when the chains are perpendicular (flat-on). From the minimization of the free energy, the equilibrium sizes can be determined:

$$a^* = \frac{3\gamma_l + \gamma_{cs} - \gamma_{ms}}{\Delta G_v} \qquad l^* = \frac{4\gamma_e}{\Delta G_v} \tag{3–4}$$

$$a^* = \frac{4\gamma_l}{\Delta G_v} \qquad l^* = \frac{2(\gamma_e + \gamma_{cs} - \gamma_{ms})}{\Delta G_v} \tag{5–6}$$

Since we know that the fold surface has a much higher free energy than the lateral surface ($\gamma_e >> \gamma_l$) and we can estimate all other terms, the conformation of the chains will be edge-on as long as $\gamma_{cs} - \gamma_{ms} < 0$. Grazing incidence x-ray diffraction (GIXRD) experiments by Cho *et al.*[40] on isotactic polypropylene (i-PP) showed that the crystallinity at the polymer-substrate interface increased as the surface energy of the substrate increased. The orientation of the chains at the interface was more parallel (edge-on) than in the bulk sample; however, as the surface energy increased, the chains took on a perpendicular orientation. This indicates the transition from edge-on to flat-on crystals.

GIXRD results show that the surface, or air-polymer interface, may also play a role in the crystal orientation.[41-46] GIXRD is sensitive to depths of tens to thousands of angstroms dependent on the incidence angle, α, as compared to the critical value, α_c. This allows the researcher to truly probe the surface and bulk of the film. Factor, Russell, and Toney[41,42] found that an aromatic polyimide forms a liquid-crystalline or a crystalline region at the surface of films with varying thicknesses and thermal treatments. They concluded that this orientation will affect the morphology throughout the bulk of the film. Similar results were found for other polymers such as polyflourene,[43] PET,[44,45] and poly(ethylene 2,6-naphthalate).[46] In all cases, the polymer segments (aromatic rings) showed preferential ordering parallel to the surface in the crystalline packing arrangement. In all cases, the ordering appeared to occur before crystallization of the bulk of the film.

Many experimental results show the effect of film thickness on the crystal morphology.[26,35] In general, as the film thickness decreases, the morphology goes from spherulites to dendrites. Taguchi *et al.*[35] studied the affect on crystals of i-PS. They found that as the thickness decreases below 14 nm, a shift from spherulites to diffusion-limited growth exists. Mareau and Prud'homme[26] have shown that PCL exhibits this transition at a film thickness around 15 nm. The GIXRD results[40,42,46] also show that the preferential arrangement at the surface does not affect the bulk orientation above a certain thickness. If the film is thick, the free energy of developing edge-on lamellae is lower due to the lateral-surface interfacial area. The material is supplied to the top of nuclei to promote growth perpendicular to the surface. In thinner films, it is energetically favorable to develop flat-on crystals because the fold-surface interfacial area wants to be maximized. Because of the thickness, it is easier to supply material parallel to the substrate. Finally, to overcome the nucleation barrier, thermal and density fluctuations are needed. Because of the thickness of the film, these fluctuations are different in the parallel and perpendicular directions, meaning that it may be easier to overcome the barrier to crystallization in one direction as opposed to the other. Both interfacial energy and film thickness play an important role in determining the crystal morphology in crystallizable thin films.

2.4. *Special Features*

Within the discussed morphologies of polycrystalline samples, there are few instances where new features arise. In one case, Duan *et al.*[47] and Wang *et al.*[48] found concentric-ringed structures, similar to banded spherulites in the bulk and thin films.[48-52] The second case was originally discovered by Keith and coworkers[13] and later revisited by Zhang *et al.*[39] This is the curved or scrolled crystals of PE. These dendritic crystals are curved along the long axis of the branches.

The concentric rings were discovered in optical microscopy and compared to those found in bulk banded crystallites. In the bulk, the banding is caused by twisting of the radial lamellae.[52] Because the banded crystals are spherulitic, they are birefringent. The crystals observed by Duan *et al.*[47] and Wang *et al.*[48] were not birefringent; therefore, they are not spherulites. According to TEM images and ED, these crystals are flat-on stacks. It is stated by both groups that the high MW molecules crystallize first; while, the low MW samples diffuse to the fold surface and branch vertically. As this occurs, a depletion zone is created, due to the vertical growth, and the concentric rings are a series of peaks and valleys. Wang *et al.*[48] verified this by changing thickness and temperature, which effectively changes the growth kinetics. By varying these parameters, they saw an increase in the periodic spacing of the rings. This phenomenon was seen over a wide temperature range for both samples. They also were seen in thickness ranging from 100 nm to 300 nm.

The second morphology of interest is the curved, or scrolled, lamellae of polyethylene. Work done by Keith *et al.*[13] showed that PE crystals grown from films exhibited a unique morphology. Branches of dendritic lamellae formed sections of curved crystal faces. In addition, these blade-like crystals had two thicknesses, one along each (200) face. It was found that the thickness difference as well as the curvature is related to the chain tilting[52] that occurs in PE lamellar crystals. The curved crystals are actually edge-on and have branched from the flat-on crystals. The edge-on crystals derive their curvature from the difference in fold-surface energy. This disparity is from the difference in fold surface topology due to chain folding.

In the work of Zhang and coworkers,[39] the curvature occurs in the flat-on, dendritic branches. These crystals also exhibit two thicknesses along the (200) direction. The curvature of the PE crystals is in the direction of the thinner side; however, branching occurs more often on the thicker side. For the bending in flat-on crystals, it is stated that the difference in growth rate for each (200) face (verified by the difference in thickness) is the main driving force for curvature.

Polycrystalline morphologies are the most common form of polymer crystallization. They are easily obtained with large ΔT, or low crystallization temperature. Polycrystalline morphologies provide insight into the general growth behavior of polymers. Growth of spherulites is easy to follow as a function of radius. Dendrites are crucial in studying the effects of diffusion in crystallization. However, in order to have a full scope of crystallization of polymers, one must also discuss the single crystalline morphology.

3. Single Crystal Morphology

The second classification of crystals is single crystals. These are crystals in which all parts can be related to the microscopic translation lattice.[2] They are a single lamella, consisting of two fold surfaces, that have a characteristic shape based on the growth planes of the translational lattice. In most cases, this shape is polygonal. The chain direction is always perpendicular to the substrate (flat-on). Due to the difference in surface free energy between the fold surface and the lateral surface, the lateral dimensions are on the order of microns; while, the thickness is on the order of nanometers. The crystal thickness is still controlled by the degree of ΔT. In order to achieve this morphology, the ΔT must be smaller compared to that of dendrites. Single crystals themselves do not have any large defects. Screw dislocations may occur normal to the lamellar surface. These crystals maintain the same order and symmetry but are not true single crystals.

In thin films, the molecular mobility and metastable crystal thickness are important in growing single crystals. The polymer molecules must crystallize in such a way that the growth of the lateral face and diffusion to that face are similar. This provides a controlled crystallization process allowing for the near-perfect crystals to form. Unlike spherulites or dendrites, there is no branching caused by secondary nucleation or disparity between diffusion and growth. Thus it is important to have a sufficient transport layer to supply molecules to the growth front. In addition, the crystallization process is strong enough to allow for crystals that are thicker than the original film itself. Studying single crystals in thin films may provide a look into the crystallization process and help provide a better understanding.

3.1. *Polygons*

Most single crystals are polygonal in shape. This is due to the structure of the crystalline unit cell. Because a single sheet of polymer crystal is created, the growth of this crystal resembles the order and symmetry of the basic building block, which is the unit cell. For instance, a cubic unit cell will produce a single crystal that has four growth faces, 90° from each other. In other words, it will have a square-shaped morphology. There are several different kinds of unit cells; therefore, there are many different morphological shapes. We will discuss a few of them.

Since we are still looking at the morphological aspects of polymer crystals, AFM and TEM techniques were the main analytical tools once again. There are only a few reports of single crystals from thin films in the literature. These are syndiotactic polypropylene (s-PP),[53,54] i-PS,[55-57] PLLA,[38,58] PEO,[59,60] and PCL.[26] Each has a unique morphology. The PP exhibits a rectangular shape. The i-PS has a hexagonal shape. Some crystals of PLA, PEO, and PCL have truncated shapes. PEO has a truncated square shape, and PLA and PCL have truncated lozenge (diamond) shapes. The truncated shapes are similar to hexagonal shapes; however, the angles are not all the same. These truncated shapes are caused by differences in the growth rate of various crystallographic planes. They will be discussed in more detail below.

Truncated crystals are those that deviate from the normal quadrilateral shape to a hexagonal shape. This change in shape is due to a change in the growth rate in the crystallographic planes.[61] For example, in the case of PE single crystals, the crystal grows with four {110} faces. When truncation occurs, the growth in the [200] direction is slower than in the [110] direction. This is manifested in the new growth faces, {200}, being observed along with four {110} faces, giving the crystal a hexagonal shape. This change in growth rate has several parameters that may affect it; however, for thin films, they are molecular weight and crystallization temperature.

The growth of the single crystals in thin films is similar regardless of shape. As the crystal grows, there is a small concave area between the growth face and the amorphous film.[58] This is caused by the difference in density between the amorphous polymer and the crystalline region. It is known that there is an amorphous layer above and below the crystal. This layer can assist in screw dislocation growth because there is an easily accessible supply of amorphous material.[55,58,60] Also, there can be overgrowth caused by substrate defects. These overgrowths originate from a large heterogeneous nucleus from which the basal, or original, lamella grows. All of these defects can cause the crystal to be thicker than the original film. The lack of material at the air-film interface can prevent the overgrowths or dislocations from developing fully in this case.

In addition to defects normal to the lamellar surface, there have been some crystals with abnormal facets, or growth faces. These were found in the PLLA[38,58] and PCL.[26] The breaking from the truncated shape in the PCL (Figures 3b and 3c) was found to be a function of the thickness. As the film became thinner than the thermodynamic crystal thickness (~ 15 nm), the {100} growth faces were ruptured. In addition, they have a lower crystallographic order as shown by ED. Because the crystal is growing thicker than the film, the molecules' diffusion to the growth face is more difficult. The amount of material supplied to the growth face is lower as well. All of these factors create the irregular shape of the lamellar crystal.

Crystallization temperature is important in the final crystal form. It controls both the crystal thickness, regardless of film thickness, and morphology. Taguchi and coworkers[56] showed the effect of crystallization temperature on thickness. They verified that the thickness of i-PS follows the relationship predicted by the Hoffman-Lauritzen theory. The crystal thickness increases with increasing T_c, or decreasing ΔT. The morphology of the single crystal transitions from a single crystal to a dendrite with an increase in undercooling. At a $\Delta T = 32°C$ ($T_m^o \approx 242°C$), the single crystal morphology is formed. At $\Delta T = 37°C$, small kinks in the growth faces indicate that a transition towards the dendritic morphology is occurring. The dendrite forms at $\Delta T = 47°C$ and eventually reaches the compact seaweed (CS) morphology at $\Delta T = 62°C$. If the crystallization temperature was decreased further, the spherulite morphology would eventually be reached.

3.2. *Helical*

One unique morphology that has not been found in many samples is helical crystals. These crystals have been found in one sample by Li and coworkers.[62-64] The sample is a chiral main-chain liquid crystalline polyester, PET(R*)-9. It consists of benzene rings, an alkyl spacer of 9 methyl groups, and an R chiral center. This polyester exhibits several different liquid crystalline as well as one crystalline phase. The crystalline phase can be grown from thin films of the polyester on a glass substrate. The crystallization of PET(R*)-9 can be found in two forms. One is a rectangular lamellar shape; the other is a helical crystal where the lamella is twisted with a regular pitch.

Although the growth of these helical crystals is not well understood, there are a few features that have been determined. In order to form the helical structure, as seen in Figure 3, the crystallites must form a double twist. This means that the crystallite "sheets" rotate around the helical axis as well as twist along the cross section. These twists are caused by defects in the crystal to give it a frustrated structure. The formation of helical crystals as opposed to flat is dependent on several parameters including film thickness. In thinner films, the flat crystal is dominant due to the restriction of the molecular diffusion. It has been shown that in slightly thicker films, the helix can grow from the flat crystal itself, further indicating that the underlying mechanism is based on defects in the crystal. Finally, the pitch of the helix may vary, even at a single crystallization temperature.

The helical crystal of PET(R*)-9 may be the link to the effects of chirality on higher level structures from the atomic level, the chiral

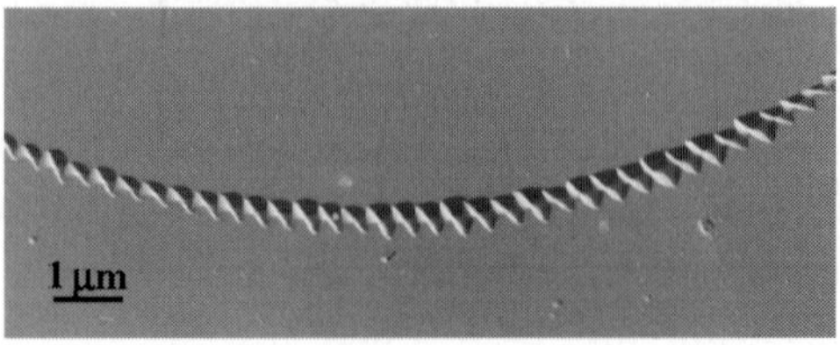

Fig. 3. TEM image of a helical crystal of PET(R*)-9.

carbon itself, to large scale, 3-dimensional helical aggregates, similar to those found in biological systems. The chirality is transferred from the atom to the helical structure of one chain. These chains can now organize themselves through crystallization into a higher lever helical structure.

3.3. *Special Features*

Similar to the polycrystalline samples, there are a few special features that may be found in single crystals as well. One such feature, truncation, was already discussed. Here we will discuss two more, undulation, or striations, and differences in sector thickness. The i-PS, PLLA, and PCL samples all exhibit striations, or ridges, on the fold surface of the crystal.[26,57,58] s-PP has been found to have a different thickness for the two adjacent sectors.[53,54]

Striations can be found on the fold surfaces of single crystals. These undulations are typically perpendicular to the growth face. They can be seen in Figure 4. There are a few explanations for the undulations which can be related. Taguchi *et al.*[57] show through dark field TEM that the striations in i-PS are not uniform. They conclude that these striations must be caused by a buckling of the lamellar surface. The buckling occurs due to surface stresses that occur. They also state that the stresses cannot be uniform on both sides because scrolling, or twisting, would be more prominent, so the stresses originate from an inhomogeneity in the folds.

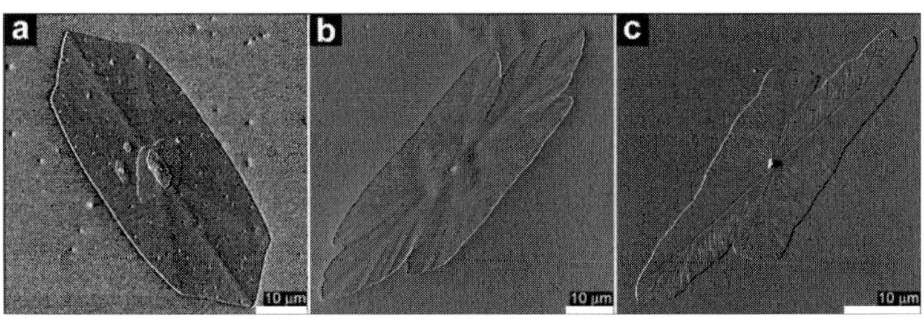

Fig. 4. AFM amplitude images of PCL with striations, ruptured {100} faces, and increasingly irregular growth. *Source*: Mareau (2005)

The idea of inhomogeneous folds was also discussed by Mareau and Prud'homme[26] as well as Kikkawa and coworkers.[58] Both have shown through melting studies that the sectors, or sections of similar fold direction that are typically parallel to the growth face, that exhibit striations melt first. This is an indication of imperfect crystallization. In fact, the folds within the sector may not all arrange in the same fashion. They may have different reentry positions. This leads to inhomogeneities and surface stresses which could cause the striations to occur.

The second feature to address is differences in height for adjacent sectors of s-PP, Figure 5. Work done in the Cheng group[53,54] has shown that the (100) sectors are thicker than the (010) sectors. These elongated rectangular crystals were grown on carbon-coated glass. The elongation indicates that the [100] growth rate (short axis, long growth face) is slower than the [010] (long axis, short growth face). Because the growth rate is slower, it allows the crystal to perfect itself during crystallization. The (010) sector must thicken at a slower rate. The difference between the two thicknesses is dependent on crystallization temperature as well as time. Both of these play a role in growth rate and thickening. Melting of the thinner sector first also provided evidence of the imperfection in the crystal. Interestingly, the bottom surface of the crystal is flat, meaning that the thickness difference only occurs on the free surface. This also indicates that the crystal grows on the substrate, unlike the other single crystals discussed earlier.

Fig. 5. TEM image of s-PP where the B sector is thicker than the A sector. *Source*: Bu (1996)

Single crystals are an important morphology to study to gain better knowledge of the crystallization of polymers. Using small supercoolings, the growth rate of the crystal is better controlled. The shape of the single crystal is the first indication of the underlying unit cell. It can also show the effective growth rates of different crystallographic planes. The unique structures such as helices, striations, and differences in sectors are a rich area of research in order to better generalize polymer crystallization. We have covered the polycrystalline and single crystalline forms of crystallizable thin films. The final crystal type comes from a specific category of thin films. Those are monolayer films.

4. Monolayer Films

Monolayer films are films that are made of a single layer of polymer molecules. These are created by an attraction between the end groups of the chain and the substrate. Adsorption of the molecules is critical in creating a monolayer film. These films exhibit similar morphologies in dendrites and single crystals; however, the growth of these films becomes more interesting since the molecular motion of the chain is limited. These films are important in studying the fundamentals of polymer crystallization.

In all of the studies of polymer monolayers, low molecular weight PEO was used.[65,71] Films of PEO were cast onto the oxide surface of silicon wafers or freshly cleaved mica to promote adsorption. This low MW PEO will crystallize into integer fold crystals with the number dependent on the T_c. In most instances, when the sample is annealed at higher temperatures, there is some pseudo-dewetting that occurs. Pseudo-dewetting is when some excess chains form small droplets on top of the monolayer. This occurs due to the self-repulsion of the chains. The monolayer is still adsorbed; therefore, it is not a true dewetting process. This is important because nucleation cannot occur in films thinner than 10 - 15 nm.[1,72] These dewetted droplets crystallize first and act as the nuclei for growth in the monolayer.

The morphology of these monolayer crystals ranges from dendrites to single crystals. In most cases, the diffusion-limited aggregates (DLA) are the dominant morphology, whether dendrites or seaweed. The reason for this will be discussed in the next section. These crystals have a thickness that is much larger than the film thickness. Temperature is still the dominant driving force for thickness determination. In addition, the molecules undergo other transformations if the temperature is changed after crystallization. Much like crystals in the bulk, the film crystals also exhibit a thickening mechanism.

Zhai and coworkers[68] studied the thickening mechanism of the DLA's. They found that by increasing the temperature from 35°C to 60°C (the melting point of PEO) the thickness, as well as the morphology, changed. At 35°C, the crystals were in the compact seaweed form. As the temperature increased, the seaweed broke into several different "islands." Further increasing the temperature led to an increase in size of the "islands" until finally an irregularly shaped single crystal was formed. Throughout this process, certain areas of the crystals became thicker. The authors found that at lower temperatures, the thickening was controlled by the breakdown of the morphology. At higher temperatures, however, the crystals underwent a type of Ostwald ripening. Thinner crystals, which are less stable, lost material through melting which then transported to the thickening crystal. This was verified by ΔV (change in volume) calculations. They showed that the ΔV was equal to zero, proving that the amount of material had not changed. In the next section, we will discuss the mechanism for material transport, crystallization, and growth of DLA's.

4.1. *Mechanism of Diffusion-Limited Aggregation (DLA)*

Reiter and Sommer[65,66] were the first researchers to study crystallization in monolayer films. Using pseudo-dewetted layers of PEO, they determined the mechanism of crystal nucleation and growth in these ultrathin films. They found that the crystallization of the dewetted droplets acts to nucleate the crystallization of the monolayer. Once the crystallization is nucleated, the crystals grow through diffusion-limited aggregation. Both seaweed and dendritic patterns are obtained based on ΔT.

The mechanism of growth was determined using AFM. Reiter and Sommer[65] found that at the layer consisted of three regions. They are the crystalline region, a depletion zone located at the edge of the crystal, and the amorphous monolayer. The depletion zone is created by a change in density between the amorphous and crystalline phases. In order to crystallize, the amorphous molecules must diffuse through the depletion zone to the growth front. Once at the growth front, there is a probability of adsorption and desorption. For adsorption, the chain crystallizes and moves up the crystal face. If it desorbs, the chain returns to the amorphous region. The combination of the depletion zone, the adsoption/desorption mechanism, and molecular diffusion causes the DLA morphology and branching to appear. These factors are controlled by the crystallization temperature, hence the variability between compact and fractal seaweeds and dendrites.

In addition to experiments in this area, Reiter and Sommer[67] also modeled the crystallization using a modified lattice model. They approach the problem with several considerations. They are: 1) chains are considered as points on a lattice, 2) crystallization starts only from a defined primary nucleus, 3) in the crystalline state, the chain is perpendicular to the surface, 4) changes into a state of higher internal order are connected to the entropy, and 5) the maximum occupation number corresponds to an extended chain conformation. For the primary nucleation, a line is used to represent the edge of the droplet where crystallization occurs. Their results exhibited the same morphologies as the experiments. They also found that the two relevant length scales, distance between the branches and the width of the branches, had very different dependences on temperature, thus leading to the different morphologies.

4.2. *Nucleation-Limited versus Diffusion-Limited Growth*

The limiting factor in growth has been discussed in various sections. One is the suppression of the nucleation due to the thickness of the film. The other is the limitation of the diffusion of the individual chains to the growth surface. Using monolayer films with dewetted droplets, it is possible to determine which of these limitations affects the growth. Zhu and coworkers[71] have used PEO with different end groups to illustrate the changes in growth rate due to each of these constraints.

Studies on two PEO samples, one with two -OH end groups (HPEO) and one with an -OH and an $-OCH_3$ (MHPEO), were done using *in situ* AFM.[71] A monolayer of the polymer was deposited onto a mica surface. During heating, pseudo-dewetting occurred and resulted in several droplets on a single layer of the PEO. It is assumed that the -OH end groups interact with the hydrophilic mica surface, meaning that the HPEO forms loops and the MHPEO forms tethers. When the sample is held above the melting point, the droplets undergo an Ostwald ripening type phenomena where the large drops get larger and the small drops get smaller. The only mechanism for this to occur is the transport of the molecules through the monolayer. This also supports the idea that the molecules that move to the crystal growth front travel through the monolayer as discussed earlier.

Zhu and coworkers[71] found that the two polymers exhibited different growth rate dependencies. The MHPEO sample showed a two-stage growth. Single crystals were formed at a ΔT of 2.5°C, and the growth along the [100] direction (r) was measured to monitor the growth. Both stages were linear with time ($r \propto t$). Due to the facetted shape and linear growth rates, it can be concluded that the growth is nucleation-limited (NL). The growth rate is dependent on the attachment of the chain at the growth front (first stage) and subsequent extension along the lamella (second stage). The second stage occurs due to an increase in the nucleation barrier because of the chain extension during crystallization.

At a supercooling, ΔT, between 2°C and 6°C, the HPEO sample does not exhibit a facetted, single crystal morphology. They are round in shape with an extended chain crystal. The first stage is extremely fast and difficult to follow. To monitor the second stage growth of this sample, the authors measured the volume change as a function of time. The volume is linear with time, which means that the growth is proportional to the square root of time ($r \propto t^{0.5}$). This indicates a diffusion-limited (DL) growth mechanism. Similar to the previous results, the diffusion of the chains is limited by the depletion zone around the crystal growth front. It would be safe to assume that with higher supercoolings, the morphology would transition to that of the DLA. In addition, the MHPEO sample shows DL growth at higher supercoolings ($\Delta T = 4.5°C$). The differences in these two samples are attributed to the end-group interaction with the surface.

Monolayer films are the best systems for studying the growth mechanisms of polymer crystals. Using hot-stage AFM, crystallization growth, kinetics, reorganization, and melting are easily studied. Confining the molecules to these quasi-2D environments induces new ways to limit the diffusion of the crystallizable chains. Now that we have discussed the different morphologies that can bc found in polymer thin films, we will address the areas of growth and kinetics. The final section of this chapter will cover changes in these with temperature, thickness, and the role of substrate surface energy.

5. Crystal Growth and Kinetics

The last consideration for crystallization in polymer thin films is the kinetics. Confining the crystallizable polymer in a thin film will affect its nucleation, growth rate, and kinetics. These issues have been addressed by studies previously discussed;[33-35] however, most of the work has been done by Frank *et al.*[1,59,60,72,73] They have taken several approaches to study the crystallization kinetics. They have used different polymers, PEO[59,60] and poly(di-*n*-hexylsilane) (PDHS).[1,72,73] Experimental techniques include AFM, FT-IR spectroscopy, and fluorescence spectroscopy for the PEO samples, and UV spectroscopy for the PDHS sample. For the spectroscopic techniques, a change in the peak associated with the crystalline fraction was correlated to the amount of crystallinity. *In situ* AFM was used to monitor morphology as well as growth. In addition, they analyzed their results using the Hoffman-Lauritzen approach and the Avrami-Evans theory, which relates the volume percent crystallinity to the crystallization time and two adjustable parameters:[2]

$$v^c = 1 - \exp(-Kt^n) \tag{7}$$

where v^c is the percent crystallinity, K depends on experimental conditions, and n depends on the dimensionality of growth.

Using the PDHS sample, the percent crystallinity was monitored over time using UV spectroscopy.[1,72,73] A low-energy peak at 364 nm corresponds to the all-trans conformation of the Si backbone. The high-energy peak at 316 nm corresponds to the disordered state. By monitoring the absorbance at each peak, it is possible to determine the percent crystallinity. After quenching to the crystallization temperature, the percent crystallinity was monitored over time. In thick films, crystallization reached the maximum shortly after quenching. As the film got thinner, the crystallinity was decreased; the time to reach the maximum crystallinity also increased. It was also found that below 10 - 15 nm there was no crystallinity because homogeneous nucleation could not occur. Similar results were obtained when the crystallinity of PEO was measured using FT-IR spectroscopy. In both cases, the behavior was modeled using the Avrami equation (Eq. 7) to determine the role of thickness on the dimensionality of the crystallization.

Fitting the data with the Avrami equation provided good agreement[72] (Figure 6). PDHS exhibited a similar K-value over all thicknesses and crystallization temperatures. The meaning of this result is complicated since K is dependent on the shape of the crystallite, the growth rate, the number of potential nuclei, and the probability of nucleation. The exponent, n, was found to decrease with decreasing thickness. As the thickness decreases, n approaches 1. This means that the growth is one-dimensional and athermal; however, by increasing the ΔT, the exponent reaches a value of 3.5 again indicating that a higher supercooling, ΔT, greatly affects the mode of crystallization.

The results for the PEO sample are quite different.[59] For PEO, the K-value decreased with decreasing thickness and the exponent stayed the same, around 4.0. This indicates that the previously mentioned parameters are quite different for PEO as compared to PDHS. Also, the crystallization of PEO is between 2D ($n = 3$) and 3D ($n = 4$). This difference may be due to the fact that PDHS crystallizes parallel to the substrate surface; while, PEO crystallizes with the chain direction perpendicular to the surface.

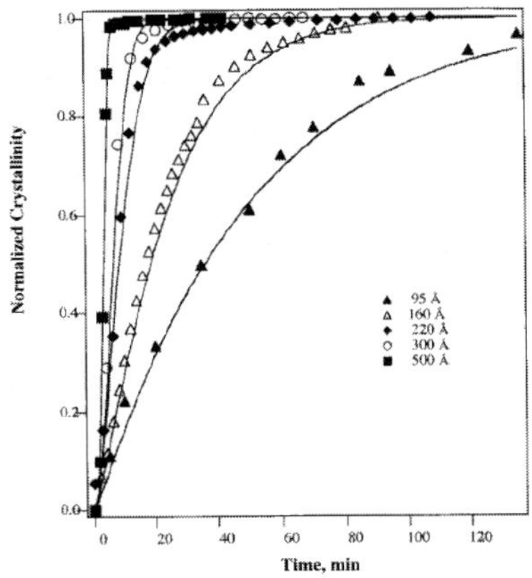

Fig. 6. Avrami fit to the crystallinity vs. time plot with varying film thickness. *Source*: Despotopoulou (1995)

5.1. *Growth Rate as a Function of Thickness*

Equally important to the amount of growth (percent crystallinity) in a polymer thin film is the growth rate of the crystallization. Studies have shown that the growth rate is highly dependent on thickness as well as temperature.[33-35,60,74] In all cases, the growth rate was monitored by optical microscopy or AFM. Typically, the radial growth of a spherulite or dendrite was measured as a function of time to determine the rate. Once the rate was found, it was possible to find the relationships with thickness and temperature.

In the work done by Sawamura *et al.*,[74] they found that the growth rate of i-PS could be modeled with the following equation:

$$G(D) = G(\propto)(1 - d/D) \tag{8}$$

where $G(D)$ is the growth rate in a film of thickness D, $G(\infty)$ is the bulk growth rate, and d is a constant. This relationship held regardless of substrate (glass or carbon-coated glass) or molecular weight. In all cases, the value of d was found to be 6 nm. Taguchi *et al.*[35] found similar results; however, their d-value was 7.2 nm, and there was deviation from the fit below a thickness of 8 nm. Sawamura and coworkers[74] hypothesized that the change in growth rates was due to the reptation of the polymer chain as it moves towards the growth face.

In 2003, Massa and coworkers[34] provided an entire growth landscape as a function of thickness for PEO thin films. Figure 5 shows the growth rate as a function of thickness from the bulk value at thicknesses near 1 μm to the growth rate at a thickness of 14 nm. Their results differ from those of Sawamura[74] and Taguchi.[35] They found that there is a local minimum at D ~ 150 nm, a peak at D ~ 100 nm, and a "kink" at D ~ 45 nm. Each of these is explained in the following way. The minimum corresponds to a change from "edge-on" crystals to "flat-on" diffusion-controlled growth. The peak at 100 nm is related to a change in the density of the DLA backbones. Finally, the "kink" is caused by a change to the CS morphology where the backbone and branch widths are comparable and tip-splitting occurs. The authors state that these changes in morphology with decreasing thickness may result in the deviations seen in the previous studies.

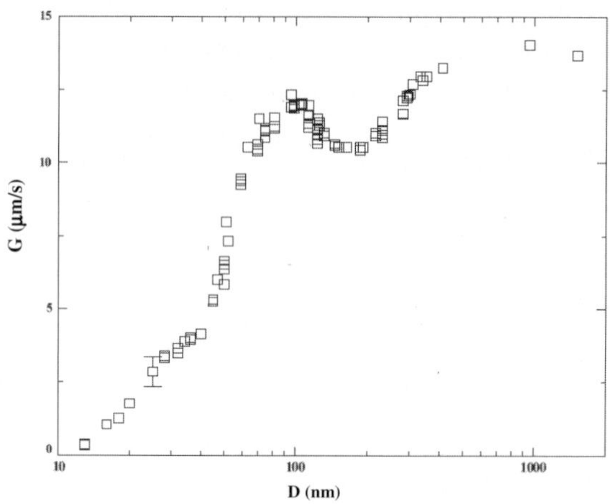

Fig. 7. Growth rate (G) vs. film thickness (D) for PEO thin films. *Source*: Massa (2003)

5.2. *Growth Rate as a Function of Temperature*

As was alluded to in the previous sections, the growth rate is also affected by the crystallization temperature. When the growth rate is fast, the spherulite morphology is apparent. The DLA morphology and the single crystal morphology appear as the growth rate decreases. This has been addressed for bulk samples in the theory of Hoffman and Lauritzen.[16] They found that the growth rate, G, is exponentially proportional to the inverse of the supercooling, $\exp\{-1/\Delta T\}$, or $\exp\{-1/(T_m^o - T_x)\}$. Schönherr and Frank[60] and Dalnoki-Varess *et al.*[33] found that the same applies for thin films.

Both groups used PEO to study the affects of T_c on the growth rate. Their results are quite similar. The data follows a $G \sim \exp[1/\Delta T]$ relationship with slight variation with changing film thickness. Schönherr and Frank[60] compared their results directly to the Hoffman-Lauritzen prediction. They found that the confinement effect creates a constant offset of the linearized plots; however, by using appropriate adjustable parameters, the data falls onto a single curve. They hypothesize that the transport term of the theory does not correct for either the activation energy for reptation across the melt-crystal interface or the temperature at which molecular motion ceases. Dalnoki-Varess and coworkers[33] found that their data follows the equation:

$$G = G' \exp[-b / T_c (T_m^o - T_c)] \quad (9)$$

where G' is independent of temperature but dependent on thickness. In contrast, b has a single value for all thicknesses. The fit of a single b-value indicates that the energy barrier for the addition of another polymer segment is unaffected by the confinement; whereas, the G' dependence is indicative of a transport difference in the films.

6. Summary

In this chapter, we have covered several areas of polymer crystallization in films with a thickness of less than 1 μm. The morphology, growth, and kinetics are all affected by confining the polymer to quasi-2-dimensional space. At first approximation, one might think that these affects may only occur once the film thickness is on the length-scale of the polymer chain (R_g); however, we find that in films that are orders of magnitude larger, there is still a change from the bulk behavior.

The morphology of the polymer crystals undergoes several transitions. Spherulites, diffusion-limited aggregates, which include dendrites and seaweed patterns, and single crystals have all been observed in thin films. The transitions from one morphology to the next are due to the degree of undercooling, or crystallization temperature, or the thickness. The changes with respect to temperature are similar to those in the bulk. By confining the polymer in thinner films, the diffusion is decreased. This change in diffusion with respect to the growth rate leads to the different morphologies. As the diffusion constant approaches the growth rate, DLA's become the dominant morphology. If the growth rate is much larger than the diffusion, then 2D spherulites can be observed.

In addition to the overall morphology, the crystal orientation can vary under the given conditions. Crystals can assume an "edge-on" orientation or a "flat-on" orientation. This orientation is defined by the chain axis of the crystallized material. If the chain axis is parallel to the surface, they are "edge-on." If it is normal to the surface, then they are "flat-on." In nearly all cases, spherulites grow with an "edge-on" orientation; while, DLA's and single crystals grow with a "flat-on" orientation. The orientation is determined in the nucleation step. The surface energy of the substrate and the availability of the material play an important role in the nucleation.

The growth rate and kinetics are also determined by the temperature and film thickness. Some films have shown that there is a limiting thickness for crystal nucleation. In general, the thickness also affects the final crystallinity of the film. The growth rate decreases with decreasing thickness as well. The effect of temperature, however, follows a similar trend as in the bulk. The differences are again attributed to the limitations on diffusion and available material.

The field of polymer crystallization has gained important observations from films a monolayer thick to several hundred nanometers. As technology moves toward smaller scales, the properties of polymeric materials must be studied to keep at the forefront of research. In addition, our knowledge of polymer crystallization is ever changing with better experimental techniques and polymer samples. These thin films provide an additional method to understand crystallization in the bulk.

Crystallization in polymer thin films will prove to be an important field for years to come.

Acknowledgment

Writing this review was supported by the National Science Foundation (DMR-0516602).

References

1. C. W. Frank, V. Rao, M. M. Despotopoulou, R. F. W. Pease, W. D. Hinsberg, R. D. Miller and J. F. Rabolt, *Science*, 912 (1996).
2. B. Wunderlich, *Macromolecular Physics* (Academic Press, New York, 1973).
3. G. R. Strobl, *The Physics of Polymers* (Springer-Verlag: Berlin, 1996).
4. P. H. Geil and D. H. Reneker, *J. Polym. Sci.*, 569 (1961).
5. E. W. Fischer, *Kolloid Z. u. Z. Polym.*, 458 (1968).
6. A. Keller, *Rep. Prog. Phys.*, 623 (1968).
7. A. Kovacs, A. Gonthier and C. Straupe, *J. Polym. Sci.: Polym. Symp.*, 283 (1975).
8. A. Kovacs, C. Straupe and A. Gonthier, *J. Polym. Sci.: Polym. Symp.*, 31 (1977).
9. A. Kovacs and C. Straupe, *J. Cryst. Growth*, 210 (1980).
10. A. Keller and S. Z. D. Cheng, *Polymer*, 4461 (1998).
11. K. Armistead and G. Goldbeck-Wood, *Adv. Polym. Sci.*, 219 (1992).
12. D. M. Sadler and G. H. Gilmer, *Polymer*, 1446 (1984).
13. H. D. Keith, F. J. Padden, B. Lotz and J. C. Wittman, *Macromolecules*, 2230 (1989).
14. A. Toda and A. Keller, *Colloid Polym. Sci.*, 328 (1993).
15. S. Z. D. Cheng and B. Lotz, *Polymer*, 8662 (2005).
16. G. R. Strobl, *Eur. Phys. J. E*, 295 (2005).
17. J. I. Lauritzen, Jr. and J. D. Hoffman, *J. Res. Natl. Bur. Stand.*, 73 (1960).
18. J. L. Keddie, R. A. L. Jones and R. A. Cory, *Europhys. Lett.*, 59 (1994).
19. J. L. Keddie, R. A. L. Jones and R. A. Cory, *Faraday Discuss.*, 219 (1994).
20. J. A. Forrest and J. Mattson, *Phys. Rev. E*, R53 (2000).
21. O. Prucker, S. Christian, H. Bock, J. Rühe, W. Frank and W. Knoll, *Macromol. Chem. Phys.*, 1435 (1998).
22. R. A. L. Jones, *Curr. Opin. Colloid Interface Sci.*, 153 (1999).
23. J. H. Kim, J. Jang and W. C. Zin, *Langmuir*, 2703 (2001).
24. J. H. van Zanten, W. E. Wallace and W. L. Wu, *Phys. Rev. E*, R2053 (1996).
25. R. Pearce and J. Vansco, *Polymer*, 1237 (1998).
26. V. H. Mareau and R. E. Prud'homme, *Macromolecules*, 398 (2005).
27. A. J. Lovinger and R. E. Cais, *Macromolecules*, 1939 (1984).
28. E. Brener, H. Müller-Krumbhaar and D. Temkin, *Phys. Rev. E*, 2714 (1996).

29. L. Li, C. M. Chan, K. L. Yeung, J. X. Li, K. M. Ng and Y. Lei, *Macromolecules*, 316 (2001).
30. Y. Sakai, M. Imai, K. Kaji and M. Tsuji, *Macromolecules*, 8830 (1996).
31. Y. Sakai, M. Imai, K. Kaji and M. Tsuji, *J. Cryst. Growth*, 244 (1999).
32. Y. K. Godovsky and S. N. Magonov, *Langmuir*, 3549 (2000).
33. K. Dalnoki-Varess, J. A. Forrest, M. V. Massa, A. Pratt and A. Williams, *J. Polym. Sci. Phys. Ed.*, 2615 (2001).
34. M. V. Massa, K. D. Dalnoki-Varess and J. A. Forrest, *Eur. Phys. J. E*, 191 (2003).
35. K. Taguchi, H. Miyaji, K. Izumi, A. Hoshino, Y. Miyamoto and R. Kokawa, *J. Macromol. Sci. Part B Physics*, 1033 (2002).
36. Y. Wang, S. Ge, M. Rafailovich, J. Sokolov, Y. Zou, H. Ade, J. Lüning, A. Lustiger and G. Maron, *Macromolecules*, 3319 (2004).
37. D. C. Bassett and A. S. Vaughan, *Polymer*, 717 (1985).
38. Y. Kikkawa, H. Abe, T. Iwata, Y. Inoue and Y. Doi, *Biomacromolecules*, 940 (2001).
39. F. Zhang, J. Liu, H. Huang, B. Du and T. He, *Eur. Phys. J. E*, 289 (2002).
40. K. Cho, D. Kim and S. Yoon, *Macromolecules*, 7652 (2003).
41. B. J. Factor, T. P. Russell and M. F. Toney, *Phys. Rev. Lett.*, 1181 (1991).
42. B. J. Factor, T. P. Russell and M. F. Toney, *Macromolecules*, 2847 (1993).
43. S. Kawana, M. Durell, J. Lu, J. E. Macdonald, M. Grell, D. D. C. Bradley, P. C. Jukes, R. A. L. Jones and S. L. Bennett, *Polymer*, 1907 (2002).
44. M. Durrell, J. E. Macdonald, D. Trolley, A. Wehrum, P. C. Jukes, R. A. L. Jones, C. J. Walker and S. Brown, *Eurphys. Lett.*, 844 (2002).
45. P .C. Jukes, A. Das, M. Durell, D. Trolley, A. M. Higgins, M. Geoghegan, J. E. Macdonald, R. A. L. Jones, S. Brown and P. Thompson, *Macromolecules*, 2315 (2005).
46. Y. Zhang, S. Mukoyama, Y. Hu, Y. Ozaki and I. Takahashi, *Macromolecules*, 4009 (2007).
47. Y. Duan, Y. Jiang, S. Jiang, L. Li, S. Yan and J. M. Shultz, *Macromolecules*, 9283 (2004).
48. Y. Wang, C. M. Chan, L. Li and K. M. Ng, *Langmuir*, 7384 (2006).
49. G. Schuur, *J. Polym. Sci.*, 385 (1953).
50. A. Shram, *Kolloid-Z.*, 18 (1957).
51. H. D. Keith and F. J. Padden, *J. Polym. Sci.*, 415 (1958).
52. H. D. Keith and F .J. Padden, *J. Polym. Sci.*, 123 (1959).
53. Z. Bu, Y. Yoon, R. M. Ho, W. Zhou, I. Jangchud, R. K. Eby, S. Z. D. Cheng, E. T. Hsieh, T. W. Johnson, R. G. Geerts, S. J. Palackal, G. R. Hawley and M. B. Welch, *Macromolecules*, 6575 (1996).
54. W. Zhou, S. Z. D. Cheng, S. Putthanarat, R. K. Eby, D. H. Reneker, B. Lotz, S. Magonov, E. T. Hsieh, R. G. Geerts, S. J. Palackal, G. R. Hawley and M. B. Welch, *Macromolecules*, 6861 (2000).

55. S. J. Sutton, K. Izumi, H. Miyaji, Y. Miyamoto and S. Miyashita, *J. Mater. Sci.*, 5621 (1997).
56. K. Taguchi, H. Miyaji, K. Izumi, A. Hoshino, Y. Miyamoto and R. Kokawa, *Polymer*, 7443 (2001).
57. K. Taguchi, Y. Miyamoto, H. Miyaji and K. Izumi, *Macromolecules*, 5208 (2003).
58. Y. Kikkawa, H. Abe, T. Iwata, Y. Inoue and Y. Doi, *Biomacromolecules*, 350 (2002).
59. H. Schönherr and C. W. Frank, *Macromolecules*, 1188 (2003).
60. H. Schönherr and C. W. Frank, *Macromolecules*, 1199 (2003).
61. E. Passaglia and F. Khoury, *Polymer*, 631 (1984).
62. C. Y. Li, D. Yan, S. Z. D. Cheng, F. Bai, T. He, L. C. Chien, F. W. Harris and B. Lotz, *Macromolecules*, 524 (1999).
63. C. Y. Li, S. Z. D Cheng, J. J. Ge, F. Bai, J. Z. Zhang, I. K. Mann, L. C. Chien, F. W. Harris and B. Lotz, *J. Am. Chem. Soc.*, 72 (2000).
64. C. Y. Li, J. J. Ge, F. Bai, B. H. Calhoun, F. W. Harris, S. Z. D. Cheng, L. C. Chien, B. Lotz and H. D. Keith, *Macromolecules*, 3634 (2001).
65. G. Reiter and J. U. Sommer, *Phys. Rev. Lett.*, 3771 (1998).
66. G. Reiter and J. U. Sommer, *J. Chem. Phys.*, 4376 (2000).
67. J. U. Sommer and G. Reiter, *J. Chem. Phys.*, 4384 (2000).
68. X. M. Zhai, W. Wang, Z. P. Ma, X. J. Wen, F. Yuan, X. F. Tang and B. L. He, *Macromolecules*, 1717 (2005).
69. X. M. Zhai, W. Wang, G. Zhang and B. L. He, *Macromolecules*, 324 (2006).
70. D. S. Zhu, Y. X. Liu, A. C. Shi and E. Q. Chen, *Polymer*, 5239 (2006).
71. D. S. Zhu, Y. X. Liu, E. Q Chen, M. Li, C. Chen, Y. H. Sun, A. C. Shi, R. M. Van Horn and S. Z. D. Cheng, *Macromolecules*, 1570 (2007).
72. M. M. Despotopoulou, C. W. Frank, R. D. Miller and J. F. Rabolt, *Macromolecules*, 6687 (1995).
73. M. M. Despotopoulou, C. W. Frank, R. D. Miller and J. F. Rabolt, *Macromolecules*, 5797 (1996).
74. S. Sawamura, H. Myaji, K. Izumi, S. J. Sutton and Y. Miyamoto, *J. Phys. Jpn.*, 3338 (1998).

CHAPTER 8

FRICTION AT SOFT POLYMER SURFACE

Manoj K. Chaudhury[1], Katherine Vorvolakos[2] and David Malotky[3]

[1]*Department of Chemical Engineering*
Lehigh University
Bethlehem, PA 18015

[2]*FDA Center for Devices and Radiological Health*
Office of Science and Engineering Laboratories
Division of Chemistry and Material Science
Silver Spring, MD 20993

[3]*The Dow Chemical Co.*
Midland, MI 48674

The modes of attachments, detachments and relaxations of molecules of rubbers and gels on solid surfaces are keys to understanding their frictional properties. An early stochastic model of polymer relxations on surfaces was given by Schallamach, which has now evolved in various ways. A review of these developments is presented along with the experimental data that elucidate the kinetic friction of smooth rubber against smooth surfaces. These soft rubbers exhibit various types of instabilities while sliding on surfaces. A few examples of these instabilities are provided.

Introduction

Friction, in both microscopic and macroscopic systems, is dominated by the processes occurring at or near surfaces, i.e. in a thin film region. The objective of this chapter is to try to understand some of the molecular level processes underlying friction by examining the progress that has taken place over many years with studies involving low modulus polymers. Aside from its tremendous technological importance, crosslinked polymer or rubber is a nice model system for tribological investigations, as its surface and mechanical properties can be easily controlled by the appropriate choice of the molecular weight, crosslinking density, and chemical composition. Another advantage of rubber comes from its low modulus, which allows it to conform well to moderately rough surfaces. A large body of early scientific research in friction was carried out with rubber.[1-9] The questions that intrigued rubber tribologists are: how much of rubber friction is controlled by interfacial and bulk deformation processes, how the dynamics of polymer chains are affected by the interface, what is the mode of relaxation of molecules at the interfaces, to name a few. The pioneering systematic studies of Grosch[10] attempted to answer some of these questions long ago, by making careful measurements of sliding friction of various rubbers against varieties of rough and optically smooth surfaces. His main observation (Figure 1) was that the friction coefficient of rubber first increases with velocity, reaches a maximum and then it decreases. In these studies, Grosch[10] discovered a remarkable correlation between friction coefficient and the rheological properties of the rubber. When the velocity corresponding to maximum friction is divided by the frequency corresponding to maximum viscoelastic dissipation, a length scale is found that correlates with the average distance between surface asperities. Based on this observation, Grosch[10] suggested that friction on rough surfaces is dominated by the energy dissipation within a zone in the rubber that is deformed by the surface asperities. As the rough surface slides over a rubber, these zones undergo cyclic compression/deformation cycles, thus dissipating energy proportional to the loss modulus of the rubber. More recently, further extending Grosch's ideas, Persson[23] developed a rather comprehensive theory of rubber friction on rough surfaces. After Grosch,[10] Ludema and Tabor[11] carried out careful experiments of rubber friction in both rolling and sliding modes. These authors came to the conclusion that while rolling friction is controlled by the viscoelastic properties of rubber, which gained support from several later studies,[12-16] the sliding friction on smooth surfaces does not necessarily correlate well with the viscoelatic properties. They suggested that the friction and slip of polymer chain segments against each other is the main mechanism contributing to sliding friction. On the other hand, Schallamach[17] proposed an adhesion-based theory of rubber friction suggesting that the cyclic processes of extension, detachment and re-attachment of the rubber molecules at the interface are the fundamental mechanisms of friction of rubber on surfaces. Below, we examine these various ideas of rubber friction in some detail.

Schallamach's Theory of Rubber Friction

Schallamach's idea[10] was that the interfacial bonds responsible for the adsorption of the polymer chains of rubber on a solid surface can always break due to thermal fluctuation. However, when a force is applied, the bond breaking probability increases exponentially with force. By solving a kinetic equation of bond rupture, Schallamach argued that the average force at which a bond breaks increases logarithmically with the rate at which a load is applied to the polymer chain. Interestingly, these basic ideas of Schallamach have also been independently discovered in the biophysics community, where they are frequently used to explain the rate dependent bond rupture forces in various protein-ligand systems.[18-20]

Schallamach's[17] theory has two main aspects. According to one aspect of the theory, the force needed to detach a polymer chain from a surface increases with velocity. However, the number of active load-bearing chains decreases with the sliding velocity as well. The force to detach the chain, according to the elementary theory of elasticity, is proportional to the extension of the polymer chain $V\bar{t}$, $\bar{t}$ being the average time a chain remains bonded to a surface before it is detached. This average time is estimated from the following equation:

$$\bar{t} = \int_0^{\infty} (\Sigma / \Sigma_o) dt \tag{1}$$

where, Σ_0 is the total number of polymer chains per unit area. Σ is the areal density of working chain at any given time, which is obtained from the following rate equation:

$$\frac{d\Sigma}{dt} = -\frac{\Sigma}{\tau} e^{-U/kT} e^{k_s Vt\lambda/kT} \tag{2}$$

where U is the depth of the adsorption potential well, k_s is the spring constant of the rubber molecule, V is the sliding velocity and λ is the width of the potential well. The last term in equation 2 suggests that the rate of dissociation of polymer chain from a surface is enhanced due to the reduction of activation energy by the applied force. Integration of equation 2 yields the areal chain density, which can be substituted in equation 1 to obtain $\bar{t}$. The average density of the active load bearing chain is obtained as:

$$\bar{\Sigma} = \Sigma_o(\bar{t}/(\bar{t}+\tau)) \tag{3}$$

where τ is the amount of time a chain spends in a detached state before re-attaching to the surface. Combining equations 1, 2 and 3, the frictional force per unit area can be obtained from the sum of all the spring forces: $\bar{\Sigma} k_s V\bar{t}$. The theory developed by Schallamach is rather successful in providing a qualitative picture of adhesion that captures many important elements of Grosch's observations. Chernyak and Leonov[21] later provided certain improvements to Schallamach's model and proposed the following equation for shear stress:

$$\sigma \approx \Sigma_o \frac{\int_0^\infty \varphi\left(\frac{r(t)}{\delta}\right) p(V,t)\,dt}{V(\bar{t}+\tau)} \tag{4}$$

where $\varphi(r(t)/\delta)$ is the elastic energy stored in the polymer chain, $p(V,t)$ is the transition probability of the polymer chain in going from the bonded to the relaxed state. The numerator of equation 4 is the work done in stretching the polymer chain to the breaking point, while its denominator represents the mean distance traveled by the chain. Using the steady state stochastic model of bonding-debonding processes, Chernyak and Leonov[21] determined the expressions for the transition probability and the time spent in the bonded state in terms of the sliding velocity. Their final expression for the shear stress is:

$$\sigma = \sigma_a (m+1) \frac{u(1+s)\left(1 - \exp\left(\frac{-1}{u}\right) - \exp\left(\frac{-1}{u}\right)\right)}{m+1-\exp\left(\frac{-1}{u}\right)} \tag{5}$$

where m is the ratio of the lifetimes of the polymer chain in the free and bound states at zero sliding velocity, s is the ratio of the viscous retardation time over the lifetime at rest, and u is the dimensionless velocity of sliding defined by $u = \frac{V\tau_\circ}{\delta a}$, where τ_o is the lifetime of the polymer chain in the bound state at rest, δ is the average distance between the polymer body and the wall and a corresponds to the cotangent of the maximum angle the polymer chain subtends on the wall at forced breakup. $\sigma_a = \frac{kT\Sigma_\circ \delta}{(1+m)R_F^2}$, R_F being the Flory radius of the polymer chain. The resultant shear stress exhibits a bell-shaped behavior with respect to the sliding velocity (Figure 1). The peak stress decreases with the molecular weight, and it exhibits a broad maximum for high molecular weight polymers.

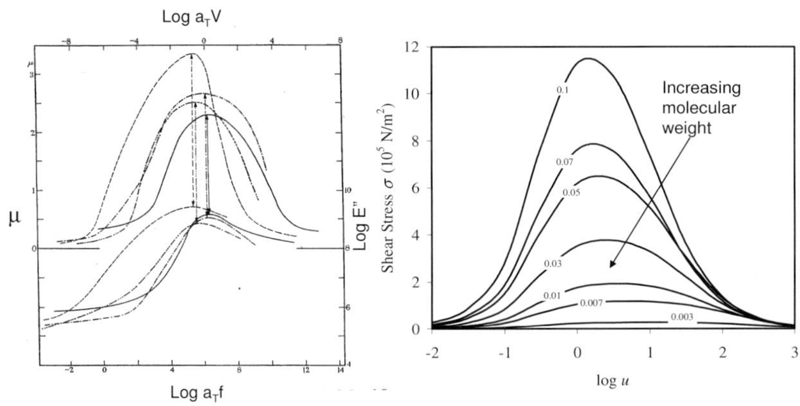

Fig. 1. (left) Friction coefficients of various rubbers against solid surfaces, as obtained by Grosch,[10] are plotted against the logarithm of speed shifted by the WLF shift factor. On the same graph, the viscoelastic loss coefficient is plotted against the logarithm of the frequency, also shifted by the WLF factor. (right) Schematic of the shear stress as function of a dimensionless velocity for rubber sliding against a solid surface as predicted by the general theories of Schallamach[17] and Chernyak and Leonov.[21]

Experiments of Rubber Friction with Molecularly Smooth Surfaces

Recently, Vorvolakos and Chaudhury[22,24] studied the friction of a PDMS (polydimethylsiloxane) rubber on a smooth solid surface in the light of the earlier studies by Grosch. The smoothest surface used by Grosch was optical quality glass, which does not guarantee molecular level smoothness. Furthermore, the solid surfaces were rather ill-defined from today's surface science standard; therefore, the effects of surface properties could not be discerned from the older studies of friction. In the new studies, Vorvolakos and Chaudhury[22] used molecularly smooth PDMS (polydimethyl siloxane) elastomers sliding against a molecularly smooth self-assembled monolayer of alkyltricholosilanes or a molecularly smooth polystyrene as a function of the crosslinking density of the elastomer, sliding speed, and temperature.

In these cases, where the interfacial interaction is dominated by van der Waals forces, the frictional force increased with sliding velocity sub-linearly up to a critical velocity, beyond which the rubber exhibited a stick slip behavior (Figure 2). It was also observed that the frictional force decreases with the molecular weight of the rubber, and with temperature, which are qualitatively consistent with the theories of Schallamach, as well as of Chernyak and Leonov.[21] There are certain differences observed with the SAM-coated silicon wafer and polystyrene, which we will discuss later in the chapter.

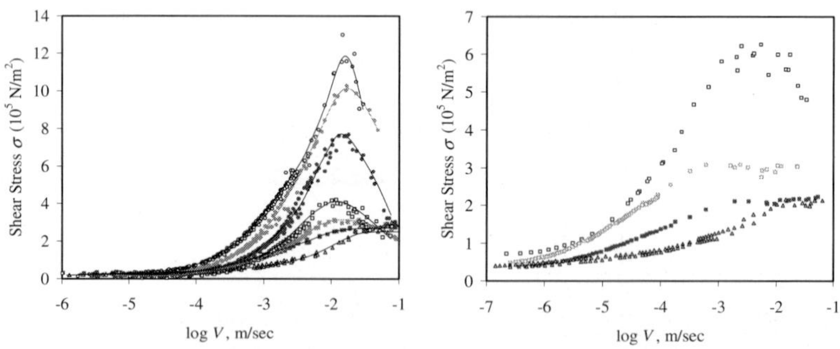

Fig. 2. (Left) Shear stress of PDMS on SAM-covered Si wafer. The molecular weights of the polymers of decreasing friction correspond to 1.3, 1.8, 2.7, 4.4, 8.9, 18.7, and 52.1 kg/mol, respectively. (Right) Shear stress of PDMS on Polystyrene: PDMS of decreasing friction corresponds to molecular weight 4.4, 8.8, 18.7, and 52.1 kg/mol, respectively.

Persson and Volikitin's Theory of Rubber Friction on Smooth Surfaces

Recently, Persson and Volikitin[23] questioned whether polymer chains can undergo the cyclic processes of attachment and detachment in a confined geometry as was suggested by Schallamach. They also questioned whether Schallamach's model is suitable for a weakly corrugated potential, where it may be much easier for the chains to slide than to break adhesively. Persson and Volikitin[23] envisioned that the sliding of rubber against a smooth surface is due to the correlated stick-slip motion of the patches of rubber elements which they call stress segments (Figure 3). Applying the method of Langevin dynamics to viscoelastic rubber, Persson and Volikitin[23] argued that the fluctuation stress in rubber segments can be as high as mega-Pascal, which is enough to depin the rubber segment from the surface. When the rubber slides, the stress at the stress segments reach a critical value when a localized slip occurs accompanied with the dissipation of the stored elastic energy in the rubber.

In their analysis, rubber friction turns out to be closely related to the frequency-dependent elastic modulus of the rubber. They recovered (Figure 3) the bell-shaped curve of friction versus sliding velocity as observed in the experiments of smooth surfaces of Grosch[17] as well as of Vorvolakos and Chaudhury.[22]

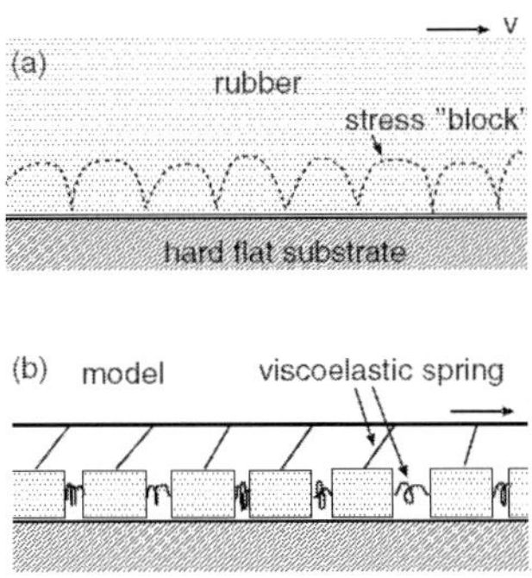

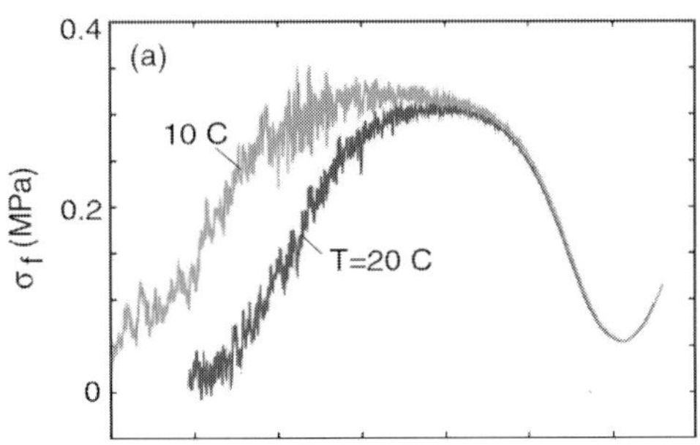

Fig. 3 Fluctuation model[23] of stress segments as proposed by Persson and Volotikin (left). Computed shear stress[23] as a function of sliding velocity is shown on the right.

Friction Resulting from Stress-Biased Diffusion

Here we describe a model that was used in reference 24, where friction of rubber was thought to be related to the biased diffusion of the polymer molecules in contact with the surface. This model is different from that of Schallamach in the sense that the chains undergo random Brownian motion in contact with the surface, which is biased when a force is applied. For small force, the corresponding Fokker-Planck equation for the dynamics of rubber molecules in contact with the surface is:

$$\frac{Dp}{Dt} = -\frac{D_o}{kT}\frac{\partial(fp)}{\partial x} + D_o\frac{\partial^2 p}{\partial x^2} \tag{6}$$

Here, p is the probability density, f is the force the chain experiences that depends on the chain extension, D_o is the diffusion coefficient. The operator D/Dt indicates a material derivative. At steady state, equation 6 becomes:

$$V\frac{dp}{dx} = -\frac{D_o}{kT}\frac{d(fp)}{dx} + D_o\frac{d^2 p}{dx^2} \tag{7}$$

Solution of equation 7 for a linear elastic polymer chain of spring force $k_s x$ is:

$$p = p_o \, exp\left(\frac{Vx}{D_o} - \frac{k_s x^2}{2kT}\right) \tag{8}$$

The physical significance of equation 8 is that the probability distribution is Gaussian in the absence of any relative sliding. However, when the sliding occurs, the distribution is shifted. The average stress at the interface is:

$$\sigma = \Sigma\left(\int_{-L}^{L} fpdx\right) / \left(\int_{-L}^{+L} pdx\right) \tag{9}$$

where L is the maximum extension of the polymer chain. In equation 9, the areal density of load bearing chains $\sum$ depends on the statistical segment length a and the number (N) of statistical segment per chain as: $\Sigma = 1/\sqrt{N}a^2$. Equation 9 in conjunction with equation 8 reduce to

$$\sigma = \sum \zeta_o V \tag{10}$$

where, $\zeta_o = kT / D_o$ is the friction coefficient. This is the low velocity limit of friction, where the frictional stress is areal density of the load bearing chains times the average frictional force a chain experiences due to its diffusive motion. The friction coefficient ζ, in general, has two contributions: one resulting from the surface interactions and the other resulting from viscous friction of the polymer chains with the surrounding chains near the surface. At the simplest level of approximation, the bulk contribution to ζ is proportional to N, while the surface contribution if proportional to $N^{1/2}$, so the net friction coefficient is:

$$\zeta \sim \zeta_o\left(\sqrt{N}e^{U/kT} + N\right) \tag{11}$$

where ζ_o is the segmental level friction coefficient in the bulk. If the first term of equation 11 is stronger than the second term, the interfacial shear stress would be $\sigma \approx \frac{\zeta_o V}{a^2} e^{U/kT}$, which is independent of the molecular weight (or modulus) of the polymer. At a simpler, but somewhat more general level, the kinetic friction force can be described by an equation[25,26] of the type:

$$f = f_0 \sinh^{-1}(V / V_0) \tag{12}$$

where, V_0 is a molecular velocity scale. When $f >> f_o$, both the sinh and cosh functions become exponentials and the force varies logarithmically with velocity, i.e.

$$f \sim f_0 \ln(V/V_0), \text{ and } \sigma \sim \Sigma f_0 \ln(V/V_0) \tag{13}$$

where, V_0 is a molecular velocity scale. Thus, while at a low sliding velocity, the shear stress increases linearly with sliding velocity, it varies logarithmically at high sliding velocity. These results are similar to Schallamach's predictions.

Friction, however, does not keep on increasing indefinitely with velocity. It should saturate when the force reaches the zero temperature limit of the force: $k_s \bar{x} \sim U/\lambda$. However, at the high velocity range, there is a subtle interplay of elastic and frictional forces experienced by the chain. This region may be analyzed using a method proposed by Persson[27] to study the behavior of adsorbed molecules in the high velocity limit using Langevin dynamics. We present here the treatment of Persson[27] that starts with the following Langevin equation:

$$m\frac{d^2x}{dt^2}+\zeta\frac{dx}{dt}=f-\frac{\partial U(x)}{\partial x}+F \tag{14}$$

Here, $U(x)$ is the depth of the potential well, f is the delta correlated stochastic force and F is the external force. By considering the coordinate of the end group of the chain as $x = Vt - x_a$, where x_a is the fluctuation of the end group position and the shape of the potential well as $U(x) = U(1 - \cos kx)$, equation 14 can be re-written as in equation 15 after expanding the potential with respect to x_a.

$$m\frac{d^2x_a}{dt^2}+\zeta\frac{dx_a}{dt}=f-kU\sin(kVt)-k^2Uu\cos(kVt)+F-\zeta V \tag{15}$$

Persson[27] presented a solution of equation 15 in the high velocity limit as follows:

$$F \approx \zeta V\left(1+\frac{U^2}{2m^2V^4}\right) \tag{16}$$

Thus, at the very high velocity limit, the frictional stress is given by $\sigma = \sum \zeta V$ with $\sum \sim 1/Na^2$ and $\zeta \sim N$; thus, the friction force increases with the molecular weight of the polymer.

Thus, frictional force has three main behaviors. In the low velocity limit, it is dominated by the surface friction coefficient and increases linearly with velocity, but being independent of the molecular weight as a first approximation (the situation may be different if specific interactions, such as H-bonding interactions act at the interface). As the velocity increases, the friction force enters a non-linear domain, where the non-linearity results from the fact that the transition probability from one potential well to the next increases exponentially with force. In this range, the frictional shear stress decreases with the molecular weight of the elastomer. Finally, when the force is much larger than the force of the potential well, friction increases linearly with velocity and molecular weight.

It should, however, be pointed out that in a real rubber, some coupling of interfacial and bulk dynamics may occur. Furthermore, at high sliding velocities, some of the polymer chains may lose contact with the counter surface, or the polymer surface may undergo some roughening, amounting to a decrease of the load-bearing chains with velocity as was envisaged by Schallamach. When this possibility is taken into account, one expects the frictional stress to exhibit an S-shaped curve, where the friction at first increases, then it decreases and finally increases again with sliding velocity. This is also the form predicted by Persson and Volotikin (Figure 3) using their fluctuation model of stress domains and by Leonov[28] after correcting the earlier theory of Chernyak and Leonov by introducing an additional time scale due to Brownian fluctuation:

$$\sigma \cong \sigma_a u \frac{1-(1-m-1/u)\exp(-m-1/u)}{1+\gamma m-\exp(-m-1/u)} \tag{17}$$

where the additional parameter γ is the ratio of the free life time of the polymer chain to its characteristic time of Brownian motion on the surface.

Examination of Experimental Data

The simple analysis presented above suggests that the shear stress is independent of the molecular weight at low sliding velocity. However, with increasing velocity the shear stress should vary as the square root of the elastic modulus. Although there were some hints that friction decreases with the increase of molecular weight in the studies of Newby and Chaudhury,[31] the effect was more clearly demonstrated in the studies reported in references 24 and 22. Figure 2 summarizes some of the important results of reference 22. When the sliding speed is in the range of 10^{-7} m/s to 10^{-4} m/s, shear stress varies very little with velocity. However, when the sliding velocity is in the range of 10^{-4} m/s to 10^{-2} m/s, shear stress increases logarithmically with the sliding velocity. Then, above V~10^{-2} m/s, sliding becomes unstable followed by stick-slip friction. Using the velocity corresponding to maximum friction ~10^{-2} s and a characteristic length scale ~ 1 nm, one estimates the relaxation time of PDMS chains on the order to be about 10^{-7}s, which is considerably larger than the viscous relaxation time of the dimethylsiloxane oligomer (~10^{-11}s). As a crude approximation, the ratio of these two time scales should be $\sim e^{U/kT}$. This leads to an estimate of *U* of about 22 kJ/mole. Interestingly, the activation energy of rubber friction from the temperature dependent[22] friction measurements is found to be about 25 kJ/mole, which is close to the above estimate of *U*.

When friction is measured with PDMS on polystyrene surfaces, it is found that the shear stress is somewhat higher than on hydrocarbon SAM. Furthermore, the onset of stick-slip transition for polystyrene occurs at a lower velocity (~10^{-3} m/s). This critical velocity would correspond to a potential well depth of 27 kJ/mole, which corresponds well to the activation energy (27 kJ/mole) of friction obtained from the temperature dependent studies. These differences in friction of PDMS on SAM and polystyrene exemplify the role of interfacial interaction in friction as was pointed out by Ludema and Tabor.[11] Another clear evidence of the role of surface properties on friction was provided by Vorvolakos et al.[32] (unpublished results). In these studies, small amounts of uncrosslinked PDMS chains of various molecular weights were mixed with a vinyl end-functional PDMS and the later was crosslinked using the well-known hydrosilation chemistry. The free chains were added in such a small amount (1 weight percent) that the modulus of the network was unaltered. However, the friction of the surface against a polystyrene coated glass sphere decreased dramatically with the molecular weight of the free polymer (Figure 4).

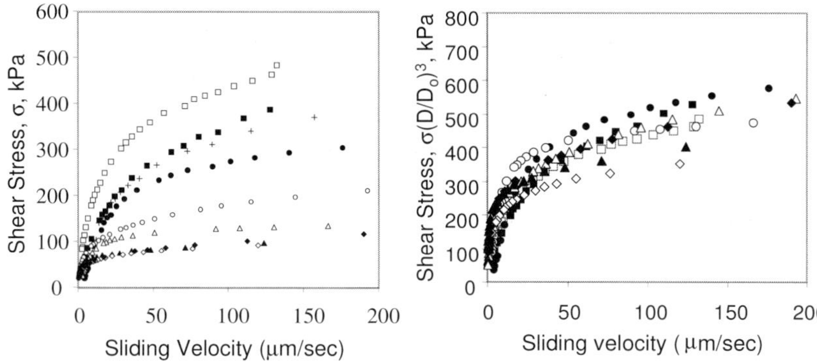

Fig. 4. (Left) Shear stress of a PDMS elastomer (Mn = 8kg/mole), which was modified by additives of unreactive PDMS of the following molecular weights 0.7 kg/mole (□), 2 kg/mole (■), 14 kg/mole (●), 49 kg/mole (Δ), 116 kg/mole (○), 204 kg/mole (▲), 308 kg/mole (◊), and 423 kg/mole (♦). Clearly, the friction decreases with the increase of molecular weight of additive. (Right) The shear stress data of the left figure are re-normalized by the thickness of the PDMS (*D*) that blooms to the surface. The thickness (D_o) of the reference polymer is that of a 0.7 kg/mole additive.

When the molecular weight of the polymer is lower than that of the network itself, a slightly increased friction is observed. It should be mentioned here that the decrease of friction for high molecular weight polymer has also been observed in the melt/solid interfaces.[29-30] Inn and Wang[29] suggested that the stress to desorb a polymer chain from a solid decreases with the molecular weight of the polymer. Hirz et al.[30] suggested that the decrease of friction at high molecular weight is related to the reduction of shear stress because of increased thickness. In order to examine the effect of the thickness of the free polymer at the surface, Dave Malotky[32] (unpublished results) developed an AFM based technique. Like the previous experiments of Mate et al.,[33] Malotky found that an AFM tip is pulled towards a substrate by capillary force when it

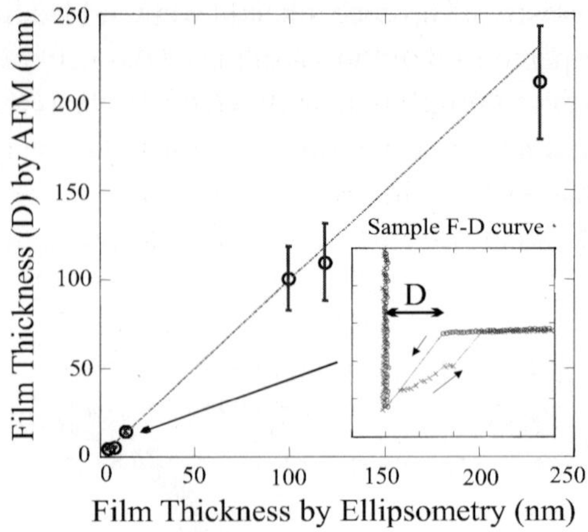

Fig. 5. Thickness of thin spin-cast PDMS films on silicon wafer as measured by Ellipsometry and AFM jump-in distance. Note that the jump-out distance is much larger than *D*.

touches the free surface of a thin liquid film. The AFM jump-in distance (*D*) correlates well with the thickness of the film obtained by ellipsometry (Figure 5). However, when the AFM tip is pulled away from the film, it is held by the liquid bridge over a distance (d) that is considerably longer than *D*. These two distances are however linearly correlated, i.e., $D \approx 0.09d + 3.46$. This is a useful relationship as it is much easier to measure *d* on a soft polymer than *D*, which was used by Malotky to estimate the thickness of liquid contamination films on the surface of a crosslinked polymer. His results show that the thickness of the layer for small molecular weight PDMS is comparable to their radius of gyration. However, for high molecular weight PDMS, the thickness is considerably smaller than R_g. It appears that the large molecular weight PDMS molecules are entangled with the network and they project only a small amount outward in the air. It is noticed that friction decreases with the thickness of this exposed layer. By making a rather crude assumption that the areal density of this exposed portion of the polymer chain varies as $\sim 1/D^2$, we expect that the shear stress of a contaminated network to be proportional to $\sim 1/D^2$. However, when the data of Figure 4 (left) is replotted by multiplying the shear stresses of the polymers with $(D/D_0)^2$

(D_0 being the thickness of a reference additive), a good collapse of data is not obtained. In fact the normalization as shown in Figure 4 (right) of the shear stress by multiplying with $(D/D_0)^3$ yields a better result. Although the physics of this scaling is not clear at present, the dramatic decrease of friction with the increase of the molecular weight of the free PDMS additives suggests the role of interface in friction quite convincingly.

Now we return to the subject of investigating the relationship between friction and elastic modulus. First of all, we notice that if the data presented in Figure 2 are normalized by dividing the friction stress with the peak stress, a reasonable master plot is obtained where all the friction data collapse (Figure 6). Furthermore, the peak stress itself increases with the shear modulus of the rubber as $\mu^{0.75}$, where the exponent is found to be somewhat larger than the expected value of 0.5. In Figure 1 as well as in Figure 6 (right) we see that the shear stress of the high molecular weight polymers reaches a plateau value at high velocity without exhibiting the friction maxima observed with the other polymers.

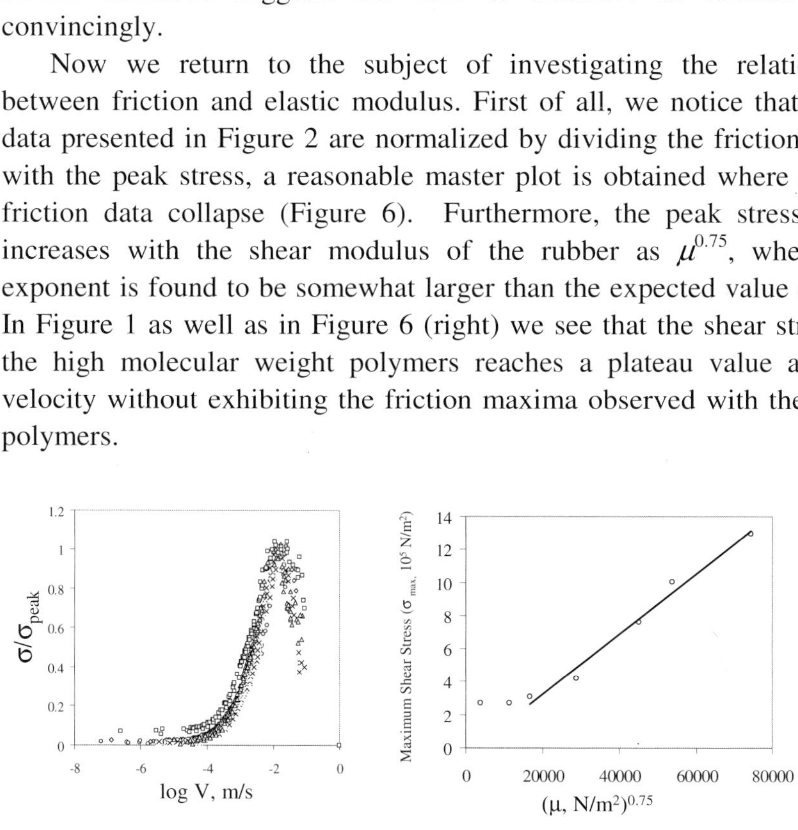

Fig. 6. (Left) Normalization of shear stress by dividing it with the peak shear stress. Here the data for the PDMS elastomers, which only exhibit peak shear stress and not plateau (see the right figure) are included. (Right) Maximum shear stress varies as $\mu^{0.75}$, expected for high molecular weight polymers.[22]

An interesting parallelism has been found in a recent experiment[34] where a rectangular prism of rigid glass was sheared against thin PDMS films of different molecular weights at different velocities (Figure 7).

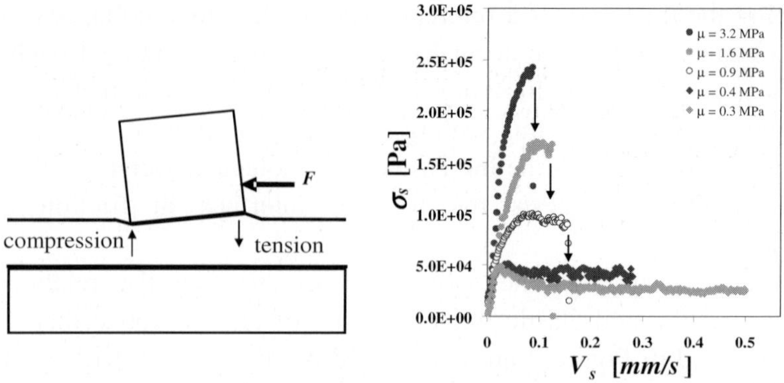

Fig. 7. Sliding friction of a rectangular prism of hydrophobic glass on 100 μm thick PDMS films of different shear modulus. In this particular configuration, as the glass prism slides, a tensile force is generated at the trailing edge and a compressive force is generated at the advancing edge. When the tensile stress reaches a critical value, the cube comes off the film. The arrows on the right figure indicate when the adhesive fracture occurs. As can be observed on the figure on the right, this kind of failure occurs for high modulus polymers, but not for the polymers of lowest modulus or highest molecular weights. For these highest molecular weight polymers, the normal stress that develops in response to the shear stress is not sufficient to cause adhesive failure. Thus the glass prism keeps on sliding even with increasing speed without coming off the PDMS film. Failure occurs when a defect is encountered. (Adapted from Ref. 34).

The rigid glass fractures adhesively from the low molecular films at a critical sliding velocity when the normal torque reaches a critical value. For high molecular weight polymers, however, sliding continues even at high velocity without showing any sign of adhesive fracture (except, of course, when a defect is reached). The results of the macroscopic experiments signify that that not enough normal stress is generated for the high molecular weight polymers to detach adhesively. It is plausible that for a related reason at the microscopic level, the areal chain density of the high molecular weight polymers does not decrease to an appreciable value at high velocity to exhibit the typical maximum observed with the low molecular weight polymers. The lack of molecular weight dependence at the high MW polymer may also mean that the entanglement crosslinking density takes over the overall behavior beyond a critical molecular weight.

Friction in Gel

The subject of gel friction has been elegantly reviewed by Baumberger and Caroli.[35] Here we discuss only a few salient points, which have some resemblance to rubber friction. Gong, Oshada et al.[36-42] considered three situations with respect to the friction of a solvated gel against a solid surface. In the first case, the solvent molecules interact more strongly with a solid surface than the polymer chains themselves causing a depletion of the latter at the interface. There is thus a thin solvent layer in between the gel and the solid substrate, the thickness of which depends on the external pressure. By estimating this effective layer thickness that decreases with the applied pressure, Gong et al.[41] estimated the viscous friction between the gel and the substrate as follows:

$$\sigma_{rep} / E = \frac{\eta V}{E^{2/3}\left(k_B T\right)^{1/3}} \frac{P / E}{1+(P / E)/(1+P / E)^{1/3}} \tag{18}$$

At low pressure, equation 18 becomes,[41]

$$\sigma \cong \frac{\eta V P}{E^{2/3}\left(k_B T\right)^{1/3}} \tag{19}$$

indicating that the shear stress increases with the applied pressure, but decreases with the elastic modulus of the gel.

In the second case, the polymer chains are attracted to the surface. Friction in this regime has two components: one is similar to that of Schallamach, in which the polymer chains undergo cyclic processes of attachment and detachment to the substrate, and the other is hydrodynamic. The shear stress in the regime of adsorption dominated friction is given by the following expression.[41]

$$\sigma_{attr} / E = \frac{\eta V}{E^{2/3}\left(k_B T\right)^{1/3}} \left[\frac{1}{2} + \left(1 + \frac{3}{2}\varepsilon \phi^{-1/2}\right)(1 + P / E)^{1/3}\right] \tag{20}$$

where ε is a measure of the adsorption energy between a monomer unit and the substrate, and φ is the volume fraction of the solvent. At high sliding velocity, Gong et al. used Schallamach's argument that the number of load-bearing polymer chains spends more time in the detached state than in the attached state causing a decrease of friction with the sliding velocity. However, as the hydrodynamic friction between the gel and the substrate picks up again, friction increases with the sliding velocity. In this limit, the friction stress is given simply in terms of the viscous shear stress. Thus the stress-velocity diagram of gel friction is, in many ways, similar to that of crosslinked rubber. Friction first increases with velocity, it decreases at the intermediate regime and then it increases again (Figure 8). However, unlike the low and the high velocity

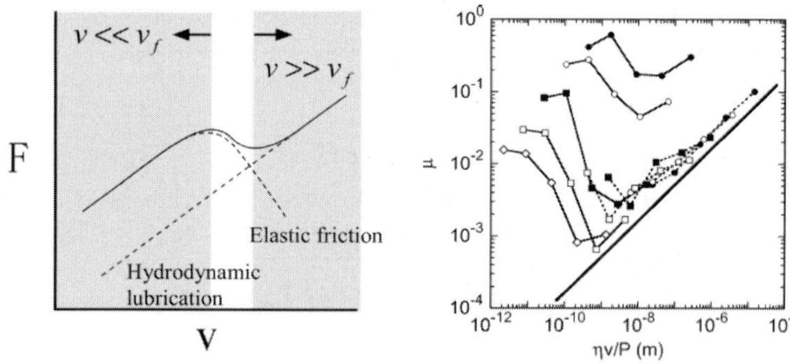

Fig. 8. Friction of gel showing an S-shaped behavior (left) as expected from the theory.[42] (right) Some high velocity results indeed show the second increase of the friction with velocity.[42]

regimes of rubber friction where the friction is dominated by surface and interchain friction, it is the solvent friction that controls the low and high velocity regimes in the case of gel friction. The case of increasing friction of gel in the high velocity region has been demonstrated recently by Gong et al.[42] (Figure 8). However, this region has not yet been clearly identified for rubber.

At this juncture, it is worthwhile to mention an interesting elastohydrodynamic effect in gel friction that was suggested by Sekimoto and Liebler[43] as well as Skotheim and Mahadevan[44] in the context of a curved body sliding against a thin soft gel coated substrate (Figure 9).

The main effect comes from the deformation of the gel layer that perturbs the parabolic profile of the gap distances close to the contact region, which affects the hydrodynamic pressure of the liquid that is sheared in the gap. When the perturbation of the profile is taken into consideration in the solution of the standard lubrication approximation of thin film flow in the gap, it is found that the pressure distribution is asymmetric. Integration of this pressure yields the normal force of the following form[43,44]:

$$F \sim \frac{\eta^2 V^2 R^2}{h_o^3}\left(\frac{H_l}{\mu}\right) \tag{21}$$

where H_l is the thickness of the gel, h_o is the gap thickness, and R is the radius of the cylinder. In the case of gel layers sliding against curved surfaces, the above normal force can further reduce friction between the surfaces.

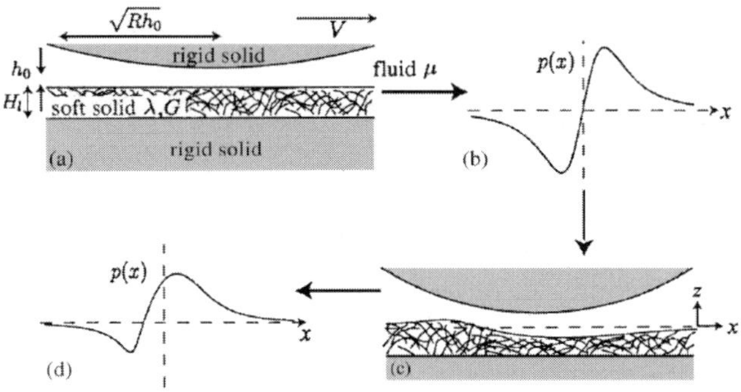

Fig. 9. This figure shows schematically how asymmetric pressure distribution develops on soft gel when a circular cylinder slides over it (Ref. 44).

Interfacial Instability and Friction

It has been known for a long time that rubbers exhibit a dynamic instability when sliding on a surface at high velocities. Systematic studies by Grosch showed that friction decreases with sliding velocity when such instabilities are encountered. A nice theoretical analysis of the problem has been provided by Ronsin and Coeuyrehourcq.[45] Recently, Baumberger et al.[46,47] made some fascinating observations about the dynamic response at the interface of a sliding gel. When a force is applied at a high rate of deformation of a gel, which is in adhesive contact with a rigid substrate, the force on the gel increases linearly till it reaches a critical state. At this critical state, the force drops to a lower value and homogeneous sliding of gel occurs on the surface. The friction force corresponding to the homogenous sliding exhibits a power law behavior with velocity with an exponent of about 0.4. At low sliding velocity, however, homogeneous sliding does not occur. Instead the force exhibits periodic stick slip dynamics (Figure 10, left). This kind of instability is not in response to the coupling of the system to a weak spring, but it reflects a series of events: breaking of certain numbers of interfacial bonds, transition to a dynamic state, reformation of the bonds and the repetition of the process. Detailed mechanisms of these events were ascribed to the propagation of self-healing slip pulses that nucleate at the trailing edge of the gel slab and propagate to the leading edge. As the slip pulse propagates, the region behind it goes to a stick region, till the pulse reaches the leading edge and another slip pulse is nucleated at the trailing edge to repeat the process (Figure 10).

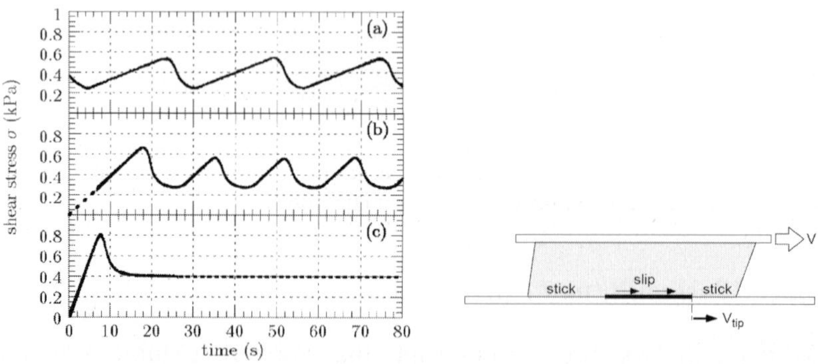

Fig. 10. (Left) Stick-slip friction in soft gel. (Right) The propagation of a slip pulse is depicted. As the pulse propagate, the surfaces behind it re-adhere.[47]

In confined systems, such as thin films of soft rubber sliding against a rigid substrate, that are prone to instability (Figure 1), bubbles are formed at the trailing edge that propagate towards the leading edge with remarkably high velocities. An approximate analysis describing the propagation of such interfacial bubbles were recently presented in reference 34. The bubble speed can be estimated by balancing the power delivered to the bubbles to the rate of dissipation of energy due to periodic viscoelastic dissipation in the rubber, Chaudhury and Kim[34] found an expression for the bubble velocity as:

$$V_b \sim V_{b0}\left(\frac{\sigma_s}{\mu}\right)^2 \tag{22}$$

where V_{bo} (~ 10 m/s) is a characteristic velocity, and σ_s is the applied shear stress. The equation was verified experimentally for the case of a rigid block sliding against thin PDMS elastomers (Figure 11). We end this chapter with the brief discussion of the role of interfacial aging in sliding instability[48] in a somewhat remote analogy to its effect in sliding friction, as discovered by Baumberger et al.[46,47] These experiments were designed to study the energetics of hydrogen bond formation at the interface. To study this effect, an oxidized PDMS rubber hemi-cylinder was brought into contact with a flat silicon wafer coated with a thin layer of grafted PDMS chains. The cylinder was rolled against the flat after different amounts of contact time. From the applied torque, the adhesive fracture energy was estimated, which starts out with a low value similar to that of van der Waals interaction, but it increases fast as contact time progresses and finally reaches to a near plateau value.

An interesting observation is made when the hemicylinder is rolled on the surface after some period of contact. For a soft cylinder in contact with a surface, the contact is rectangular with two lines parallel to the long axis of the cylinder. During rolling, one of these lines moves forward before the trailing line starts to move. When the trailing line reaches the boundary of the old and new contact, i.e. at the intersection of stronger adhesion due to longer contact time and lower adhesion due to shorter contact time, an elastic instability occurs and the hemi-cylinder rolls forward with considerable speed.

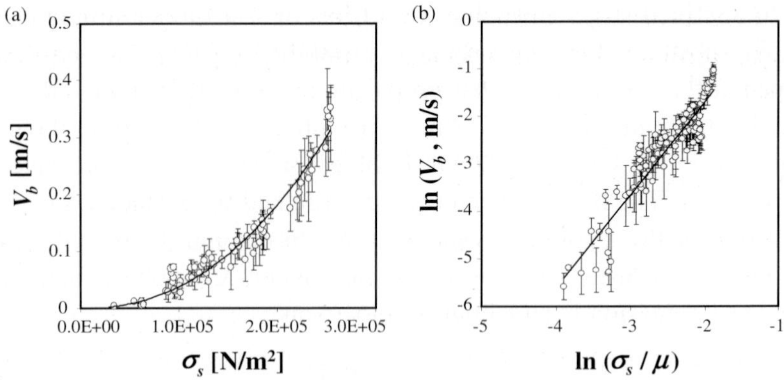

Fig. 11. Bubble velocity[34] in thin soft elastomers increase with the applied shear stress, but decreases with modulus as predicted by equation 26.

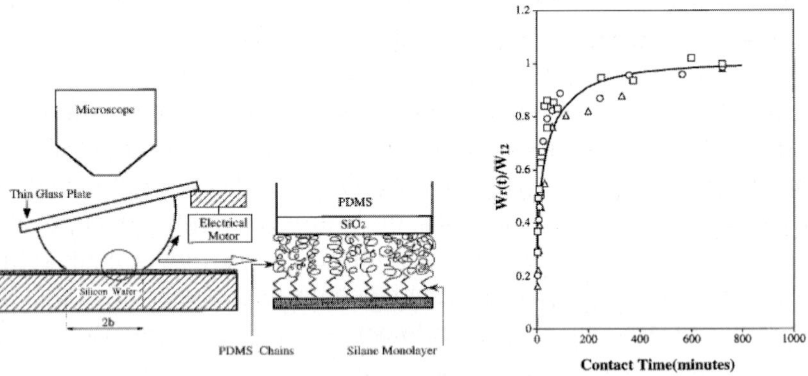

Fig. 12. (Left) Schematic of the measurement of rolling torque of a PDMS hemicylinder against the grafted brush of PDMS.[48] (Right) The fracture energy (which is expressed as the ratio of the fracture energy at a given time $W_r(t)$ to that (W_{12}) after 12 hours) obtained from the rolling torque measurements.[48]

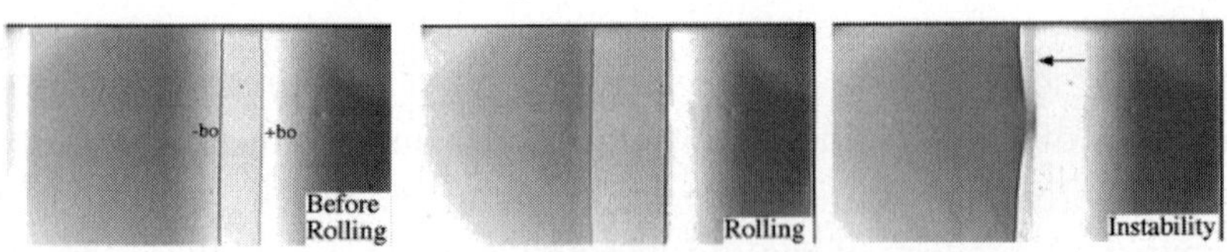

Fig. 13. Elastic instability[48] observed with the rolling of cylinder on a surface that shows age dependent adhesion. On the figure marked "Instability", the trailing edge, which is at the junction of high and low adhesion regions, undergoes an instability and the hemicylinder exhibits fast rolling before the trailing edge passes to the region of lower adhesion.

These experiments provide evidence that not only sliding friction, but also rolling friction is affected by interfacial processes and time of contact.

Summary and Perspective

We have tried to present a brief review of the old and new concepts of rubber friction that show certain parallelism with the sliding friction of gels. These results should also be relevant to various thin film studies involving grafted polymers. Because of somewhat limited focus of the review, many important studies could not be reviewed here, such as the studies involving the friction of polymer brush in solvents[49] or the chain pull-out effects[50-52] that play important roles in the friction of grafted brush against rubbery networks. However, the brief review of the friction of crosslinked networks against impenetrable surfaces raises several important questions that have not yet found clear answers. The various models used to describe rubber friction differ considerably from each other in their essential physical foundation. For example, the model of Schallamach considers the cyclic processes of attachment and detachment of the polymer chains at the interface, whereas Persson's model is based on thermally activated fluctuation of the stress segments that never de-adheres from the surface. Another model[22] describes friction as a stress-biased diffusion of polymer chains in contact with the surface. Definitive experiments are needed to examine the interface and friction measurements simultaneously, to address the detailed questions of dynamics of polymer chains as well as its interactions with surfaces. Some important developments in this direction are taking place, especially with the advent of sum-frequency generation spectroscopy.[53] However, more developments to interrogate the interface are needed. An important question that needs to be resolved is the aging behavior of polymer chains at the interface and its relationship to the adhesive and tribological interactions that prevail at the interface.

References

1. Ariano, R. *India Rubber J.*, 79, 56 (1930).
2. Derieux, J. B. *J. Elisha Mitchell Sci. Soc.*, 50, 53 (1934).
3. Dawson, T. R. *Rubber: Physical and Chemical Properties*; Dawson, T. R., Porritt, B. D., Eds.; The Research Association of British Rubber Manufacturers: Croydon, U.K., 1935; pp 381-386 (1935).
4. Roth, F. L.; Driscoll, R. L.; Holt, W. L. *J. Res. Natl. Bur. Stand.*, 28, 439 (1942).
5. Thirion, P. *Rev. Gen. Caoutch.*, 23, 101 (1946).
6. Papenhuyzen *Ingenieur, Utrecht,* 75, 53 (1938).
7. Bueche, A. M.; Flom, D. G. *Wear*, 2, 168 (1959).
8. Bowden, F. P.; Tabor, D. *The Friction and Lubrication of Solids* (Clarendon Press: Oxford, 1950).
9. Greenwood, J. A.; Tabor, D. *Proc. Phys. Soc., London*, 71, 989 (1958).
10. Grosch, K. A. *Proc. R. Soc. London, Ser. A*, 274, 21 (1963).
11. Ludema K. C. and Tabor D. *Wear*, 9, 329 (1966).
12. Maugis, D. Barquins, M. *J. Phys. Lett.*, L95, 42 (1981).
13. Roberts, A. D. *Rubber Chem. Tech.*, 52, 23 (1979).
14. Kendall, K. *Wear*, 33, 351 (1975).
15. Charmet, J.; Verjus, C.; Barquins, M. *J. Adhesion*, 57, 5 (1996).
16. Barquins, M. *J. Natl. Rubber Res.*, 5, 199 (1990).
17. Schallamach, A. *Proc. Phys. Soc., London, Sect. B*, 65, 657 (1952).
18. Bell, G. I. *Science*, 200, 618 (1978).
19. Evans, E.; Ritchie, K. *Biophys. J.*, 72, 1541 (1997).
20. Merkel, R.; Nassoy, P.; Leung, A.; Ritchie, K.; Evans, E. *Nature*, 397, 50 (1999).
21. Chernyak, Y. B.; Leonov, A. I. *Wear*, 108, 105 (1986).
22. Vorvolakos, K.; Chaudhury, M. K. *Langmuir*, 19, 6778 (2003).
23. Persson, B. N. J.; Volokitin, A. I. *Eur. Phys. J. E*, 21, 69 (2006).
24. Ghatak, A.; Vorvolakos, K.; She, H.; Malotky, D.; Chaudhury, M. K. *J. Phys. Chem. London, Sect. B*, 104, 4018 (2000).
25. Blake, T. D.; Haynes, J. M. J. *Colloid Interface Sci.* 30, 421 (1969).
26. Kendall, K. J. *J. Adhes.* 5, 179 (1973).
27. Persson, B. N. G. Sliding Friction: Physical Properties and Applications, 2nd ed. (Springer: Heidelberg, 2000).
28. Leonov, A. I. *Wear*, 131, 137 (1990).
29. Inn, Y.; Wang, S.-Q. *Phys. Rev. Lett.*, 76, 467 (1996).
30. Hirz, S.; Subbotin, A.; Frank, C.; Hadziioannou, G. *Macromolecules*, 29, 3970 (1996).
31. Newby, B. Z., Chuadhury, M. K., *Langmuir*, 13, 1805 (1997).
32. Vorvolakos K, Malotky D and Chaudhury M. K., (*unpublished results*). Although, experimental data at higher velocities are available for the higher molecular weight polymers, they are not included here as the corresponding data for the lower

molecular polymers could not be obtained at corresponding high velocities because of the abrasion of the rubber by polystyrene. A polystyrene coated glass was used instead of a SAM coated glass, because the former surface resulted in a greater friction force so that any changes of the surface friction of PDMS could be easily discerned.

33. Mate, C. M., *Phys. Rev. Lett.*, 68, 3323 (1992).
34. M. K. Chaudhury and K. H. Kim, *Eur. Phys. J. E.*, 23, 175 (2007).
35. Baumberger, T.; Caroli, C., *Adv. in Phys.*, 55, 279 (2006).
36. J. Gong, M. Higa, Y. Iwasaki, Y. Katsuyama, Y. Osada, *J. Phys. Chem. B*, 101, 5487 (1997).
37. J. Gong, Y. Iwasaki, Y. Osada, K. Kurihara, Y. Hamai, *J. Phys. Chem. B*, 103, 6001 (1999).
38. J. Gong, G. Kagata, Y. Osada, *J. Phys. Chem. B*, 103, 6007 (1999).
39. J. Gong, Y. Iwasaki, Y. Osada, *J. Phys. Chem. B*, 104, 3423 (2000).
40. G. Kagata, J. Gong, Y. Osada, *J. Phys. Chem. B*, 106, 4596 (2002).
41. J. Gong, Y. Osada, *J. Chem. Phys. B*, 109, 8062 (1998).
42. Kurokawa T, Tominaga T, Katsuyama Y, Kuwabara R, Furukawa H, Osada Y. and Gong, J. P *Langmuir*, 21, 8643 (2005).
43. K. Sekimoto and L. Leibler, *Europhys. Lett.*, 23, 113 (1993).
44. J. M. Skotheim and L. Mahadevan, *Phys. Rev. Lett.*, 92, 245509 (2004).
45. Ronsin, O.; Coeuyrehourcq, K. L. *Proc. R. Soc. London, Ser. A*, 457, 1277 (2001).
46. Baumberger, T.; Caroli, C.; Ronsin, O. *Phys. Rev. Lett.*, 88, 075509 (2002).
47. T. Baumberger, C. Caroli, and O. Ronsin, *Eur. Phys. J. E*, 11, 85 (2003).
48. H. She, D. Malotky, and M. K. Chaudhury, *Langmuir* , 14, 3090 (1998).
49. Klein J., Perahia D., and Warburg S. *Nature*, 352, 143 (1991).
50. Brown, H. R. *Science*, 263, 1411 (1994).
51. Casoli, A.; Brendle, M.; Schultz, J.; Philippe, A.; Reiter, G. *Langmuir*, 17, 388 (2001).
52. Leger, L, Hervet, H., Massey, G., Durliat E., *J. Phys. Condens. Matter*, 9, 7719 (1997).
53. Yurdumakan B, Nanjundiah K, Dhinojwala A. *J. Phys Chem*, 111, 960 (2007).

CHAPTER 9

RELATIONSHIPS BETWEEN MOLECULAR ARCHITECTURE, LARGE-STRAIN MECHANICAL RESPONSE AND ADHESIVE PERFORMANCE OF MODEL, BLOCK COPOLYMER-BASED PRESSURE SENSITIVE ADHESIVES

Costantino Creton

Laboratoire de Physico-Chimie des Polymères et Milieux Dispersés
CNRS-ESPCI-UPMC
10 rue Vauquelin
75231 Paris Cedex 05
France

Kenneth R. Shull

Department of Materials Science and Engineering
Northwestern University
2220 Campus Dr.
Evanston, IL USA, 60208-3108

Pressure sensitive adhesives represent one of the most important commercial applications of block copolymers. The large-strain tensile properties of pressure sensitive adhesives play a dominant role in determining the adhesive properties of these materials. In many cases changes in the polymer architecture significantly affect the large-strain properties and the resultant adhesive performance, with little or no effect on the linear viscoelastic properties. In this chapter we provide several examples, using model systems based on diblock, triblock, tetrablock and star-block copolymers of polystyrene and polyisoprene. We also discuss some examples using model acrylic diblock and triblock copolymers.

Introduction

Pressure sensitive adhesives (PSAs) consist of a thin viscoelastic layer (typically 20-100 μm) of material with a modulus at the experimentally relevant frequencies in the range of 10^5 Pa. PSA performance is determined by an interplay between adhesive/substrate interactions, and the bulk mechanical properties of the adhesive material.[1] The role of the interfacial interactions is to support a stress that is sufficiently large so that the bulk of the adhesive material can be deformed to large strains. Optimization of a PSA therefore requires that two separate materials questions be addressed:

1. What controls the stress that the interface can support?
2. What controls the large-strain deformation behavior of the adhesive material?

These questions are equally important in the overall design of a PSA. The first question is important in the design of release coatings for PSAs, for example. These coatings must be designed so that stress required to remove the PSA from the backing is less than the stress required to substantially deform the bulk of the material. This review, however, is concerned primarily with the second question, because this is the issue where block copolymers have played the greatest role.

The ability to create an adhesive material that sticks well to a variety of surfaces, but not to the backing with its release liner, arises from the nonlinearity of the overall response of a PSA. For a given adhesive, the detailed characteristics of the substrate control the stress that can be applied to the adhesive prior to detachment. Large strain deformation mechanisms responsible for most of the energy dissipating ability of a PSA are invoked when the interfacial stress exceeds a threshold value that is comparable to the elastic modulus of the adhesive layer. At this stress level cavitation and the subsequent formation of a fibrillar structure within the adhesive material take place, without substantially increasing the driving force for adhesive failure at the adhesive/substrate interface. Release liners are designed so that the maximum interfacial stress falls below the threshold value at which these bulk deformation mechanisms are invoked. Once fibrillation has taken place, the relevant properties of the material are related to its behavior under tensile loading conditions. At sufficiently large tensile strains, the adhesive strain hardens, resulting in an increased driving force for failure at the interface with the substrate, allowing a detachment of the PSA from the surface without leaving any residue. The overall adhesion energy is determined by the work expended in deforming the adhesive prior to this point of failure.

Given the obvious importance of the large-strain tensile behavior of the materials in determining the adhesive performance, it is useful to understand how this behavior is affected by the underlying materials structure. Block copolymers, used widely as the basis for commercial PSAs, are ideally suited for these types of fundamental investigations. In this chapter we consider two types of block copolymers that are often used in adhesive applications. The system for which we have the most information is a styrenic system consisting of styrene/isoprene block copolymers, although some data is included here for acrylic systems as well. The mechanical characterization of these materials is discussed in the following section, followed by a discussion of their adhesive performance in probe tack experiments. This section is followed by a brief discussion of the adhesive behavior of these materials, followed by some concluding remarks.

1. Mechanical Characterization

1.1. *Effect of Chain Connectivity on Large Strain Behavior*

One of the most interesting features of ordered block copolymer structures is their ability to introduce a dilute but highly regular density of crosslink points. We focus here on the type of ordered structure obtained when the volume fraction of the minority phase is less than 15% and the block copolymers self-assemble in a structure of spheres in a matrix. From the point of view of mechanical properties and adhesive applications, the most interesting situation is that of block copolymers where the A block is glassy and the B block is elastomeric. Extensive investigations have been carried out for the commercially important case where A is polystyrene, B is cis1,4 polyisoprene and the blend is further diluted with a high T_g, low molecular weight molecule that is miscible with the elastomeric domains and immiscible with the glassy domains. These low molecular weight diluents are generally referred to as tackifying resins in the industrial community and are necessary to dilute the entanglement network, lower the elastic modulus, and transform the rubbery and non adhesive pure block copolymers in an actual PSA. Although the microstructure of the blend is mainly determined by the volume fraction of PS, its mechanical properties are greatly dependent on the architectures of the block copolymers that are being used. It is the purpose of this part to explore the effect of copolymer architecture.

The easiest comparison to make is between a pure A-B-A triblock copolymer and a 50/50 blend of a triblock copolymer with an A-B diblock with the same polystyrene content but one half the total molecular weight. Triblock copolymers have the ability to bridge adjacent PS domains while diblocks do not. In the rest of the chapter we will refer to bridging chains and pendant chains for these two situations. For the pure triblock copolymers, mean-field simulations have shown that about 80% of the midblocks act as bridging chains while the rest act as loops.[2,3] This result remains true when the blend is diluted with a tackifying resin to overall concentration of 40 wt%. If diblocks with one half the molecular weight of the triblock are added to the blend, the concentration of bridging chains scales linearly with the triblock volume fraction in the diblock/triblock blend.[3] We also discuss results using S-I-S-I tetrablocks, and four-arm star- block copolymers, described in more detail below. The characteristics of each of the four copolymers (diblocks, triblocks, tetrablock and 4-arm star block) are listed below in Table I.

Figure 1 shows the small strain complex shear modulus measured in a parallel plate rheometer as a function of frequency for the SIS material and a SI/SIS blend with 54 wt. % SI, relative to the total amount of copolymer. Each material also contains 60 wt. % resin. The resin (T_g = 40°C) is a hydrogenated polycyclopentadiene produced by Exxon Mobil Chemical under the trade name of Escorez 5380. The introduction of a diblock copolymer results in a slight decrease in G' at low frequency and

Table I. Molecular characteristics of the styrene (S) – isoprene (I) copolymers used in the experiments. The star block can be viewed as 4 diblock copolymers with the molecular weights indicated in the table, joined together at the polyisoprene ends.

	S mol. wt. (kg/mol)	I mol. wt. (kg/mol)	S mol. wt. (kg/mol)	I mol. wt. (kg/mol)	Wt. % S	Wt. % bridgeable isoprene
SI Diblock	10.8	61	------------	------------	15	0
SIS Triblock	11.6	131	11.6	-------------	15	100
SISI Tetrablock	12.4	60	12.4	70	16	46
$(SI)_4$ 4-arm star block	10.8	61	------------	------------	15	100

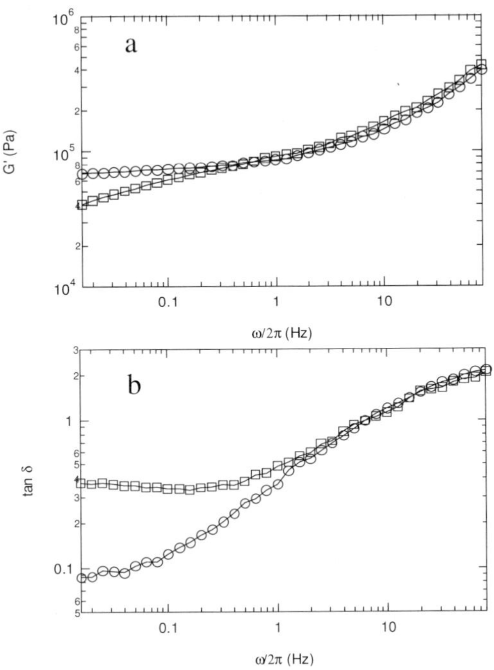

Fig. 1. (a) Storage modulus G' and (b) loss tangent tan(δ) as a function of frequency at T = 22°C, for SIS/SI blends with 60 wt. % resin. ❍: 0 wt. % SI, ❑: 54 wt. % SI.

increases the value of the loss tangent. Both effects are only apparent at frequencies lower than about 1 Hz. If tensile tests are performed up to several hundreds of percent strains on the same materials, a relatively weak strain rate dependence is observed (see Figure 2(a)) for a given blend, and a strong effect of the presence of the diblock on the softening behavior is observed at a given strain rate (see Figure 2(b)). Because these experiments are conducted with a constant crosshead velocity, V, the actual true strain rate decreases throughout the test. The reported strain rates correspond to the initial strain rates, $\dot{\varepsilon}_{init}$, obtained by dividing the crosshead velocity by the undeformed sample length. The primary effect of the presence of the diblock is to modify the shape of the stress-strain curve, giving a more pronounced softening followed by strain hardening (upturn in stress) at higher strains.

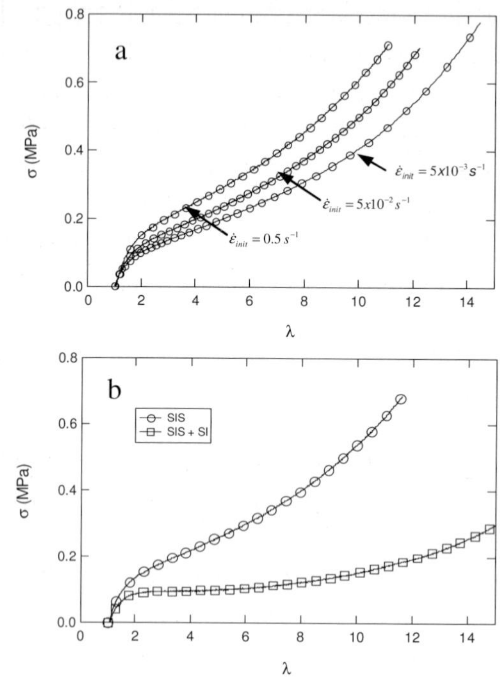

Fig. 2. Nominal stress as a function of strain for SIS/SI blends with 60 wt.% resin. (a) tensile tests for the 0 wt.% SI blend at initial strain rates of $5x10^{-3}$, $5x10^{-2}$ and 0.5 s^{-1}. (b) tensile tests at $\dot{\varepsilon}_{init}$ = 0.5 s^{-1} for blends with 0 and 54 wt.% SI.

From the molecular point of view these differences can be interpreted as follows: the small strain modulus is controlled by the density of entanglements which act as crosslink points, but progressively slip and orient in the direction of traction.[4] The type of softening observed suggests that the final number of effective physical crosslink points is significantly different between the pure triblock system and the 50/50 blend.[5] A quantitative comparison with mean-field predictions can then be used to further refine the analysis: the softening and hardening observed with this entangled triblock is consistent with the presence of a large fraction of trapped entanglements which actually act as permanent crosslinks. These trapped entanglements are not present in the 50/50 blend, but strongly contribute to the modulus of the A-B-A blend.

1.2. *Effect of Copolymer Architecture: Triblock/Diblock Blend versus Tetrablock*

A more striking example of the role played by the copolymer architecture is provided by the comparison between a diblock/triblock blend and an A-B-A-B tetrablock copolymer. In a recently proposed molecular model of the linear viscoelastic behavior, Gibert *et al.* reduce the triblock/diblock blends to a blend of bridging chains and pendant chains.[6,7] Their model suggests that a properly designed A-B-A-B tetrablock copolymer could have identical linear viscoelastic properties to a blend of A-B diblock and A-B-A triblock. Three conditions are necessary in order for this equivalence to be achieved:

1. The A content needs to be constant in order to maintain the same thermodynamically stable structure (spheres of A in a matrix of B).
2. The molecular weight of the B free end-block in the ABAB chains has to be the same as that of the B sequence of diblock chains in the ABA + AB blend.
3. The length of the B blocks trapped between two A blocks in the ABAB chains has to be chosen in order to respect the same "pendant B chains/bridging B chains" ratio. Therefore, the length of the equivalent triblock in the ABAB chains could not be equal to the length of the triblock chains in ABA + AB blends.

In order to test these ideas, a SISI tetrablock copolymer was synthesized by ExxonMobil Chemical. The polymer had an overall styrene fraction of 16 wt. % with 46.2 wt. % of the polyisoprene corresponding to the 'bridgeable' block between styrene blocks. The molecular weight of the non-bridgeable block is just slightly higher than the polyisoprene block of the SI chains. As expected, the linear viscoelastic properties of a SISI tetrablock, and a SIS/SI blend are nearly identical, provided that they are formulated according to the 3 criteria listed above. Room temperature data for the specific blend used by Roos[8] is shown in Figure 3. Gibert *et al.* have obtained similar results over a much broader temperature and frequency range, even with the addition of substantial resin fractions.[9]

An important difference between the SISI and tetrablock and the SIS/SI blends is that the tetrablock has a covalent bond linking the equivalent triblock and diblock chains. Therefore, while no difference is seen in the linear viscoelastic response, we would expect to see differences in the large strain mechanical properties, and consequently on their adhesive properties. The expected differences in the large strain tensile properties are illustrated by the comparison shown in Figure 4. The tensile curves of the tetrablock based adhesive and the corresponding based adhesive are extremely different. We also have represented for comparison the pure SIS blend data of Figure 2. In both

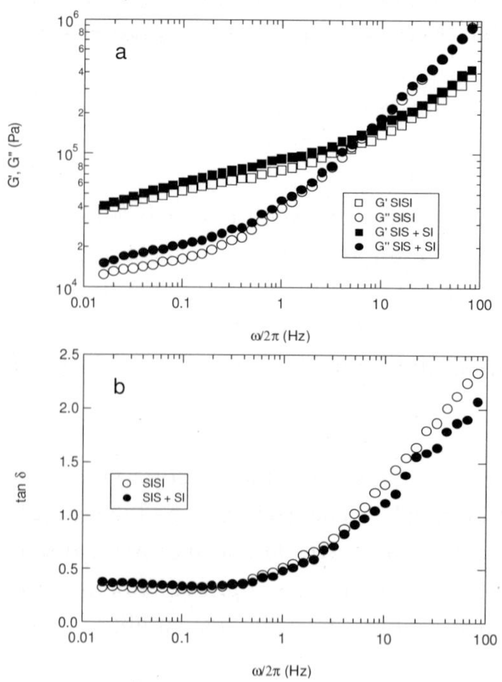

Fig. 3. Linear visco-elastic properties for the adhesive blends based on SIS + SI with 54% SI and on SISI. (a) G' and G" and (b) tan δ.

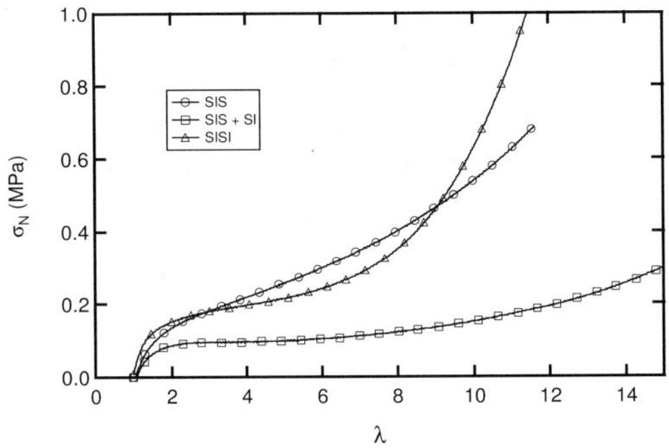

Fig. 4. Nominal tensile stress as a function of strain for the adhesive blends based on SISI, on SIS + SI and on SIS at $\dot{\varepsilon}_{init}$ = 0.5 s^{-1} (T = 22°C). The data is that of Figure 2(b) with the addition of the SISI data. Resin content for all blends is 60 wt. %.

cases there is a linear increase in the stress at low strains, followed by strain softening and then strain hardening regimes. Reduction in the bridging fraction by the addition of the additional polyisoprene block or by the addition of the SI diblock copolymer increases the strain softening, but the details depend on the way the decrease in bridging fraction is achieved. Use of tetrablock chains causes the hardening to occur at much lower strains (~600% as compared to 1200% for the SIS/SI blend). A useful way to compare large strain data from elastomers is to use a reference, neo-Hookean elastic model rather than a linearly elastic model. For a neo-hookean rubber the stress-strain curve in uniaxial tension is given by:

$$\sigma_N = G\left(\lambda - \frac{1}{\lambda^2}\right), \qquad (1)$$

so that one can define the reduced or Mooney stress, σ_R, as follows:

$$\sigma_R = \frac{c_N}{\left(\lambda - \frac{1}{\lambda^2}\right)}, \qquad (2)$$

which reduces to a constant value G for a neo-Hookean material. The difference between the two adhesives is clearly apparent in the Mooney stress representation shown in Figure 5 where σ_R is plotted as a function

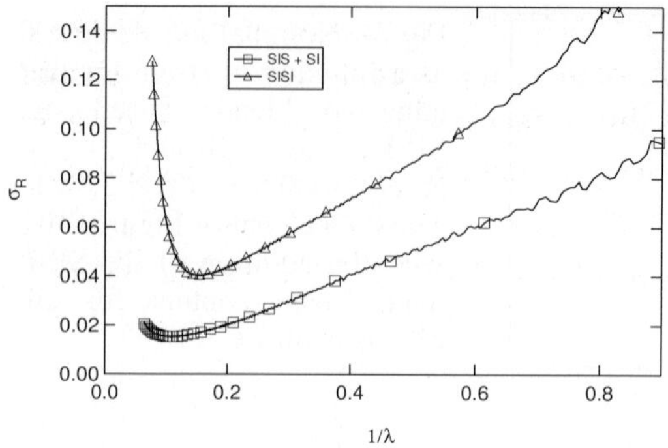

Fig. 5. Comparison of the shape of the reduced data from the tensile experiments for the SISI and the SIS + SI blends, both with 60 wt. % resin. $\dot{\varepsilon}_{init} = 0.5\ s^{-1}$ (T = 22°C). The data is that of Figure 4.

of 1/λ. The reduced stress is significantly higher for the SISI block copolymer at all deformations and the softening is clearly more pronounced, with a very sharp upturn for the strain hardening.

Although an unambiguous molecular interpretation of these results is difficult to make, the results clearly show that the architecture-dependent connectivity between polystyrene domains can only be discerned from large strain measurements, and not from measurements of linear viscoelasticity. Given the large strain involved in the debonding processes of adhesives, this result implies that it is also difficult to predict quantitatively the adhesive performance from the linear viscoelastic properties.

1.3. *Effect of Copolymer Architecture: Star versus Linear*

Another example of the effect of copolymer architecture is obtained from a comparison involving a four-arm star-block copolymer, $(SI)_4$, which has the branch point at the ends of the polyisoprene blocks of the constituent arms. Each arm of the star corresponds to one SI diblock or to one half of the SIS triblock. Hence the total molecular weight of the star is twice that of a triblock. Comparison between a $(SI)_4$/SI blend and a SIS/SI blend shows two important differences. First, the $(SI)_4$/SI blend softens for higher levels of stress, indicating a higher density of crosslinks or trapped entanglements, and the finite extensibility of the SI_4/SI blend is apparent for lower values of λ. The difference in the finite extensibility and in the density of crosslink points is very clear from the tensile tests (see Figure 6(a)). If we say that the finite extensibility of the blend should scale as the square root of the molecular weight between chemical or physical crosslink points, the star tensile curve should be moved to lower extensions by a factor $\sqrt{2}$, which is the case in Figure 6(b). This result is illustrated by renormalizing the deformation by the theoretical maximum extension λ_{max}. Based on the value of R_g in the melt,[10] λ_{max} is equal to 19 and 26 respectively for the SI_4 and SIS copolymers.

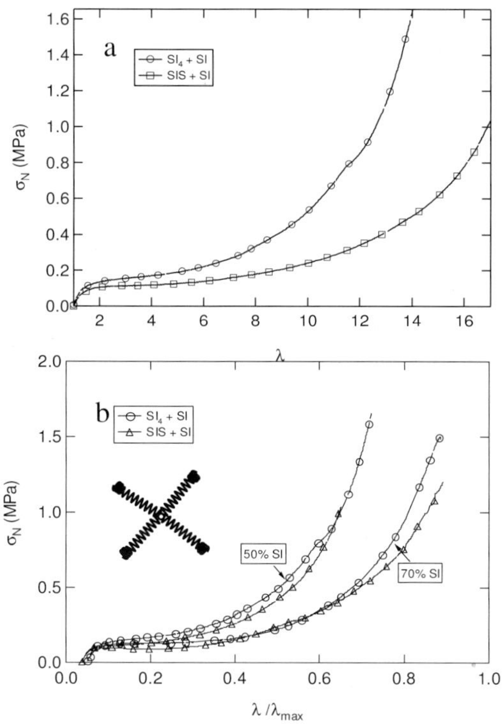

Fig. 6. Tensile data for the SIS/SI and SI4/SI blends for $\dot{\varepsilon}_{init}$ = 0.5 s^{-1} and T = 22°C. (a) Nominal stress vs. λ for the SI_4/SI blends and the SIS/SI blends. In both cases the percentage of diblocks in the copolymer blend is 50%. (b) Same data plotted as a function of λ/λ_{max}. The percentages on the figure are the percentages of diblock in the blends. Values of λ_{max} are 19 and 26 respectively for the SI_4 and SIS.

1.4. *Large Strain Hysteresis*

Because the adhesion energy is determined by the energy expended in deforming the adhesive material, experiments designed to quantify the mechanical hysteresis are instructive. Figure 7 and Figure 8 show results obtained for blends without resin and with 60 wt. % resin at room temperature. On each graph we show results for successive tensile tests

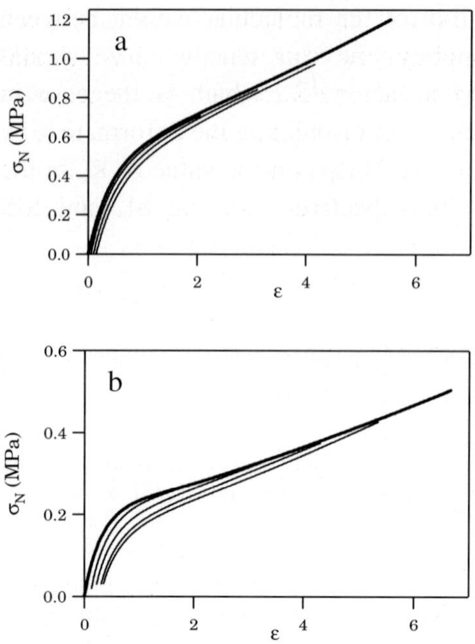

Fig. 7. Mullins type tests for materials without resin: SIS (a) and a SIS/SI blend with 54% SI (b).

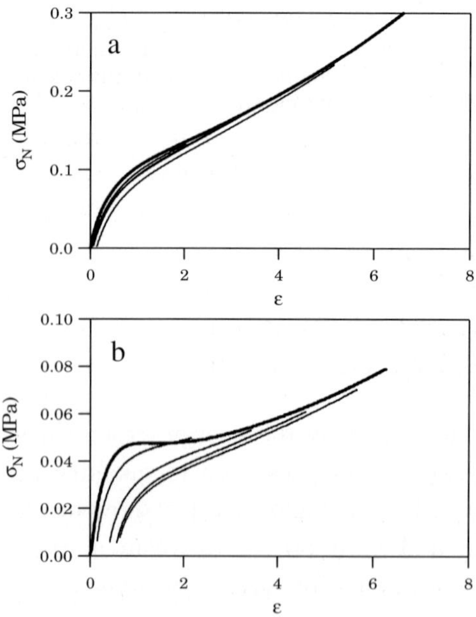

Fig. 8. Mullins type tests for blends with 60% resin: SIS (a) and SIS/SI blend with 54% SI (b).

applied to a given sample. Values for increasing values of the maximum extension, λ_{max}, are plotted for each material. For the pure SIS material all the curves can be nearly superimposed: the deformation is practically reversible. Increasing the diblock fraction increases the dissipation when successive tractions are performed. Similar results are obtained for blends containing 60 wt. % resin. At large strains, the parameter controlling dissipation appears to be the diblock content and not the resin content. Indeed the blends without diblock are quite elastic regardless the amount of resin. However the presence of resin in the blend enhances the dissipative losses when the diblock content is increased.

While the experimental protocol used in these experiments is similar to the standard tests used to detect the Mullins effect in filled elastomers,[11] no real Mullins effect was identified in our materials. Indeed the energy dissipation we measured does not seem to be due to the breakage of an organized structure but more to classic viscoelastic dissipation at the macromolecular scale. It is interesting however to note that while the resin substantially increases the glass transition temperature and the monomer friction coefficient at room temperature,[12] it does not greatly affect the measured hysteresis. This result can be attributed to the fact that the test temperature remains significantly higher than the T_g of the elastomeric domains.

Overall, we observe that, within this range of molecular weights that are much larger than the average molecular weight between entanglements, the molecular weight of the midblock has an important effect on the maximum extensibility and on the softening at intermediate strains, but it has no discernible effect on the linear viscoelastic properties. On the other hand the ratio of pendant chains to bridging chains has a very important effect on the dissipative properties on the blend, in addition to its effect on the large strain properties.

1.5. *Acrylic Systems*

An advantage of acrylic systems, including poly(2-ethylhexyl acrylate) and poly(n-butyl acrylate) (PnBA) is that effective adhesion is often obtained even without the addition of a tackifying resin, a result that can be partially attributed to the high entanglement molecular weight of these systems. Poly(n-butyl acrylate), for example, has an entanglement molecular weight of 22,000 g/mol,[13] compared to 7,000 g/mole for cis 1,4 polyisoprene.[14] The molecular characteristics of the different acrylic copolymers are listed in Table II. Because of the way these polymers are synthesized, the PnBA blocks contain 2-4 mole % acrylic acid, and the

Table II. Molecular characteristics of the acrylic block copolymer adhesives.

	A (PMMA) mol. wt. (kg/mol)	B (PnBA) mol. wt. (kg/mol)	PMMA (A) mol. wt. (kg/mol)	Wt. % PMMA
Triblock (TB1)	15	136	15	18
Triblock (TB2)	22	180	22	20
Triblock (TB3)	15	330	15	8
Diblock (DB)	18	90	18	17

triblock copolymers are contaminated with small amounts of diblock copolymers with half the total molecular weight of the triblock copolymers.[15] The large-strain properties of these acrylic polymers have not been measured, but the small-strain viscoelastic properties are shown in Figure 9. These results were obtained by bringing a hemispherical glass indenter into contact with the adhesive, and oscillating the indenter in order to obtain the storage and loss moduli *in situ* for each adhesive layer.[16] Each of the polymers has a predominantly elastic character, which is expected for the ordered microphase morphologies that are obtained for these molecular weights. Triblock TB3 has a very low PMMA weight fraction, and has the mechanical properties that one would expect for a matrix of PMMA spheres in a PnBA homopolymer matrix. The modulus of this material is actually less than the modulus of $\approx 10^5$ Pa that corresponds to the plateau modulus of PnBA. We attribute this result to the fact that adhesive layers were formed by forming the phase-separated morphology in a solvent-rich gel state, where entanglements are diluted by out by the presence of the solvent. Solvent is slowly evaporated at room temperature, where the relaxation time associated with molecular exchange is much too long for the structure of

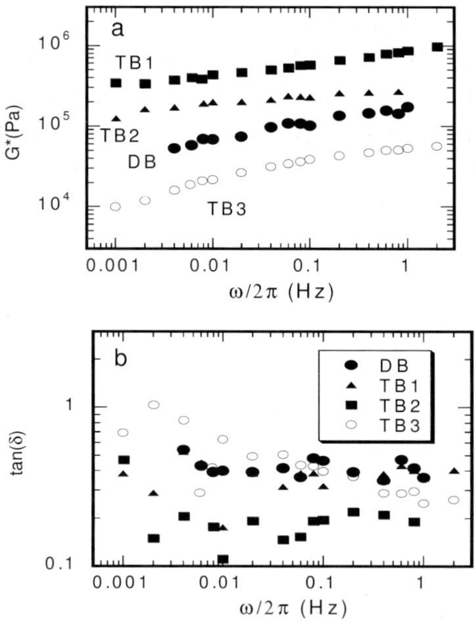

Fig. 9. Frequency dependence of the magnitude of the complex shear modulus (a) and loss tangent (b) for each of the model acrylic adhesives.

the dried material to equilibrate.[17] In this way we are able to control the microstructure of the polymer in a controlled way, giving reproducible mechanical and adhesive properties.[18]

2. Adhesive Behavior

As mentioned in the introduction, the block copolymer blends described in section 2 find interesting applications as pressure-sensitive-adhesives. An important requirement of a good PSA is that interfacial crack propagation be limited, even when the adhesive forces are due to relatively weak van der Waals forces. The material must also maintain a distinctly solid character, exhibiting minimal creep under a constant low level of stress. The capability to resist interfacial crack propagation is typically tested by bringing a flat ended or spherical probe in contact with the adhesive film and subsequently removing it at a constant velocity. An example of such a debonding curve for a flat-ended probe removed from a 100 μm thick layer of SIS/SI adhesive blend is shown on Figure 10. As it can be readily seen, the force displacement curve extends with a non- zero force up to distances several times the initial thickness of the film. The images taken during the debonding process show that the first initial peak is due to the nucleation of a series of cavities, with a size comparable to the thickness of the film, and the subsequent plateau is due to the extension of a fibrillar foam in the direction of traction until the fibrils detach cleanly from the surface. This general scenario occurs for all the styrene/isoprene block copolymer blends discussed in section 2. However the detailed shape of the force displacement curve varies.

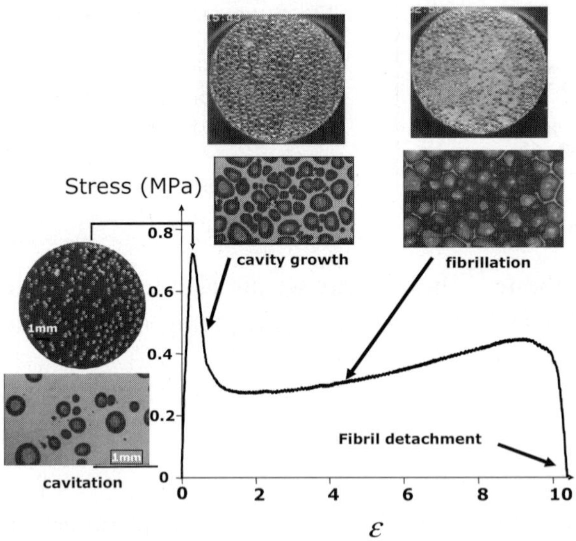

Fig. 10. Schematic of the debonding of an SIS/SI PSA blend from a steel surface. The stress represents the force normalized by the initial area of contact and the strain is the displacement of the probe normalized by the initial thickness of the PSA film.

Typical debonding curves are shown on Figure 11 for the different SI block copolymer blends discussed in section 2. The level of the plateau in stress during the extension of the fibrillar foam is clearly related to the large strain tensile test as can be readily seen by comparing Figure 11(a) with Figure 2(b), Figure 11(b) with Figure 4 and Figure 11(c) with Figure 6. It is particularly noteworthy to see that while the small strain moduli of the SIS/SI, SISI and SI_4/SI materials are all very similar at the typical strain rate applied during these probe tests, the level of stress required to extend the fibrillar foam varies significantly with the copolymer architecture. Such differences in the resistance to fibril

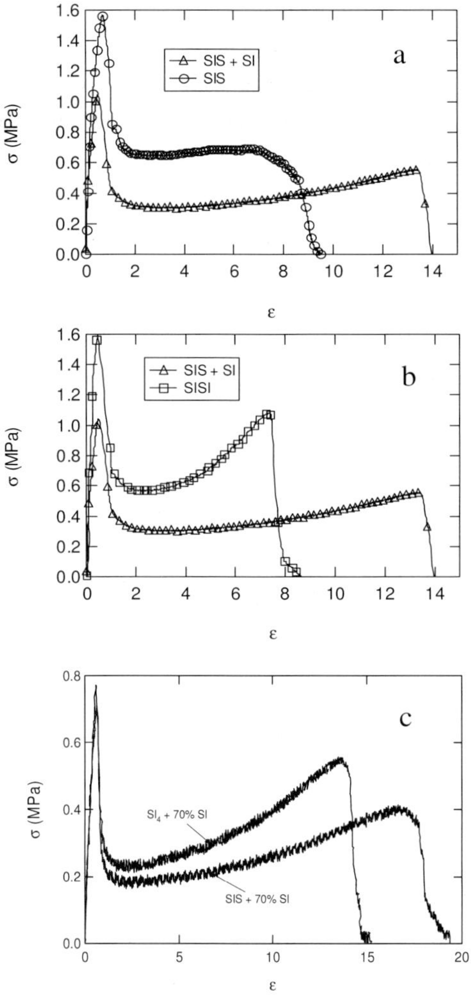

Fig. 11. Tensile part of the nominal stress vs. strain curves obtained on 100 μm thick layers of PSA with a cylindrical flat–ended steel probe. (a) Comparison of the SIS and SIS/SI blends (V = 100 μm/s). (b) Comparison of the SISI tetrablock and the SIS/SI blends (V = 100 μm/s). (c) Comparison of the SI_4 and SIS/SI blends (V = 10 μm/s). Each material contains 60% resin.

extension during debonding can lead to important differences in the adhesive properties evaluated by industry-standard methods.[9]

Figure 1 and Figure 4 show that both in small strain at low strain rate and at intermediate strain, the presence of pendant chains increases the level of dissipated energy during stretching of the material. This different level of hysteresis has an important effect on the adhesive properties. These differences become increasingly obvious when the adhesive blends are detached from surfaces where van der Waals forces are weaker, such as a polyolefin surface.[19]

Tack curves obtained for the acrylic systems are shown in Figure 12, along with representative images of the contact area. These curves were obtained by bringing a glass hemispherical indenter with a radius of 6mm into contact with the adhesive in order to establish a well-defined circular contact, and then retracting the indenter at a fixed velocity of 2.5 μm/s.[15] The confinement ratio, given by the ratio of the maximum contact radius to the undeformed adhesive thickness, is between 2 and 3 for the experiments performed on the acrylic systems, a value that is sufficiently high so that internal cavitation is just beginning to become important.[20] The stress reduction after the initial maximum in Figures 10 and 11 is associated with the release of confinement associated with cavitation. Because the initial confinement is much lower in the experiments done with acrylic systems (Figure 12) than for the experiments done with the styrene/isoprene systems, (Figures 10 and 11), the stress reduction associated with cavitation in the acrylic systems is much less pronounced. The findings for the different acrylic polymers can be summarized as follows:

- Cavities form near the point of maximum nominal stress for the DB1, TB2 and TB3 systems.

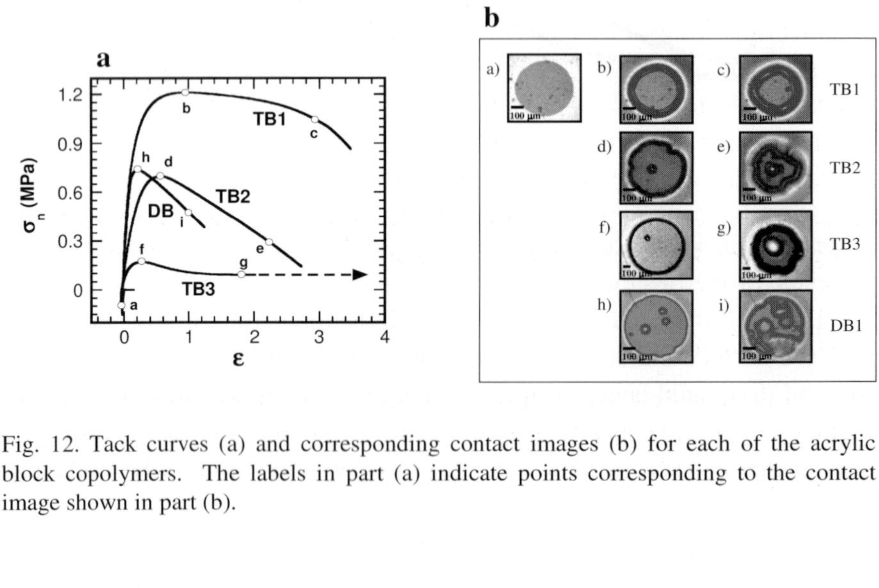

Fig. 12. Tack curves (a) and corresponding contact images (b) for each of the acrylic block copolymers. The labels in part (a) indicate points corresponding to the contact image shown in part (b).

- The cavitation stress is comparable to Young's modulus for the adhesive in each case.
- Each of the triblock materials fails adhesively at the copolymer/glass interface.
- Cavity growth is quite rapid for the diblock copolymer. Failure corresponds to cavity coalescence, and the failure of the adhesive itself is partially cohesive, results that are all consistent with the relatively low strength of a block copolymer material with no PnBA midblocks able to bridge the PMMA domains.
- Cavitation is not observed for the TB1 copolymer, which has the highest elastic modulus. For this material contact area shrinks uniformly as the glass indenter is pulled away from the adhesive layer.

Because the contact geometry for the TB1 triblock copolymer is relatively simple throughout the tack test, this sample is an excellent candidate for studies of the adhesive failure process. It was shown for this material that a single relationship existed between the contact radius and the applied displacement, independent of the load that was actually being applied to the adhesive material[15]. This surprising result illustrates the complexity of the adhesive failure process for these viscoelastic materials at large strains. While some important insights have been obtained about the relationship between the molecular properties of the adhesive and substrate and the adhesive properties of the interface, there is no well-defined detachment criteria that can be applied for adhesives that have been deformed to high strains. The concepts of linear elastic fracture mechanics, which work well for bulk elastomers and for thin adhesive layers that detach at low applied strains, cannot be applied in a straightforward manner to these situations.

3. Conclusions

The focus of this chapter has been on the relationships between the molecular architecture of block copolymers and block copolymer blends, the large strain behavior of these materials, and their adhesive performance. Examples have been provided to highlight the importance of factors affecting the large-strain response and the related tack behavior, but that have little or no effect on the linear viscoelastic properties. Specific examples for spherical domain morphologies, with spherical glassy domains in a rubber matrix, are as follows:

- Reducing the bridging fraction by the addition of diblock copolymer substantially enhances the observed strain softening, and delays the onset of strain hardening to much larger strains (Figure 2(b)). The non-recoverable deformation is also increased (Figures 7 and 8).
- The overall shape of the stress-strain curve is greatly modified if the molecular weight of the bridging chains is reduced, giving a more pronounced strain softening at higher stress levels and a strain hardening at lower strains (Figure 4 and Figure 5).
- Glassy domains behave in a way that is similar to chemical crosslinks (Figure 6 and accompanying discussion).
- The stress/strain curves measured during a tack curve (deformation of a thin film along its thickness direction by adhesive contact with an indenter) mimic the results obtained from standard tensile tests (comparison of Figure 11 and Figure 4).

The results obtained with the acrylic systems were more qualitative. The important point here is that the mechanical response and adhesive properties can be tuned over a large range from relatively subtle changes in the molecular architecture of the copolymer, even in situations where the linear viscoelastic properties do not differ substantially.

Acknowledgement

CC gratefully acknowledges the financial support of the European Commission under the GROWTH program of the 5th framework program: Project N° G5RD-CT2000-00202 DEFSAM, as well as the collaboration with Jacques Lechat, Nicolas Kappes and Ken Lewtas from

ExxonMobilChemical Europe within the framework of this project. KS acknowledges funding from the National Science Foundation, Division of Material Research.

References

1. Creton, C. *MRS Bull.*, **28**, 434, (2003).
2. Matsen, M. W.; Thompson, R. B. *J. Chem. Phys.*, **111**, 7139, (1999).
3. Daoulas, K. C.; Theodorou, D. N.; Roos, A.; Creton, C. *Macromolecules*, **37**, 5093, (2004).
4. Rubinstein, M.; Panyukov, S. *Macromolecules*, **35**, 6670, (2002).
5. Roos, A.; Creton, C. *Macromolecules*, **38**, 7807, (2005).
6. Gibert, F. X.; Marin, G.; Derail, C.; Allal, A.; Lechat, J. *J. Adhes.*, **79**, 825, (2003).
7. Derail, C.; Cazenave, M. N.; Gibert, F. X.; Marin, G.; Kappes, N.; Lechat, J. *The Journal of Adhesion*, **80**, 1131 (2004).
8. Roos, A. Sticky Block copolymers: Structure, Rheology and Adhesive Properties. Ph.D. Thesis, Université Paris VI, 2004.
9. Gibert, F.-X., Ph.D. Thesis. In Université de Pau et des Pays de l'Adour, 2001.
10. Fetters, L. J.; Lohse, D. J.; Colby, R. H., In *Physical Properties of Polymers Handbook*, Mark, J. E., Ed. American Institute of Physics: New York, 1996; pp 335.
11. Mullins, L.; Tobin, N. *Rubber Chem. Tech.*, **30**, 555, (1957).
12. Nakajima, N.; Babrowicz, R.; Harrell, E. R. *J. Appl. Polym. Sci.*, **44**, 1437, (1992).
13. Ahn, D.; Shull, K. R. *Langmuir*, **14**, 3637, (1998).
14. Fetters, L. J.; Lohse, D. J.; Milner, S. T.; Graessley, W. W. *Macromolecules*, **32**, 6847, (1999).
15. Drzal, P. L.; Shull, K. R. *J. Adhesion*, **81**, 397 (2005).
16. Crosby, A. J.; Shull, K. R.; Lin, Y. Y.; Hui, C.-Y. *J. Rheology*, **46**, 273, (2002).
17. Seitz, M. E.; Burghardt, W. R.; Faber, K. T.; Shull, K. R. *Macromolecules*, **40**, 1218, (2007).
18. Flanigan, C. M.; Crosby, A. J.; Shull, K. R. *Macromolecules*, **32**, 7251, (1999).
19. Creton, C.; Roos, A.; Chiche, A., In *Adhesion: Current Research and Applications*, Possart, W. G., Ed. Wiley-VCH: Weinheim, 2005; pp 337.
20. Crosby, A. J.; Shull, K. R.; Lakrout, H.; Creton, C. *J. Appl. Phys.*, **88**, 2956, (2000).

CHAPTER 10

STABILITY AND DEWETTING OF THIN LIQUID FILMS

Karin Jacobs

Dept. of Experimental Physics, Saarland University, D-66123 Saarbrücken, Germany

Ralf Seemann

Dept. of Experimental Physics, Saarland University, D-66123 Saarbrücken, Germany and Max-Planck-Institute for Dynamics and Self-Organization, D-37073 Göttingen, Germany

Stephan Herminghaus

Max-Planck-Institute for Dynamics and Self-Organization, D-37073 Göttingen, Germany

The stability of thin liquid coatings is of fundamental interest in everyday life. Homogeneous and non-volatile liquid coatings may dewet either by heterogeneous nucleation, thermal nucleation, or spinodal dewetting. Wetting and dewetting is explained on a fundamental level, including a discussion of relevant interactions. This chapter will also address the various dewetting scenarios and explain how the effective interface potential governs the behavior obtained for various stratified substrates and film thicknesses.

1. Introduction

'To wet or not to wet'[1] is the question this chapter tries to answer. It concerns the fundamental aspects of the stability of thin liquid films or coatings. The liquid in question can be any homogeneous liquid,[a] such as, e.g., an aqueous solution, an oil, or a polymer melt. Whether or not

[a]Homogeneous on the length scale set by the thickness of the liquid film under consideration. In particular, demixing or segregation effects are thereby assumed to be negligible.

a liquid wets a given surface is not only of fundamental interest, but also of substantial technical importance. The authors have been approached by numerous companies with interesting problems concerning wettability, ranging from the automotive industry, the pharmaceutical and chemical industry, to the food processing, printing, and textile industry. In principle, the general answer to all wettability problems is simple: a liquid wets a surface if it can gain energy by enlarging the interface with that substrate. To provide this answer for a specific system under study, however, may be quite cumbersome. For the sake of clarity, we restrict our considerations to liquids on solid, homogeneous substrates.

2. Experimental Model Systems for Simple Liquids

In order to study the basic mechanisms of a certain class of phenomena, it is of central importance to identify suitable model systems. Since the main problems in the controlled preparation of thin films and surfaces lies in their inherent propensity to become contaminated by dust or other impurities, the first quantitative studies of surface forces in wetting films have been performed with cryogenic systems, in which impurities are naturally frozen out.[2–4] However, it has been realized that these systems involve strong exchange of material between liquid and vapor phase and thus are not well suited for studies of dynamic aspects, such as dewetting and structure formation phenomena.

Thin laser-annealed metal films were candidates,[5,6] yet at the same time, techniques of preparing systems which involve more complex liquids, such as polymer melts, in a well-defined and clean way became progressively available. As a result, the numerous benchmark studies of the basic mechanisms of dewetting have been performed with polymer melts (see e.g. the review articles of D.G. Bucknall[7] and P. Müller-Buschbaum[8]). Systematic experimental studies were begun around the 1990s[9–13] inspired by P.-G. de Gennes' theoretical work.[14,15] Polymer melts are on the one hand close to application (coatings, photo resist), yet on the other hand easily controllable in the experiments. Polymers such as polystyrene (PS) are very suitable model liquids since they have a very low vapor pressure in the melt, and mass conservation can safely be assumed. Moreover, they are chemically inert, non-polar, and their dynamics can be tailored by choosing different chain lengths and annealing temperatures. For molecular weights below the entanglement length ($\sim$17 kg/mol), the melt can safely be treated as simple (Newtonian) liquid for the low shear rates in dewetting experiments.[16,17]

Below the glass transition temperature T_g, the films are glassy and can be stored for subsequent analysis.[b] PS in atactic steric configuration moreover does not exhibit any tendency for crystallization.

As substrates, Si wafers with their natural amorphous Si dioxide layer (SiOx) have been widely used. The present chapter thus concentrates on Si wafers as substrates, which can be purchased with a very small surface roughness (rms roughness smaller than 0.2 nm). The wettability of these substrates can be tuned by chemically grafting monomolecular functional layers onto the native silicon oxide overlayer. For instance, hydrophobization is often achieved by grafting a self-assembled monolayer of octadecyltrichlorosilane[25,26] onto the wafer.

For dewetting experiments it is necessary to prepare a thin liquid film in a non-equilibrium state on a substrate. Usually, a thin polymer coating is prepared from a solvent solution by one of the following standard techniques: spin coating, dip coating, or spraying.[c] The solvent evaporates during the preparation procedure, leaving behind a smooth, glassy polymer layer. By spin coating a solution of polystyrene in toluene one can easily achieve a PS layer of thickness in the range of a few nanometers up to several micrometers. The roughness of the polystyrene layer is then similar to the underlying substrate. For experimental details and caveats of preparation, the reader is referred to the pertinent literature.[10,13,16,27–30] The thickness of the films can be determined by standard techniques, such as ellipsometry.[31] The dewetting process can be studied by optical microscopy and/or atomic force microscopy (AFM). Most studies use AFM in non-contact mode operation to avoid any damage of the soft surface, which even allows to image the dewetting scenario *in situ*.[32,33]

To induce dewetting, the films are heated above the glass transition temperature. Figure 1 shows a series of optical micrographs of an 80 nm thick PS film of molecular weight of 65 kg/mol ('PS(65k)') dewetting a silanized Si wafer. The series depicts a pattern formation process that is typical for most dewetting films, whether for a water film on a waxed surface, or a coating on a dusty substrate.

The process of dewetting can be divided into three stages: in the early stage, holes are generated by a rupture process, c.f. Fig. 1(a); in the in-

[b]It is important to note that T_g of thin films can be substantially different (mostly lower) than the bulk value.[18–24]

[c]For all these techniques, the surface to be coated must be wettable by the polymer solution. Otherwise, the film must be prepared on another substrate, e.g. mica, and then transferred to the surface of interest.

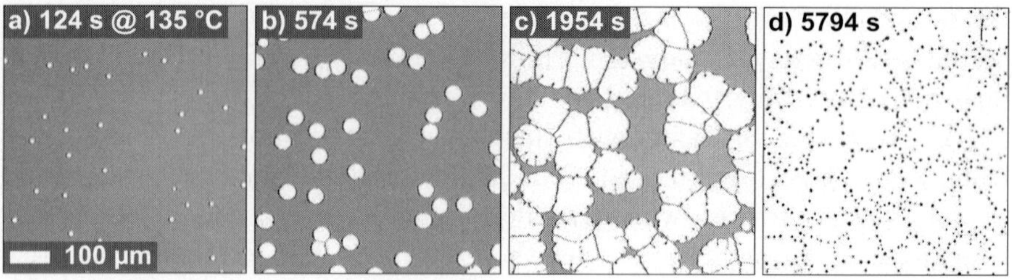

Fig. 1. Pictures series taken by a light microscope: a 80 nm thick polystyrene film of 65 kg/mol molecular weight is dewetting at 135°C from a hydrophobized silicon substrate for approximately (from left to right) 2 min, 10 min, 30 min, and 100 min.

termediate stage, the radius of the holes increases, leading to hole coalescence, c.f. Fig. 1(b)–(c). In the intermediate stage, the focus is on the dynamics involved in the dewetting process, its impact on the hole profiles and on its influence on dewetting patterns. From the dynamics of hole growth[27,28,34–40] as well as from the shape of the liquid rim surrounding the hole one can get information about the slip or no-slip boundary condition of the liquid close to the solid substrate.[17,41,42] In the late stage, the straight ribbons that separate two coalescing holes decay into droplets due to the Rayleigh-Plateau instability,[43] c.f. Fig. 1(c)–(d). Slight differences in the size of the droplets cause slight pressure differences, leading small droplets to shrink and large droplets to grow (Ostwald ripening[44]). Due to the small vapor pressure and the low mobility of the polymer molecules, the ripening process is extremely slow. Most experiments are stopped before the 'final state', i.e. one single drop on the surface, is reached. In what follows, we focus on the initial stage of dewetting: why does the polystyrene film dewet the hydrophobized Si wafer at all? The next section will be dedicated to the discussion of the various energy contributions involved and their mutual balance.

3. The Energy Balance

In a Teflon© coated frying pan, a stable oil layer can only be achieved if one pours enough oil into the pan. Gravity then stabilizes the thick oil film. If the oil film thins well below the so-called capillary length λ_{cap}, capillary (intermolecular) forces dominate over gravitational forces and the film may dewet.[45] The capillary length is given by $\lambda_{cap}=\sqrt{\sigma_{lv}/\rho g}$ (for olive oil: $\lambda_{cap} \approx 1.7$ mm, for water: $\lambda_{cap} \approx 2.7$ mm), where σ_{lv} is the liquid/vapor surface tension and ρ the density of the liquid. Therefore, dewetting of liquid films driven by intermolecular forces can only occur for 'thin' films, i.e. films thinner than λ_{cap}.

A liquid film can be metastable and then needs a nucleus to induce dewetting. In the following we will focus on liquid films much thinner than λ_{cap} and discuss under which conditions a thin liquid layer will be stable, metastable or unstable on top of a substrate.

3.1. *The Young equation*

Dewetting is a dynamic process that begins in a non-equilibrium situation,[14,46–52] namely the flat film on the surface and ends when reaching an equilibrium state: one droplet or a set of droplets.[d] Thus, let us first start with a droplet, c.f. Fig. 2:

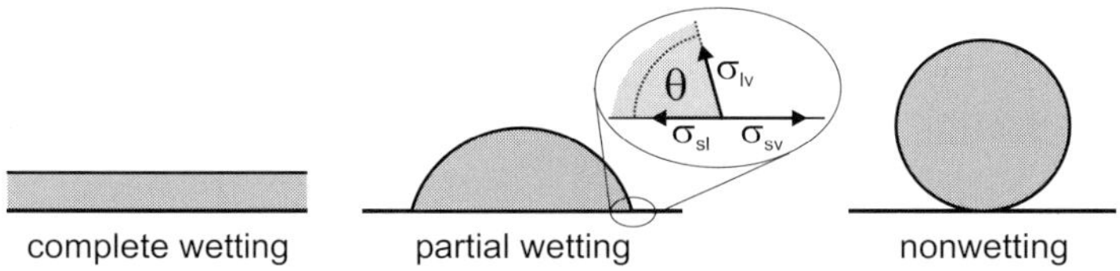

Fig. 2. Sketch of a liquid drop atop of a solid substrate. Complete wetting is characterized by a contact angle $\theta = 0$, partial wetting by $0 < \theta < \pi$, and nonwetting by $\theta = \pi$ (from left to right)

A droplet on a homogeneous surface usually exhibits the form of a spherical cap and the tangent to the droplet at the three phase contact line includes an angle θ with the substrate. In equilibrium, the angle is given by a balance of macroscopic forces:

$$\cos\theta = \frac{\sigma_{sv} - \sigma_{sl}}{\sigma_{lv}} \tag{1}$$

This is the Young equation[46] of 1805, where σ_{sv} and σ_{sl} are the solid/vapor and solid/liquid interfacial free energies, and σ_{lv} is the liquid/vapor interfacial free energy (or tension).

For $\theta = 0°$, i.e. $\sigma_{sv} - \sigma_{sl} \geq \sigma_{lv}$ the droplet will spread and will completely wet the substrate. For $0° < \theta < 180°$ one speaks of partial wetting, and for $\theta = 180°$ of non-wetting. All three cases are sketched in Fig. 2. In

[d]Equilibrium situation actually is one single droplet, yet it normally takes too long to reach this state. Before, a network of droplets is formed.

case of water (resp. oil), a surface is termed 'hydrophilic' ('oleophilic') if $0° \leq \theta < 90°$ and 'hydrophobic' ('oleophobic') for $90° \leq \theta \leq 180°$.[e] In other words, the contact angle in Young's equation is determined by the free energies of interfaces between semi-infinite media.

This is, however, not the full story. Although a first glance at Fig. 2 suggests that the free energy of a homogeneous film should be written as $\sigma_{\text{film}} = \sigma_{sl} + \sigma_{lv}$, this approach neglects possible interactions of the two interfaces (solid/liquid and liquid/vapor) with each other, across the liquid film. Such interactions come up e.g. by virtue of van der Waals forces between the involved materials, which may span several tens of nanometers. In the vicinity of the three phase contact line, the interfaces can thus deviate significantly from a straight intersection at Young's angle (cf. Fig. 3). We will come back to this point later, after having discussed in some detail the long-range forces of a stratified system.

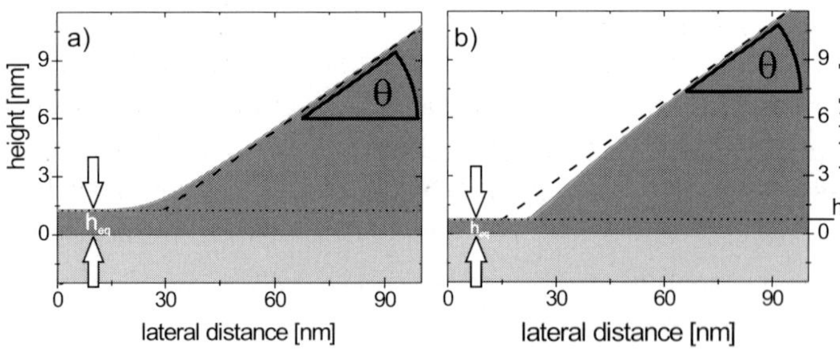

Fig. 3. Droplet profiles in the vicinity of the three phase contact line as expected for (a) an unstable and (b) a metastable situation. The profiles have been calculated for the case of (a) a $d_{SiOx} = 191\ nm$ on Si and (b) a $d_{SiOx} = 1.7\ nm$ SiOx layer on Si, the potentials of which are shown in Fig. 7. The dashed line marks the macroscopic contact angle $\theta = 7.5°$ for a PS drop and the arrows mark the equilibrium wetting layer thickness h_{eq}.

These forces are of similar importance for predicting the stability of a liquid film. Their dependence on film thickness determines not only the linear stability against small perturbations of the free surface, but also the scenario by which the initially uniform film is transformed into its equilibrium state, which consists of droplets of contact angle θ on the surface.[53–55] For example, a droplet of photo resist on a semiconductor may exhibit an equilibrium contact angle $\theta = 20°$ and therefore wets the surface only par-

[e]It should be noted that non-wetting has never been observed on flat surfaces, although some systems (like, e.g., mercury on glass) come quite close.

tially. What does this mean for a photo resist film that was prepared by a non-equilibrium technique (spin coating) on top of the semiconductor? According to Young's equation, the film is not stable. How will the film decay into droplets and at what speed will this process take place at a given temperature? Will there be time to dry or cure the photo resist before it dewets?

Depending on the thickness of the photo resist, the answers will be different, though Young's contact angle θ is always the same! Macroscopic and molecular terms describing stability conditions often were mixed up and have led to confusion. In what follows, we shall discuss which forces are to be considered, and what typical length scales are involved.

3.2. *The effective interface potential*

Following the preceding section, we *define* the so-called effective interface potential $\phi(h) \equiv \sigma_{\text{film}} - \sigma_{sl} - \sigma_{lv}$. It comprises both short-range and long-range interactions and is defined as the excess free energy per unit area which is necessary to bring two interfaces from infinity to a certain distance. From its very definition, it is clear that $\phi(h) \to 0$ as h tends to infinity. The excess free energy of an infinitely thick film is thus given by the sum of the free energies of its two interfaces, in accordance with intuition.

For dielectric systems, there are only two relevant types of interactions, steric repulsion and van der Waals forces.[f]

$$\phi(h) = \phi(h)_{steric} + \phi(h)_{vdW} \tag{2}$$

In Fig. 4 three typical curves of $\phi(h)$ are sketched to illustrate the general principle. Line (1) characterizes a film that is stable on the substrate, since energy would be necessary to thin the film. The equilibrium film thickness is infinite. The two other curves exhibit a global minimum of $\phi(h)$ at $h = h_{eq}$: Curve (2) characterizes a film that is unstable, whereas line (3) describes a metastable film.

It is readily shown by a linear stability analysis[56–58] that if the second derivative of $\phi(h)$ with respect to h is negative ($\phi''(h) < 0$), unstable modes exist whose amplitudes grow exponentially according to $\exp(t/\tau)$, where τ is the growth time that is characteristic for the respective mode. Furthermore, there is a characteristic wavelength λ_s of these modes the amplitude of

---

[f] A detailed description for numerous different situations involving e.g. polar molecules or hydrogen-bonds or the interactions of a colloidal sphere interacting with a metallic surface can be found in the textbook of J. Israelachvili.[48]

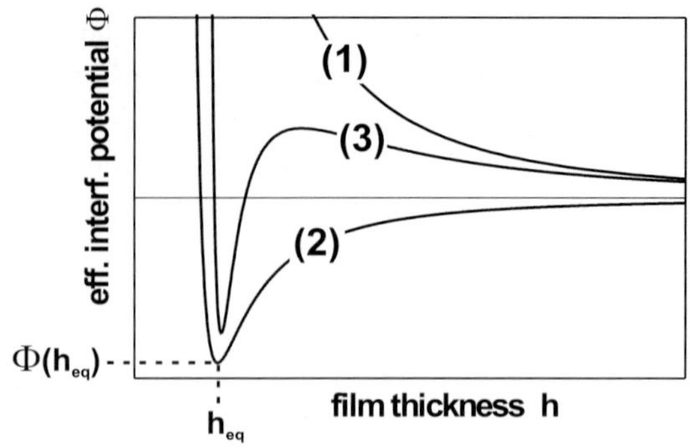

Fig. 4. Sketch of the effective interface potential ϕ as a function of film thickness. Line (1) denotes the stable case, line (2) the unstable one and curve (3) the metastable case. The thickness of the stable wetting layer is termed $h = h_{eq}$, and is typically in the order of some nm, the depth of the global minimum of ϕ is named $\phi(h_{eq})$.

which grows fastest and will therefore dominate the emerging dewetting pattern.

The linear stability analysis also reveals that the spinodal wavelength λ_s is linked to the (second derivative of the) effective interface potential:[56,57]

$$\lambda_s(h) = \sqrt{\frac{-8\pi^2\sigma_{lv}}{\phi''(h)}} \,. \tag{3}$$

The spinodal wavelength λ_s is the key to clearly identify a spinodal dewetting process in experimental systems.[g]

One question is now to be answered: What links the effective interface potential to Young's contact angle? The effective interface potential can describe a non-equilibrium situation, yet Young's contact angle is only defined for the equilibrium. A. Frumkin has published in 1938 that[60]

$$\frac{\phi(h_{eq})}{\sigma_{lv}} = \cos\theta - 1 \,. \tag{4}$$

What is the consequence of Eq. (3) and Eq. (4)? Determining the spinodal wavelength λ_s as a function of film thickness h enables us to gain insight

[g]The name 'spinodal dewetting' has been coined due to the analogy to spinodal decomposition of a blend of incompatible liquids, which occurs if the second derivative of the free energy with respect to the composition is negative.[59] As in dewetting, this mechanism is clearly distinct from heterogeneous nucleation, in which randomly distributed impurities dominate the emerging pattern.[13]

into the course of $\phi''(h)$. By additionally measuring the equilibrium layer thickness h_{eq} and the contact angle θ, it is possible to reconstruct the complete effective interface potential,[6,29,30] c.f. Fig. 7, as will be shown later.

The steric and the van der Waals part of the effective interface potential are characterized by different exponents and different interaction constants, therefore we will discuss the interactions separately:

Steric repulsion and chemical interactions are relevant only within a few Ångstroms of film thickness, and the resulting force is therefore termed 'short-range force'. The repulsion is due to overlapping electron shells and is typically described by a higher-order polynomial function and varies as $1/h^{12}$, where h is the distance between the interacting bodies. Considering two planar surfaces, this repulsion yields an interaction energy varying as

$$\phi(h)_{steric} = \frac{C}{h^8}\,, \tag{5}$$

where C is a constant characterizing the interaction strength.[h] The reason for the lower exponent in the flat film geometry lies in fact that mutual interaction between all involved atoms or molecules have to be considered. For an extended derivation of the equations we like to refer to J. Israelachvili's approach.[i]

Van der Waals interactions characterize attractive intermolecular forces of quantum-electrodynamic origin. They arise from the variations of the zero-point energies of the collective electromagnetic modes of the system under study. The van der Waals energy between two molecules turns out to vary as $1/h^6$ in the non-retarded approximation. This approximation holds as long as the lateral dimensions of the system are much smaller than the wavelength of the electromagnetic fields at the dominant excitation energies. Considering retardation effects, the interaction falls off as $1/h^7$. In the following, we use only non-retarded potentials since in the experimental systems, effects due to retardation are mostly camouflaged by experimental errors. Nevertheless, in some special cases including cryogenic systems, retardation effects have been reported[2] and a recent theoretical study[61] discusses their possible influence on thin films in the context of the full theory of Dzyaloshinskii, Lifshitz, and Pitaevskii.[62]

[h] This definition of $\phi(h)_{steric}$ here is such that C is positive and thus the interaction always repulsive.

[i] Here, chapter 10 in the textbook of J. Israelachvili[48] is very helpful.

Considering again two planar surfaces, the non-retarded interaction yields[48,63]

$$\phi(h)_{vdW} = -\frac{A}{12\pi h^2}\,, \tag{6}$$

where A is the Hamaker constant[64] and $\phi(h)$ is the energy per unit area. A retarded interaction is described as $\phi(h) \sim A/h^3$.

The van der Waals forces of two media interacting through vacuum are always attractive, and A is positive. If the vacuum yet is replaced by a third medium (e.g. a liquid film), things get more involved: In a system consisting of three media $m_1/m_3/m_2$, it may happen that m_1 attracts m_3 stronger than it attracts m_2. Thus m_2 is effectively 'repelled' by m_3, this means that the Hamaker constant is negative. To describe the entire system, the Lifshitz theory[65] has to be applied. Here, the interacting bodies are treated as continuous media and the atomic structure is ignored.[48,62,65] Instead, bulk properties as the dielectric constants and the refractive indices are used to calculate the Hamaker constant.

For the sake of completeness, we cite here the formula[j] for the Hamaker constants from the book of J. Israelachvili,[48] which has proven to be very useful. It is valid for two media 1 and 2, interacting through medium 3. All media are taken as being dielectric with a single electronic absorption frequency ν_e, which is typically in the range of $3 \cdot 10^{15}$ Hz; n is the refractive index of the medium in the visible and ϵ the dielectric constant (i.e. $n_i^2 = \epsilon_i$, taken in the visible spectral range)

$$\begin{aligned} A &= A_{\nu=0} + A_{\nu>0} \\ &\approx \frac{3}{4}kT\left(\frac{\epsilon_1-\epsilon_3}{\epsilon_1+\epsilon_3}\right)\left(\frac{\epsilon_2-\epsilon_3}{\epsilon_2+\epsilon_3}\right) \\ &\quad + \frac{3h\nu_e}{8\sqrt{2}}\frac{(n_1^2-n_3^2)(n_2^2-n_3^2)}{\sqrt{(n_1^2+n_3^2)}\sqrt{(n_2^2+n_3^2)}\{\sqrt{(n_1^2+n_3^2)}+\sqrt{(n_2^2+n_3^2)}\}} \end{aligned} \tag{7}$$

The Hamaker constant A will then give the strength of the van der Waals forces between the two interfaces solid/liquid and liquid/air.

For air as medium 1, Si as medium 2 and polystyrene as medium 3, Eq. 7 gives[k] $A_{Si} = -2.2(5) \cdot 10^{-19}$ J, replacing Si as medium 2 by silicon dioxide, the Hamaker constant is $A_{SiOx} = 1.8(4) \cdot 10^{-20}$ J. For a silane

[j] Confer chapter 11, Eq. 11.13 of Ref. 48.

[k] The error of the Hamaker constant is a result of the uncertainties of the optical properties of the involved materials as found in literature. The error for the last digit is given in brackets.

(OTS) covered Si wafer, the Hamaker constant for the interaction with PS is calculated to be $A_{OTS} = 1.9(3) \cdot 10^{-20}$ J. For $A < 0$ ($A > 0$), the system can gain energy by enlarging (reducing) the distance h between the surfaces, in other words, a polystyrene film of thickness h is stable on Si, since $A < 0$.

In Fig. 5, the dotted line represents the van der Waals potential, $\phi(h)_{vdW}$, as given by Eq. 6 for a polystyrene film on a Si wafer, where A_{Si} is negative. The potential therefore is positive, purely repulsive[1] and the PS film will be stable. However, $\phi(h)_{vdW}$ for PS on an infinitely thick SiOx layer (solid line) is always negative, purely attractive and the PS film will be unstable, since the Hamaker constant A_{SiOx} is positive.

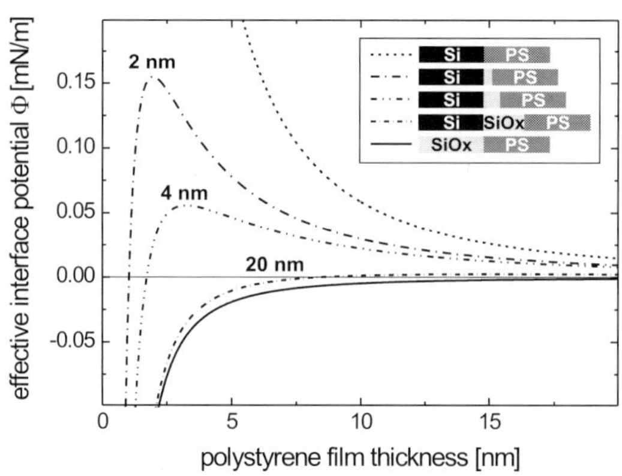

Fig. 5. Long-range part of the effective interface potential $\phi(h)$ as function of PS film thickness h for different SiOx layer thickness ranging from 0 nm (dotted line) to infinity (solid line), calculated with the formula given in Eq. (8). The Hamaker constants were calculated from the optical constants of the involved materials.[48]

Due to the $1/h^2$ dependence of the potential, the van der Waals forces are long-range forces and act significantly on distances up to about 100 nm. In stratified systems with more than one layer between the two half spaces of media 1 and 2, all mutual interactions have to be considered. The exact calculation of the van der Waals potential may therefore be quite cumber-

[1]The terms 'attractive' and 'repulsive' can be misleading if thinking of the thin film, yet the terms are chosen for a system of two media (in our case solid and air) interacting through a third one, here the liquid layer. Hence attractive and repulsive are meant for air being attracted to the solid surface or 'repelled'. In case it is repelled, the effective interface potential is repulsive for the respective thickness of the liquid layer and the liquid wets the solid.

some. This is an important point, since there is a number of publications that disregard this aspect and claim that theory and experiment do not match. A system like a PS film in air, spun on a thin silicon oxide layer on top of a Si wafer involves two thin layers (three interfaces, abbreviated as air/PS/SiOx/Si) and can be described by two Hamaker constants. So far, no discrepancies between theory and experiment have ever been observed when the formulas given above had been used. One of the most impressive successes is the quite accurate prediction of the critical wetting temperature of pentane on water.[66]

Assuming additivity of forces, the system air/PS/SiOx/Si can be characterized by a summation of the van der Waals contributions of each of the single layers with thickness h of the PS and d_{SiOx} of the SiOx layer:

$$\phi_{vdW}(h) = -\frac{A_{SiOx}}{12\pi h^2} + \frac{A_{SiOx} - A_{Si}}{12\pi(h + d_{SiOx})^2}\,, \tag{8}$$

where A_{SiOx} and A_{Si} are the Hamaker constants of the respective system air/PS/SiOx and air/PS/Si. Although not exact,[m] the concept allows to calculate the interaction energies even for stratified systems. With the help of Eqs. (7) and (8), the van der Waals potential of the experimental system is accessible, if the SiOx layer thickness is known and the Hamaker constants are calculated as shown before. Figure 5 depicts the van der Waals potential as gained from Eq. (8). Clearly, the potential is influenced by the thickness of the silicon dioxide layer.

So the widely used system air/PS/OTS/SiOx/Si can be described by

$$\phi_{vdW}(h) = -\frac{A_{OTS}}{12\pi h^2} + \frac{A_{OTS} - A_{SiOx}}{12\pi(h + d_{OTS})^2} + \frac{A_{SiOx} - A_{Si}}{12\pi(h + d_{SiOx} + d_{OTS})^2}\,, \tag{9}$$

yet the long-range potential ϕ_{vdW} is not much different to a system with an SiOx layer of thickness $d_{OTS} + d_{SiOx}$, since the optical properties of the OTS and the SiOx layer are very similar. The difference in wetting behavior will only be obvious if the short-range potential is added.

To conclude so far, $\phi(h)$ combines short- **and** long-range interactions. Short-range interactions are difficult to quantify, yet van der Waals interactions can be captured by the optical properties of the involved materials,

[m] It should noted here that numerous studies have tried to use only one so-called 'effective Hamaker constant' for stratified systems using combining rules of the form $A_{132} \approx (\sqrt{A_{22}} - \sqrt{A_{11}})(\sqrt{A_{22}} - \sqrt{A_{33}})$. However, these rules can only achieve good results under severe limitations, e.g., if the zero-frequency contribution $A_{\nu=0}$ in Eq. (7) is negligible.[48] Strictly speaking, they are plain wrong, and should be used, if at all, with utmost care. It is not surprising that many of the studies which have used them could not reconcile the theoretical expectation with the experimental results.

even for stratified systems, where the suitable van der Waals potential must be taken. If $\phi''(h) < 0$, spinodal dewetting is possible and a characteristic wavelength λ_s should be found in the dewetting pattern.

4. Experiments: Linking the Effective Interface Potential to Macroscopic Properties

In this section, experimental results shall be compared to the predictions made by the effective interface potential. Typical experimental dewetting patterns in the system air/PS/SiOx/Si are shown in Fig. 6. What can we learn from analyzing the dewetting films? I) By AFM, the contact angle θ of droplets in the late stage of dewetting can be measured. II) The equilibrium film thickness h_{eq} can be determined, e.g. by ellipsometry. III) The spinodal wavelength λ and IV) the growth time τ of the spinodal pattern can be determined. V) The morphology of the pattern. VI) The growth of the size of the holes and VII) the shape of the liquid front as function of time.

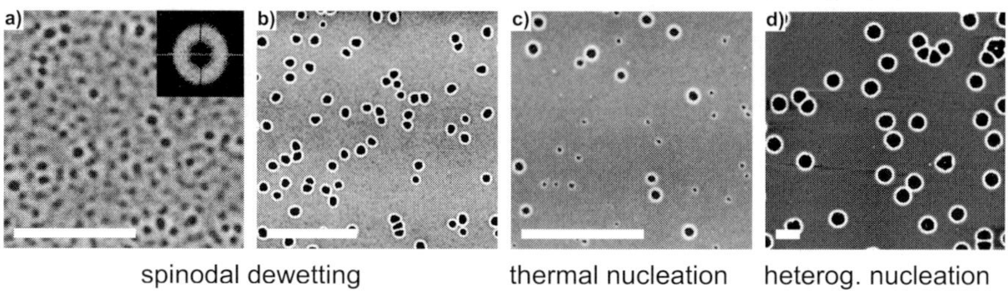

Fig. 6. (a)–(d) AFM images of dewetting PS(2k) films on Si wafers with variable Si dioxide layer and PS film thickness. Scale bars indicate 5 μm, z-scale ranges from 0 (black) to 20 nm (white): (a) 3.9 nm PS film on a Si wafer with d_{SiOx} =191 nm. The inset shows a Fourier transform of the image. (b) 3.9(2) nm PS, (c) 4.1 nm PS and (d) 6.6 nm PS on Si wafers with d_{SiOx} =2.4 nm. (The statistical analysis of the distribution of hole sites in cases (b) to (d) was performed on larger sample areas.)

Dewetting patterns like the one in Fig. 6(a) clearly exhibit the spinodal wavelength. Figure 7(a) comprises experiments with different PS film thicknesses on Si wafers with d_{SiOx} = 2.4 nm (open symbols) and with d_{SiOx} = 191 nm (filled symbols). From the data of $\lambda(h)$ data points for $\phi''(h)$ can be gained. Now, fitting the second derivative of Eq. (8) to the data, the Hamaker constants and the short-range interaction constants can be obtained as fit parameters and the entire effective interface potential can be inferred. It is plotted in Fig. 7(b).

The best fits[29,30] are achieved for $A_{Si,fit} = -1.3(6) \cdot 10^{-19}$ J and $A_{SiOx,fit} = 2.2(4) \cdot 10^{-20}$ J. The values for A nicely match the values calculated from optical properties of the media involved, as described by Eq. 7. For C we find $C_{SiOx,fit} = 6.3(1) \cdot 10^{-76}$ Jm^6 and $C_{OTS,fit} = 2.1(1) \cdot 10^{-81}$ Jm^6. In the cited references, a detailed description of the fitting and reconstruction procedure can be found.

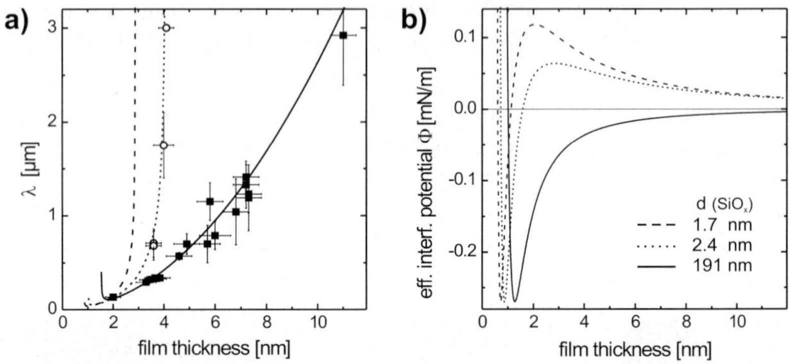

Fig. 7. (a) Spinodal wavelength as function of PS film thickness on Si wafers with $d_{SiOx} = 2.4$ nm (open symbols) and with $d_{SiOx} = 191$ nm (filled symbols). (b) The effective interface potential $\phi(h)$ for three different SiOx layer thicknesses.

The second derivative $\phi''(h)$ is plotted in in Fig. 8(a) for three SiOx layer thicknesses. An arrow marks the zero in the respective second derivative, corresponding to the relevant inflection points of the curves displayed in Fig. 7(b). Spinodal dewetting is possible only for films thinner than indicated by the arrows: PS films on Si wafers with an 1.7 nm (2.4 nm) thick oxide layer are unstable for PS film thicknesses below 3 nm (4 nm). For larger thicknesses, the films are metastable and need to overcome a potential barrier in order to finally reach the global minimum of ϕ. The maximum in ϕ, which is visible in Fig. 7(b) for the broken curves, represents only part of this barrier, since a nucleus for dewetting as it is formed, e.g., by thermal activation, is a localized structure and thus involves excess surface energy as well.[68] It is readily checked that the energy required to form such nucleus is in almost all cases large as compared to kT, such that thermal nucleation plays no role in systems of practical interest. This can be illustrated nicely

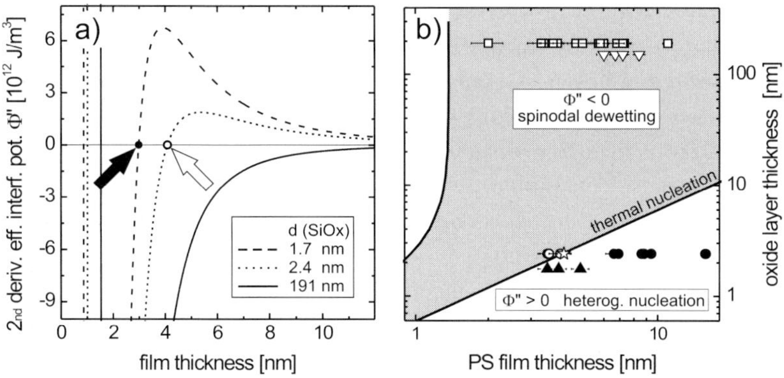

Fig. 8. (a) Second derivative of the effective interface potential as function of film thickness for three different SiOx layer thicknesses as fitted to the data of Fig. 7. (b) Stability diagram of PS films on top of Si wafers with variable oxide layer thickness.[29]

with the PS films, as shown in Fig. 8(b). PS films on thick 191 nm SiOx (solid line in Fig. 7(a) are unstable for all relevant film thicknesses. On 2.4 nm of oxide, however, a sign reversal is observed at a film thickness of about 4 nm (white arrow in Fig. 8(a). Only very close to this point, the height of the potential barrier vanishes, and homogeneous nucleation by thermal activation is possible.[45,68,69] However, nucleation can as well, and usually does, proceed by means of localized defects in the film. A defect, be it in the molecular texture of a polymeric film or just a small dust particle, may remove the potential barrier locally and thus induce dewetting. This rupture mechanism is termed 'heterogeneous nucleation',[59,67] in analogy to defect-mediated nucleation in demixing scenarios.

The fact that the effective interface potential can be directly inferred from macroscopic quantities, such as the spinodal wavelength, is one reason why experimentalists are seeking for regimes of spinodal dewetting in thin liquid film systems. However, if a company is asking why their coating does not stay stable on a surface, it is convenient to rather have a stability diagram which allows to look up where the system will be in a metastable or an unstable state. A stability diagram for the system air/PS/SiOx/Si is shown in Fig. 8(b). The solid lines of the diagram are based on Eq. 2 with a long-range potential as in Eq. 8 and separate the spinodal dewetting (unstable) regime, where $\phi''(h) < 0$, from in the regime of heterogeneous nucleation, characterized by $\phi''(h) > 0$ (metastable regime). Thermal nucleation is possible for $\phi''(h) = 0$, this line separates the two regimes. Experiments that exhibit spinodal dewetting patterns are indicated by open symbols, those of heterogeneous nucleation (randomly distributed holes) are indicated by solid symbols.[n] A star marks the set of parameters where thermal nucleation was observed. Note that in the unstable regime, heterogeneous nucleation from localized defects is also possible and indeed is sometimes observed. In fact, it usually dominates if the experiments are not performed in extremely clean conditions. Spinodal dewetting, however, can only take place in the unstable regime.

We can state here that the effective interface potential indeed is the key for wetting properties of liquids on surfaces. It can easily be obtained via the optical properties of the involved media and by taking into account the adapted van der Waals potential for stratified systems. It is worth noting that metastable potentials can only be present if there is more than one Hamaker constant involved, one with negative and one with positive sign.

Coming back to the company — a stability diagram for their specific system will reveal the way to a solution of their wettability problem: If their coating system is located in the metastable regime (in the example shown, it depends on the silicon dioxide thickness and the polystyrene thickness), the advice is to reduce the number of possible nucleation centers, e.g. by creating a cleaner environment. If the coating system yet is in the unstable regime, the entire composition of the substrate needs to be reconsidered.

As mentioned at the beginning, the shape of a droplet close to the three phase region is influenced by the long-range intermolecular interactions. By AFM, this region can be characterized. For a detailed discussion we like to refer to the work of T. Pompe et al.[71,72] In Fig. 3 two profiles are sketched that were calculated based on the knowledge of the effective interface potential. Figure 3(a) is a characteristic curve for an unstable potential as shown for $d_{SiOx} = 191\ nm$ in Fig. 7(a), whereas a profile like Fig. 3(b) is expected for a metastable potential like for $d_{SiOx} = 1.7\ nm$. The experimental AFM results (not shown here) corroborate the theoretical expectations.[73]

One question is open up to now: are spinodal dewetting patterns always as easy to detect as in Fig. 6(a)? The answer is 'no': patterns which at first glance appear like the one shown in Fig. 6(b) can be generated due to a spinodal process as well.[6] The next section we will briefly explain how

[n] As explained above, thermal nucleation can be only observed if the nucleation barrier is of order kT. On the scale of Fig. 8(b), this is fulfilled only in a region smaller than the width of the line denoted 'thermal nucleation'. Alleged observations of thermal nucleation in a wider range[70] appear very questionable.

to distinguish a dewetting pattern of a system in the metastable state from one in the unstable regime.

5. Characterizing Experimental Dewetting Patterns

The three rupture scenarios **spinodal dewetting, homogeneous nucleation**, and **heterogeneous nucleation** give rise to specific dewetting patterns. Vice versa, characterizing the experimental dewetting pattern can help to identify the rupture mechanism and to infer the effective interface potential. Figure 1 and Fig. 6 show experimental examples of films dewetting via different rupture mechanisms. Theoretically, the distinction between nucleation and spinodal dewetting appears quite clear: Vrij[56] proposed already in 1966 that a **spinodal rupture** of a free liquid film results in a dewetting pattern of 'hills and gullies' with a preferred distance λ_s after a certain rupture time τ.

Experimentally, the rupture time τ is difficult to measure since the hole which forms as the film has thinned to zero thickness must have a certain size to be observable. Experimentalists thus concentrated instead on the evidence of a preferred wavelength λ_s observable in their systems.[o] If, however, the holes are randomly (Poisson) distributed, they are assumed to stem from **heterogeneous nucleation**, reflecting the fact that defects typically exhibit random statistics. Although it was generally accepted later that (heterogeneous) nucleation from localized defects is the reason for the dewetting scenario shown in Fig. 1(a), the very nature of the nucleation defect mostly remains unclear: it might be dust particles or any other chemical or physical inhomogeneities. In some holes, by light microscopy or atomic force microscopy (AFM), a nanoscopic object could be observed right in the center of each hole. Assuming the object to be a dust particle, the first trial was to reduce the number of nucleation centers by improving the preparation conditions. The number of holes, however, could not be reduced below a certain level, which suggests that the physics of hole nucleation in polymer films may be deeper than a mere effect of 'dirt'. It has been shown experimentally that stress inside the thin films (stretched

[o]Note that a preferred hole distance λ_s has to be distinguished from a mean hole distance: In some studies it was found that the mean hole distance scaled with the film thickness as expected for spinodal dewetting.[10,74] Therefore, the scenario shown in Fig. 1(a) was often regarded as a typical spinodal dewetting scenario. However, it turned out later that this was not correct:[13] first of all, the system was metastable for the film thicknesses studied, as an analysis similar to Eq. 8 would have shown. Secondly, the typical time scale for spinodal dewetting did not fit the theoretical expectation (rupture time $\tau \propto h^5$).

entanglements[75]) caused by the preparation of the film out of solution plays a significant role and can cause holes. Details can be found in the studies of Seemann et al.[73] and Reiter et al.[39]

The experimental distinction between a hole pattern with preferred hole distance and a pattern with randomly distributed hole sites is far from being obvious. The thicker the films are, the weaker is the driving force, and the longer is the growth time τ of the spinodal mode[30,56,57] (typically, $\tau \propto h^5$) and can easily exceed experimental time scales. For thicker films, dewetting by heterogeneous nucleation may therefore be quicker and can suppress a spinodal pattern.[76,77] Moreover, chemical heterogeneities can locally cause a change in ϕ and therefore the rupture conditions of the sample may vary from spot to spot leading to a less-ordered dewetting pattern. This effect is more pronounced in thicker films due to the small driving forces and the large growth time τ. Hence a two-point correlation function (a radial pair correlation function $g(r)$ or a Fourier transform) might not be sensitive enough to detect the correlations between the hole sites, or the statistics is too poor. In this case, more powerful tools have to be applied. Minkowski functionals — based on integral geometrical methods — have shown to be a versatile method to track down higher order correlations. There is no room here to dwell on this method in any reasonable detail. The reader is referred to the pertinent literature.[6,13,32,33,78,79]

Let us come back to thermal nucleation once more. As discussed above, the potential barrier for nucleation of a dry spot may in principle be overcome by thermal activation, provided the system is sufficiently close to the sign reversal of $\phi''(h)$. The characteristic feature of this scenario is a continuous breakup of more and more holes, whereas nucleation from defects causes holes that emerge only within a sharp time window.[28] Figure 6(c) depicts an example for a 4.1 nm thick PS films on a wafer with a 2.4 nm SiOx layer: Holes of different sizes are observable. For that system, the stability diagram of Fig. 8(b) reads out accordingly that $\phi''(4.1\ nm) \approx 0$. This is another strong indication that the theoretical predictions corroborate the experimental observations and are able to capture the wettability of dielectric systems.

6. Dynamics of Spinodal Dewetting

At the beginning of spinodal dewetting studies, only one system was known to dewet spinodally: thin gold films on top of quartz substrates.[5] Metal films, however, are not so easy to deal with as compared to polymer films,

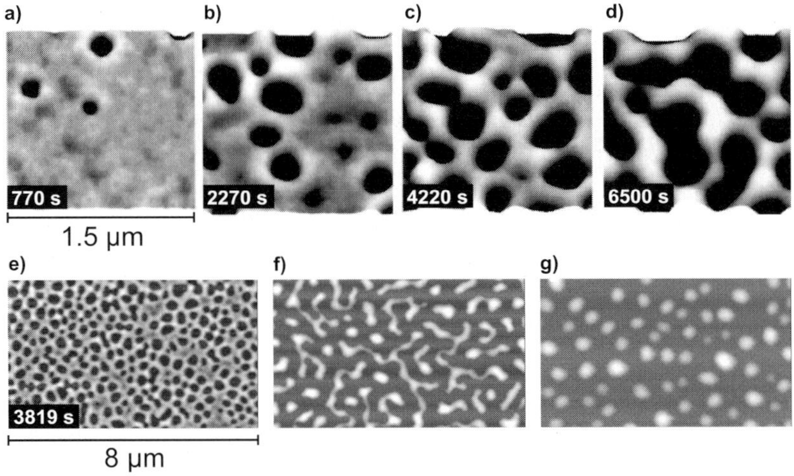

Fig. 9. Dewetting morphology of a 3.9(2) nm PS(2k) film on a Si wafer with a 191 nm thick SiOx layer as recorded by in situ AFM. Up to about 5000 s (e) the temperature was held constant at 53° and annealing times are given in the pictures. Afterwards, the temperature was successively increased to 100°. The scan size in (a)-(d) is 1.5 m. In (e)-(g) larger scans were made to check possible damage of the sample by the AFM tip. Scan (e) was taken some minutes before scan (d), whereas scan (f) was recorded some minutes after scan (d). Scan (g) characterizes the end of the dewetting process and Ostwald ripening of the droplets slowly sets in.

since the films need to be annealed by a laser. Hence, the time scale is in the nanosecond range, crystallization plays a role and the Hamaker constants are not so easy to calculate. Polymer films such as polystyrene (PS), however, are dielectric, and Hamaker constants can easily be determined via Eq. 7.

In situ AFM studies on a dewetting PS film on a Si wafer with a thick SiOx layer reveal the dynamics of the structure formation process, c.f. Fig. 9. Figure 10 depicts results of Fourier transforms of the AFM scans. Clearly, one mode is amplified fastest. The amplitude of the growing unstable mode, shown in the inset, grows exponentially with time, as expected from theory. Further investigations show that at the onset of the dewetting, no mode is specifically selected, rather, the amplitude reflects the roughness of the film. Comparing the results to deterministic simulations,[32] the pattern formation process and the morphology of the dewetting structures match very well, yet the temporal evolution of the morphology slightly differs. It turned out that thermal fluctuations accelerate the dewetting dynamics in the experiments.[33,69]

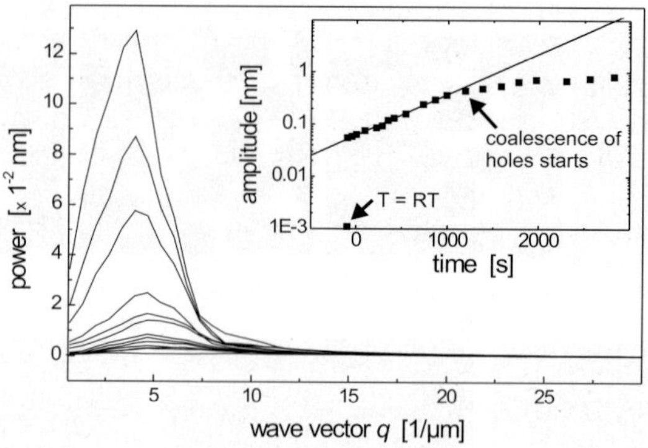

Fig. 10. Results of Fourier transforms giving the power spectral density of in situ AFM pictures at $T = 53°C$, on of which is shown in Fig. 6(a). Note that the time intervals between the curves are not constant. The inset depicts the amplitude of the undulation as a function of annealing time. The first data point at $t < 0$ s gives the roughness of the PS film surface at room temperature (RT) as revealed from a Fourier transform. The solid line is a fit of an exponential growth to the data, as expected from theory.[57]

7. Conclusion

The wettability of a substrate is a delicate interplay of forces. For dielectric systems, the wettability can be described successfully by the effective interface potential, which is a sum of short- and long-range interactions. The latter are dominated by van der Waals interactions. Their strength can be obtained by calculating the Hamaker constant via the optical properties of the involved media. For stratified systems, the additivity of forces can be assumed.

Rupture mechanism, pattern formation, morphology and dynamics of dewetting are all governed by the effective interface potential and the experiments can corroborate the theoretical expectations. The knowledge of the effective interface potential therefore allows for tailoring the wettability on demand and is a great tool for further fundamental or applied studies.

References

1. B. Evans and M. Chan, Physics World **9**, 48-52, (1996).
2. E. S. Sabisky, C. H. Anderson, Phys. Rev. A **7**, 790-806 (1973).
3. J. E. Rutledge, P. Taborek, Phys. Rev. Lett. **69**, 937-940 (1992).
4. S. Herminghaus et al., Ann. Physik **6**, 425-447 (1997), and references therein.

5. J. Bischof, D. Scherer, S. Herminghaus, and P. Leiderer, Phys. Rev. Lett. **77**, 1536 (1996).
6. S. Herminghaus, K. Jacobs, K. Mecke, J. Bischof, A. Fery, M. Ibn-Elhaj, and S. Schlagowski, Science **282**, 916 (1998).
7. D. G. Bucknall, Progress in Materials Sci. **49**, 713 (2004) and references therein
8. P. Müller-Buschbaum, J. Condens. Matter **15**, R1549 (2003) and references therein
9. C. Redon, F. Brochard-Wyart, and F. Rondelez, Phys. Rev. Lett. **66**, 715 (1991)
10. G. Reiter, Phys. Rev. Lett. **68**, 75 (1992).
11. R. Yerushalmi-Rozen, J. Klein and L. Fetters, Science **236**, 792 (1994)
12. R. Xie, A. Karim, J. F. Douglas, C. C. Han, R. R. Weiss, Phys. Rev. Lett. **81**, 1251 (1998)
13. K. Jacobs, K. Mecke, and S. Herminghaus, Langmuir **14**, 965 (1998).
14. P.-G. de Gennes, Rev. Mod. Phys. **57** 827 (1985).
15. P.-G. de Gennes, C. R. Acad. Sci. Paris **B 228**, 219 (1979).
16. R. Seemann, S. Herminghaus, K. Jacobs, Phys. Rev. Lett. **87**, 196101 (2001).
17. R. Fetzer, K. Jacobs, A. Münch, B. Wagner, T. P. Witelski, Phys. Rev. Lett. **95**, 127801 (2005).
18. J. L. Keddie, R. A. L. Jones, R. A. Cory, Europhys. Lett. **27**, 59 (1994).
19. P.-G. de Gennes, Europ. Phys. J. E **2**, 201 (2000) and C. R. Acad. Sci. **1/IV**, 1179 (2000)
20. J. A. Forrest, K. Dalnoki-Veress, Adv. Colloid. Interface Sci. **94**, 167 (2001) and references therein.
21. S. Herminghaus, K. Jacobs and R. Seemann, Eur. Phys. J. E **5**, 531 (2001).
22. S. Herminghaus, K. Landfester, and R. Seemann, Phys. Rev. Lett. **93**, 017801 (2004)
23. J. Baschnagel and F. Varnik, J. Phys.: Condens. Mater. **17** R851 (2005) and references therein.
24. M. Alcoutlabi and G. B. McKenna, J. Phys.: Condens. Mater. **17** R561 (2005) and refernces therein.
25. S. R. Wassermann, Y. Tao, and G. M. Whitesides, Langmuir **5**, 1075 (1989).
26. J. B. Brzoska, I. Ben Azouz, and F. Rondelez, Langmuir **10**, 4367 (1994).
27. F. Brochard, J.-M. di Meglio, and D. Quéré, C. R. Acad. Sci II **304** 553 (1987).
28. K. Jacobs, R. Seemann, G. Schatz, and S. Herminghaus, Langmuir **14**, 4961 (1998).
29. R. Seemann, S. Herminghaus and K. Jacobs, Phys. Rev. Lett. **86**, 5534 (2001).
30. R. Seemann, S. Herminghaus and K. Jacobs, J. Phys.: Cond. Mat. **13**, 4925 (2001).
31. R. M. A. Azzam, N. M. Bashara, *Ellipsometry and Polarized Light* (Elsevier, Amsterdam 1977).
32. J. Becker, G. Grün, R. Seemann, H. Mantz, K. Jacobs, K. R. Mecke, R. Blossey, Nature Materials **2**, 59 (2003).

33. R. Fetzer, M. Rauscher, S. Seemann, K. Jacobs, and K. Mecke, Phys. Rev. Lett. **99**, 114503 (2007)
34. C. Redon, J. B. Brzoka, and F. Brochard-Wyart, Macromolecules **27**, 468 (1994).
35. F. Brochard, P. G. de Gennes, H. Hervert, C. Redon, Langmuir **10**, 1566 (1994).
36. F. Brochard, C. Redon, C. Sykes, C. R. Acad. Sci II **314**, 19 (1992).
37. G. Reiter, P. Auroy, and L. Auvray, Macromolecules **29**, 2150 (1996)
38. G. Reiter and R. Khanna, Langmuir **16**, 6351 (2000).
39. G. Reiter, M. Hamieh, P. Dammann, S. Sclavons, S. Gabriele, T. Vilmin, and E. Raphael, Nature Materials **4**, 754 (2005)
40. R. Fetzer and K. Jacobs, Langmuir **23**, 11617 (2007).
41. R. Fetzer, M. Rauscher, A. Münch, B. Wagner, and K. Jacobs, Europhys. Lett. **75**, 638 (2006).
42. R. Fetzer, A. Münch, B. Wagner, M. Rauscher, and K. Jacobs, Langmuir **23**, 10559 (2007).
43. Lord Rayleigh, F. R. S., Proc. London Math. Soc. **10**, 4 (1878).
44. W. Ostwald, 'Lehrbuch der Allgemeinen Chemie', vol. 2, part 1. Leipzig, Germany (1896).
45. S. Herminghaus, F. Brochard, C. R. Physique, **7**, 1073-1081 (2006); see also C. R. Physique **8**, 86 (2007).
46. T. Young, Phil. Trans. Royal Soc. part I, 65 (1805).
47. S. Safran, *Statistical Thermodynamics of Surfaces, Interfaces, And Membranes*, (Addison-Wesley Publishing Company: New York, 1994).
48. J. Israelachvili, *Intermolecular Surface Forces* (Academic Press Inc.: New York, 1992), 2nd ed.
49. M. Schick in *Liquids at Interfaces*, J. Charvolin *et al*, Eds (Elsevier Science: Amsterdam, 1989).
50. S. Dietrich in *Phase Transition and Critical Phenomena*, C. Domb and J. L. Lebowitz, Eds (Academic Press: London, 1988) Vol. 12.
51. A. Oron, S. H. Davis, and S. G. Bankoff, Rev. Mod. Phys. **69**, 931 (1997).
52. A. Oron, Phys. Rev. Lett. **85**, 2108 (2000).
53. R. Seemann, R. Blossey, and K. Jacobs, J. Phys.: Condens. Matter **13** 4915 (2001).
54. A. Sharma and R. Khanna, Phys. Rev. Lett. **81**, 3463 (1998), and A. Sharma and R. Khanna, J. Chem. Phys. **110**, 4929 (1999).
55. D. Quéré, Nature Materials **3**, 79 (2004).
56. A. Vrij, Disc. Faraday Soc. **42**, 23 (1966).
57. E. Ruckenstein and R. K. Jain, J. Chem. Soc. Faraday Trans. II **70**, 132 (1974).
58. M. B. Williams and S. H. Davis, J. Colloid Interface Sci. **90**, 1 (1982).
59. V. S. Mitlin, J. Colloid Interface Sci. **156**, 491 (1993).
60. A. N. Frumkin, J. Phys. Chem. USSR **12**, 337 (1938).
61. H. Zhao, Y. J. Wang and O. K. C. Tsui, Langmuir **21**, 5817 (2005).
62. I. E. Dzyaloshinskii, E. M. Lifshitz, and L. P. Pitaevskii, Adv. Phys. **10**, 165 (1961).

63. A. W. Adamson, *Physical Chemistry of Surfaces, J. Wiley & Sons, Inc., New York, 1990.*
64. H. C. Hamaker, Physica **4**, 1058 (1937).
65. E. M. Lifshitz, Soviet Phys. JETP (Engl. Transl.) **2**, 73 (1956).
66. K. Ragil et al., Phys. Rev. Lett. **77**, 1532-1535 (1996).
67. V. S. Mitlin, Colloid Surf. A **89**, 97 (1994).
68. R. Blossey, Int. J. Mod. Phys. B **9**, 3489 (1995).
69. Y. J. Wang, O. K. C. Tsui, J. Non-Crystalline Solids **352**, 4977 (2006).
70. J. P. DeSilva, M. Geoghegan, A. M. Higgins, G. Krausch, M.-O. David and G. Reiter, Phys. Rev. Lett. **98**, 267802 (2007).
71. T. Pompe and S. Herminghaus, Phys. Rev. Lett. **85**, 1930 (2000).
72. T. Pompe, Phys. Rev. Lett. **89**, 076102 (2002).
73. R. Seemann, S. Herminghaus, C. Neto, S. Schlagowski, D. Podzimek, R. Konrad, H. Mantz, and K. Jacobs, J. Phys.: Condens. Matter **17**, S267 (2005).
74. G. Reiter, Langmuir **9**, 1344 (1993).
75. S. G. Croll, J. Appl. Polymer Sci. **23**, 847 (1979).
76. R. Konnur, K. Kargupta, and A. Sharma, Phys. Rev. Lett. **84**, 931 (2000), and references therein.
77. C. Neto, K. Jacobs, R. Seemann, R. Blossey, J. Becker, G. Grün, J. Phys.: Condens. Matter **15**, 3355 (2003) and **15**, S421 (2003).
78. K. R. Mecke, *Integralgeometrie in der Statistischen Physik - Perkolation, komplexe Flüssigkeiten und die Struktur des Universums, Reihe Physik Bd. 25* (Verlag Harri Deutsch, Frankfurt a.M., 1994).
79. K. Jacobs, R. Seemann and K. Mecke in *Statistical Physics and Spatial Statistics*, K. Mecke and D. Stoyan, Eds. (Springer: Heidelberg 2000).

CHAPTER 11

ANOMALOUS DYNAMICS OF POLYMER FILMS

Ophelia K. C. Tsui

Department of Physics, Boston University
Boston, MA 02215, U.S.A.
E-mail: okctsui@bu.edu

This chapter reviews the recent experiments involving the dynamics of polymer films with thickness comparable to the gyration radius of the polymer, which overwhelmingly show that their glass transition temperature, T_g, depends on the film thickness and can sometimes differ significant from that of the bulk. The greatest mystery is the onset film thickness of this phenomenon being ~50 nm, much larger than the typical cooperativity size (~3 nm) of glass transition. While the sliding chain model [De Gennes, *Eur. Phys. J. E* (2000)] can qualitatively explain the results of the high molecular weight (M_w) freely-standing films, it fails to account for those of the low-M_w freely-standing films and the supported films. For the latter two, at least two theories, *i.e.*, the percolation theory [Long *et al.*, *ibid* (2001)] and surface capillary wave theory [Herminghaus *et al.*, *ibid* (2001)] have been proposed that can account for the observed thickness dependence of thin film T_g. However, experimental data available at this time do not allow the theories to be distinguished. We briefly outline the physical ideas of these theories, and delineate how dynamical measurements of nanometer thick films may provide important insights about the problem.

1. Introduction

A fundamental understanding of the dynamical and mechanical properties of polymer thin films is important in many applications including organic light emitting devices, protective encapsulations in microelectronics, and lubricant coatings, *etc*. There has been mounting evidence over the past two decades showing that the properties we know well about polymers in bulk often do not apply when they are made into thin films, especially when the film thickness is comparable to the gyration radius of the polymer (typically 2 to 50 nm).[1–13] Nonetheless, a large number of these observations remain unexplained, and understanding the origins of the anomalous dynamics of polymer films has become one of the most challenging problems of contemporary polymer physics. The purpose of this chapter is to give a brief account of this rapidly evolving field of research. At the time when this chapter is written, several excellent reviews on the subject have been published.[1–4] Due to limitation of space, the selection of materials in this review may be skewed to the author's personal interest. The readers are referred to the previous reviews[1–4] for a more thorough view about the subject.

2. Experimental Observations on the T_g of Polymer Films

The study of dynamics of glass forming liquids under confinement at the nanoscale was first carried out by Jackson and McKenna[14]. In that work, a depression of T_g of 8.8K was found in *o*-terphenyl confined in

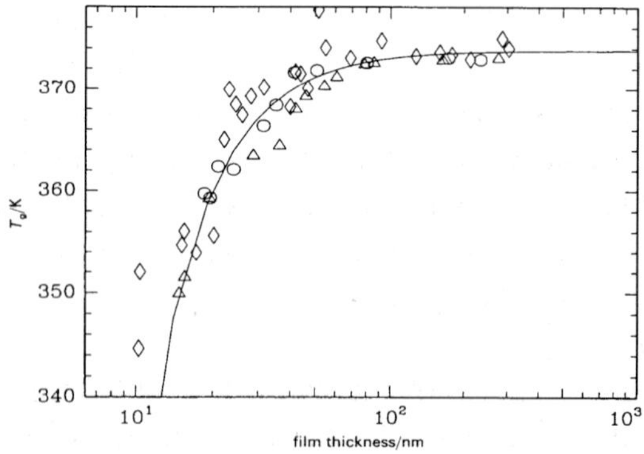

Fig. 1. T_g (K) *vs.* film thickness (nm) of PS supported on silicon with different molecular weights, M_w of 120K (circles), 500.8K (triangles) and 2,900K g/mol (diamonds). The solid line is the best fit to Eq. 1. Data by Keddie et al. reproduced from Ref. 19 (copyright RSC Publishing).

controlled pore glasses treated with hexamethyldisilazane and have an average pore diameter of 8.5 nm. Soon afterwards, Reiter found that polystyrene (PS) films with thickness < 8 nm supported by a glass substrate could dewet at temperatures below the bulk T_g, providing evidence that the T_g of these films can be smaller than the bulk value.[15–17] Keddie and coworkers[18,19] were the first to systematically measure the T_g of polymer films as a function of the film thickness, h. Figure 1 shows their result on PS supported by silicon. The most surprising finding from this result is the absence of any dependence on the molecular weight, M_w, of the polymer (120K to 2900K g/mol) in the T_g *vs.* h dependence. The data for $T_g(h)$ was found to empirically fit the following:

$$T_g(h) = T_g(\infty)[1-(\delta/h)^{\nu}], \tag{1}$$

where δ and ν are fitting parameters with values 3.2 nm and 1.8, respectively, found by Keddie *et al.*[18,19] for PS on silicon. These authors postulated that Eq. 1 could arise from a low-density, highly mobile layer at the free surface of the film having an intrinsic thickness of δ that diverges like $(1 - T/T_g(\infty))^{-1/\nu}$ as the temperature approaches the bulk T_g from below. On the other hand, qualitatively different behaviors were reported by them in the same papers.[18,19] It was observed that the T_g of poly(methyl methacrylate) (PMMA) films could increase or decrease with decreasing film thickness depending on the substrate material and surface condition as illustrated in Fig. 2. Specifically, PMMA films on gold coated silicon showed a decrease in T_g with decreasing film thickness, but those deposited on silicon covered by a native oxide layer showed an increase in T_g with decreasing film thickness. Later, other supported polymer films were also found to demonstrate an increase in T_g with decreasing film thickness, such as poly(2-vinyl pyridine) on acid-cleaned silicon oxide.[20]

A layered model is one of the earliest models proposed to explain these observations,[18-25] and by far the most accepted for the thickness dependence of the T_g of polymer films. The model presupposes that the molecular motions near the polymer-air interface are much faster than those in the bulk polymer, which can be due to segregation of chain ends

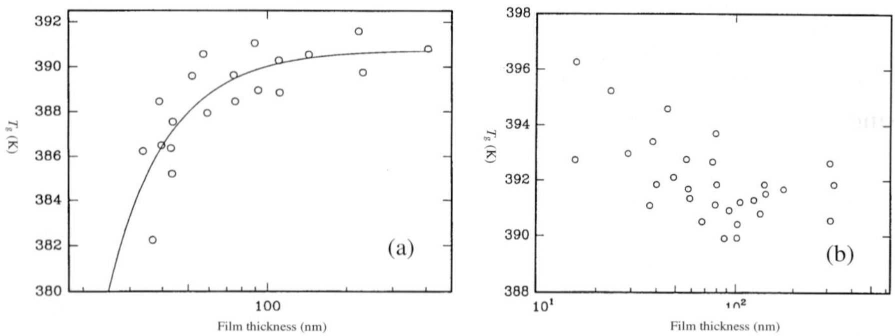

Fig. 2. T_g (K) vs film thickness (nm) of PMMA films deposited on (a) gold coated silicon and (b) silicon covered by a native oxide layer. Data by Keddie *et al.*, reproduced from Ref. 19 (copyright RSC Publishing).

to the surface[24,26] or a reduction in the chain entanglement near the polymer free surface.[24,27] On the other hand, the molecular motions at the polymer-substrate interface can be faster or slower than the polymer motions in bulk depending on whether the polymer-substrate interactions are sufficiently weak or strong, respectively. The T_g of a film is a result of the interplay between the effects of the two interfaces. The x-ray photoelectron spectroscopic (TDXPS) and angle-dependent XPS measurements of Kajiyama *et al.*[28] on poly(styrene-*b*-methyl methacrylate) showed that the T_g of the polymer got progressively smaller than that of the bulk towards the free surface. By using fluorescence labeling, Torkelson's group[29] showed that the local T_g of PS was smaller than the bulk value at the free surface, but continuously approached the bulk T_g over several tens of nm into the film. Numerous computer simulation results also support this picture.[30–35] These findings provide important evidence to the existence of dynamical heterogeneity — which is indeed a gradient — in the film, and validates the use of the layered model.

2.1. *Experimental Search for Enhanced Mobility at the Polymer-Air Interface*

The idea that there can be a mobile layer at the free surface of a polymer has fascinated numerous researchers who have devised some most ingenious experiments to search for its existence.[22,28,29,36–50] These experiments involved a large number of measurement techniques, which include angle-dependent x-ray photoelectron spectroscopy,[28] various kinds of atomic force microscopy (AFM),[37–42] near-edge x-ray absorption fine structure,[43,44] positron annihilation lifetime spectroscopy,[22,45] optical birefringence measurement,[46] melting of a topographic structure,[49,50] and local T_g probe by fluorescence labeling.[29] Although some of the results are conflicting, the majority holds that the mobility of PS, for which the thin film T_g shows the biggest reduction, is enhanced near the free surface. And between two conflicting results obtained by the same technique, the more recent one indicates that the surface mobility is enhanced. For AFM experiments showing a conflicting result, one can usually find attributes to the specificity of the polymer, inadequacy of the surface sensitivity of the technique and/or alternative interpretations of the results. I postpone further discussions until Sec. 5.2. For experiments involved rubbed PS, studies show that the relaxation phenomenon is rich and may not be simply related to the glass transition of the polymer.[47,48]

The search for an enhanced mobility surface layer is connected with the search for the origin of a reduction in the thin film T_g. In contrast, the search for the origin of an increase in the thin film T_g has not aroused nearly as much interest. It is attributable to the fact that an increase in T_g can always be ascribed to the friction between the polymer chains and the substrate. Furthermore, far more examples have been found for decreases than increases in the T_g of polymer films, which also makes the former more appealing.

2.2. *Significance of the Polymer-Air Interface and Confinement Effect Revealed by Freely-Standing Films*

The importance of the polymer-air interface in bringing about a reduction in the T_g of a polymer film can also be perceived from the results of freely-standing films,[51–57] which possess two polymer-air interfaces (Fig. 3). From the measured T_g of PS freely-standing films (Fig. 4),[51–55] one may readily notice that the magnitude of the shift in $T_g(h)$ is twice as large as that found in supported PS films (Fig. 1). In addition, the data

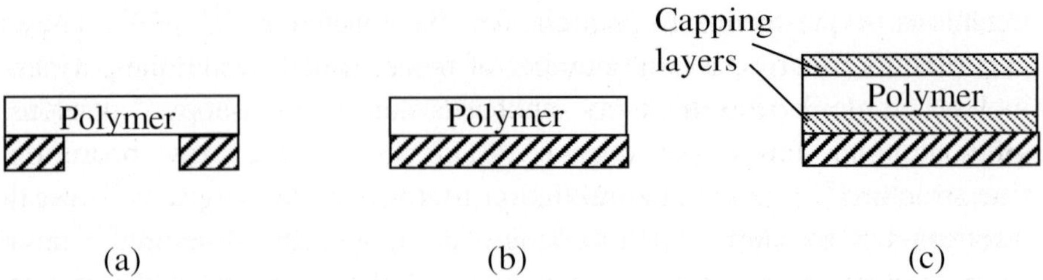

Fig. 3. Three configurations commonly employed in the study of polymer films: (a) freely-standing, (b) supported and (c) capped films. These configurations differ by the number of polymer-air and polymer-substrate interface, and allow systematic control of the effect of these two kinds of interfaces on the dynamics of polymer films.

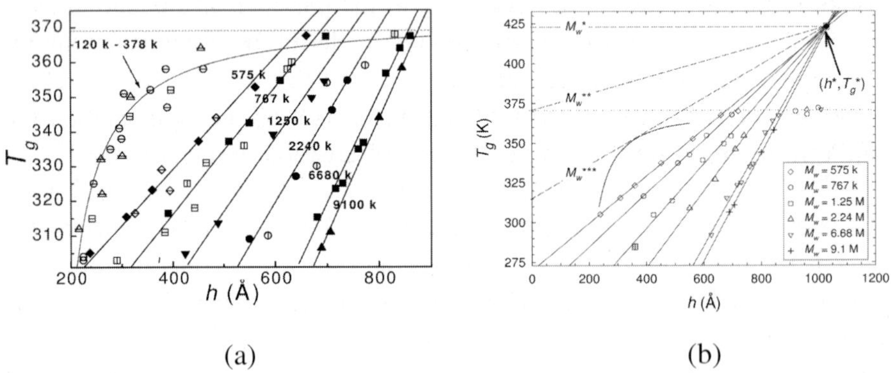

Fig. 4. (a) T_g of freely-standing PS films with M_w from 120 K to 9100 K g/mol obtained by different techniques. The solid symbols are obtained by ellipsometry. The hollow symbols are obtained by Brillouin light scattering. The solid lines are fits of the data to Eq. 2. (b) The same data shown in (a), but for $M_w > 378$ K g/mol only. The fitted lines are extrapolated above the bulk T_g and are found to meet at a point denoted by (h*,T*) as shown. The physical meaning of the convergence at (h*,T*) is not known. (Both figures are reproduced from Ref. 4, copyright Elsevier.)

behave quite differently between the $M_w \geq 575$ kg/mol films and the $M_w \leq 378$ kg/mol films, as a result of which they were referred to as the high- and low-M_w films, respectively. For the high-M_w films, $T_g(h)$ is equal to the bulk T_g at large film thicknesses, but starts to decrease linearly with the film thickness below some threshold thickness. From Fig. 4, as M_w decreases the threshold thickness decreases and the slope of

$T_g(h)$ below the threshold gets smaller. The fact that the shift in $T_g(h)$ depends on the comparison between the unperturbed size of the polymer ($\sim M_w^{1/2}$) and the film thickness evidences that chain confinement effect of some sort exists. For the low-M_w films, on the other hand, the variation of $T_g(h)$ is gradual, showing no M_w dependence as was found in supported films (Fig. 1). In fact, Eq. 1 that generally fits the data of supported films provides a good description of the low-M_w data as well (Please see the solid, curved line in Fig. 4(a)). The fitted value of δ, with ν fixed at 1.8, is 7.8 nm.[3] It is noteworthy that this value is twice the value found for the supported films. Since freely-standing films possess twice the number of polymer-air interfaces as supported films do, this result reinforces the layered model that the reduction of T_g in polymer films is caused by a mobile layer at the polymer-air surface. As for the high-M_w data (Fig. 4(a)), it was found[55] that the linear regression lines through the data actually extrapolate to meet at a single point, (h^*,T^*) = (103 nm, 423 K) above the bulk T_g (Fig. 4(b)). This suggests that the $T_g(h)$ of high-M_w freely-standing films can be written as[55]:

$$T_g - T_g^* = \alpha(M_w)(h - h^*). \tag{2}$$

Empirically,

$$\alpha(M_w) = b\ln(M_w/ M_w^*), \tag{3}$$

where $b = (0.70 \pm 0.02)$ K/nm and $M_w^* = (69 \pm 4)$ kg/mol. Forrest and Dalnoki-Veress argued that the $T_g(h)$ dependence described by Eq. 2 should yield to the low-M_w behavior (i.e., the curved line in Fig. 4(a)) when the slope $\alpha(M_w)$ is decreased to such a point that the corresponding $T_g(h,M_w)$ line touches the low-M_w curve only at a single point. This condition is illustrated by the straight line marked by M^{***} in Fig. 4(b) (the short curved line depicts a segment of the low-M_w curve). The value of M^{***} thus estimated is 300 K g/mol, which agrees well with the crossover at M_w = 378 K g/mol seen in Fig. 4(a).

Qualitatively the same results were found in freely-standing PMMA films by the Dutcher group recently[56,57] although the magnitude of the T_g reduction is roughly a factor of three less than that found in the PS films. It points to the importance of the specificity of the polymer on its thin film properties, and is consistent with the lack of enhanced surface mobility often found in acrylate polymers.[39,41,42]

2.3. *Effect of Chain Ends*

Local enrichment of the chain ends at the free surface of a polymer has been one of the most cited origins for the surface mobile layer. The phenomenon was confirmed by Kaijiyma *et al.*[38] who found that the concentration of surface chain ends diminished as the M_w of the studied PS was increased and became negligibly small when the M_w is bigger than the entanglement molecular weight (~30 kDa). Motivated by these results, we investigated[24] the thickness dependence of T_g of monodispersed PS thin films with M_w = 13.7 K and 550 K Da ($M_w/M_n \leq$ 1.06 for both polymers) to seek for any effect on the $T_g(h)$ dependence due to the higher concentration of surface chain ends expected in the 13.7 kDa than the 550 kDa PS films. As seen in Fig. 5 (main panel), the measured $T_g(h)$ demonstrate excellent fits to Eq. 1. In the inset is shown the same data normalized by the corresponding bulk T_g. Evidently, the data of both molecular weights almost overlap, indicating that the $T_g(h)$

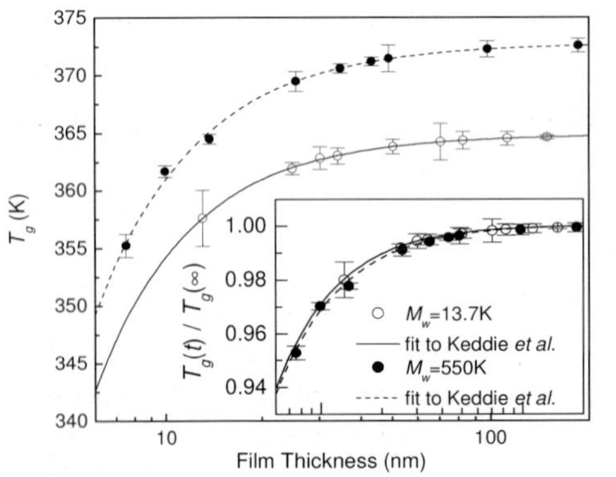

Fig. 5. (main panel) T_g *vs.* h for PS films with M_w = 550 kDa (solid circles) and 13.7 kDa (open circles). The solid and dashed lines are fits to Eq. 1. (inset) The same data but normalized by the corresponding bulk T_g of the polymer, $T_g(\infty)$, determined from the fits shown in the main panel (reproduced from Ref. 24, copyright American Chemical Society.)

dependence is independent of M_w even below the entanglement molecular weight[58] and hence the surface chain ends cannot constitute an essential cause for the reduction of T_g in PS supported films. This result is in keeping with the simulation work of Doruker and Mattice[59] who found that the segmental mobility at the free surface was enhanced for both linear and cyclic polymer chains. To investigate if there is at all any influence of the surface chain ends on the T_g of polymer films, we studied the M_w dependence (from 13.7 K to 2.3M Da) of T_g of PS films with fixed thicknesses of 15 and 50 nm, respectively.[24] Shown in Fig. 6 are the results plotted as $T_g(M_w)/T_g(M_w = \infty)$ *vs.* M_w for the two thicknesses as well as the bulk polymer, where $T_g(M_w = 2.3M)$ was taken to be $T_g(M_w = \infty)$. According to Fox and Flory (FF),[60]

$$T_g\,(M_w)/T_g\,(M_w = \infty) = 1 - m_0/M_w, \tag{4}$$

where $m_0 \sim (\rho_s - \rho_e)$ with ρ_s and ρ_e being the mass density of a chain segment and chain end, respectively. For PS, $T_g(M_w = \infty) = 373$ K and $m_0 = 455.8$.[60] As seen in Fig. 6, Eq. 4 provides fairly good fits to the data (solid lines), and the M_w dependence of T_g weakens as the film thickness decreases. The fitted value of m_0 for the 15 nm films is 40% less than that of PS in bulk. A reduction in m_0 may either be due to an increase in ρ_e or a decrease in ρ_s. Since the inset of Fig. 5 shows that the reduction in the T_g of a 15 nm film is only ~2% of $T_g(\infty)$, which puts an upper limit to the possible reduction in ρ_s of the polymer under confinement in a 15 nm thickness. A 2% reduction in ρ_s would be too small to account for the 40% reduction in $\rho_s - \rho_e$ deduced above. Therefore, the apparent reduction in m_0 is most likely due to an increase in ρ_e as the film thickness is reduced. A probable explanation to this observation would

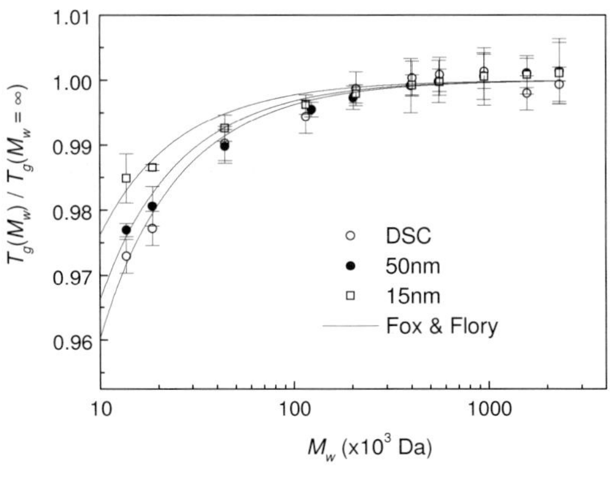

Fig. 6. $T_g\,(M_w)$ normalized by the T_g at M_w = 2.3M Da as a function of M_w for samples in bulk (open circles) and in thin films with h = 50 nm (solid circles) and 15 nm (open squares). Solid lines are fits to Eq. 4.

be an enhanced segregation of the chain ends to the surface as the polymer thickness decreases, resulting in a reduction in the number of chain ends remaining in the film body that ultimately determines the T_g of the film. To confirm that the change in T_g *vs.* M_w shown in Fig. 6 is indeed a reflection of the change in the number fraction of chain ends in the film body, we studied the $T_g(h)$ dependence of a binary blend of 13.7 K and 550 K PS (in 1:1 wt. ratio).[24] For a binary blend of polymers with disparate molecular weights, Hariharan *et al.*[61] showed that entropic effect would drive the low-M_w component to the film surface, producing a local enrichment of the low-M_w component. Thus, one expects the 13.7 K constituent in the blend films to segregate to the surface. By using Fox and Flory's (FF) model, the T_g of the binary blend in bulk should be[60]:

$$1/T_g = ½[1/T_g(M_w{=}13.7\ \mathrm{K}) + 1/T_g(M_w{=}550\ \mathrm{K})]. \tag{5}$$

In Fig. 7 (solid line) is displayed this dependence assuming the measured T_g of the respectively monodispersed films. The measured T_g's of the blend films are shown by solid squares. As seen, the data agree with this model line quite well for $h >$ ~30 nm, but approaches the T_g of the 550 K PS films when the film gets thinner. Our result strongly suggests that those chain ends segregated to the surface do not contribute to the T_g of the film, but instead only those remaining in the film body do. This reinforces our above observation that the surface chain ends are not directly related to the reduction of T_g observed in polymer films.

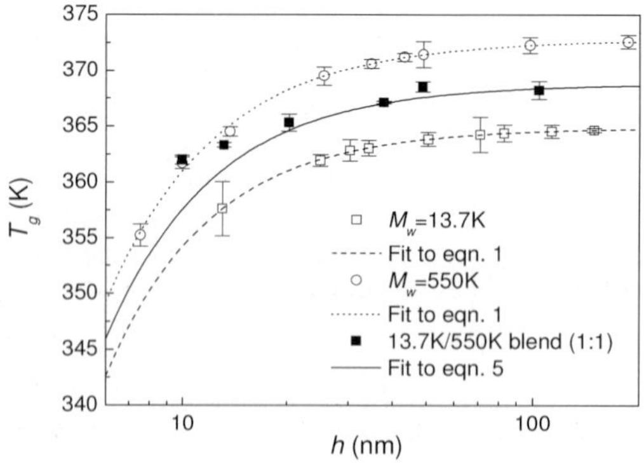

Fig. 7. Comparison between the T_g of the 13.7 K/550 K (1:1 in wt) blend films (solid squares) and that of the pure constituents reproduced from Fig. 5. The continuous lines passing through individual data sets are fits to Eq. 5.

2.4. *Effect of the Polymer-Substrate Interface*

We have examined yet another popular perception about the thickness dependence of polymer film T_g, i.e., the amount of T_g reduction or increase of a polymer supported film depends on, respectively, how repulsive or attractive the polymer-substrate interaction is. The T_g of 33 nm thick PS (M_w = 96 kDa, M_w/M_n = 1.04, bulk T_g = 373 K) coated on random copolymer of PS and PMMA (P(S-*r*-MMA)) brushes with M_w ~ 10 kDa, M_w/M_n ~ 1.1 to 1.8 was studied as a function of the styrene fraction, f, of the brush.[25] The result, plotted as T_g *vs.* f is shown in Fig. 8. As seen, as f is decreased from 1, the T_g of the films decreases and reaches ~351 K at f ~ 0.7, showing that the thin film T_g decreases as the polymer-substrate interaction gets more unfavorable. We examine in what way the polymer-substrate interaction energy may manifest in this effect. The transition of a liquid into a glass corresponds to a kinetic arrest of the liquid as a result of the necessity of cooperative molecular rearrangement in order for any motion (involving configurational changes) to be possible.[62] Because cooperativity is involved in the dynamics at the glass transition, T_g should be determined by the total

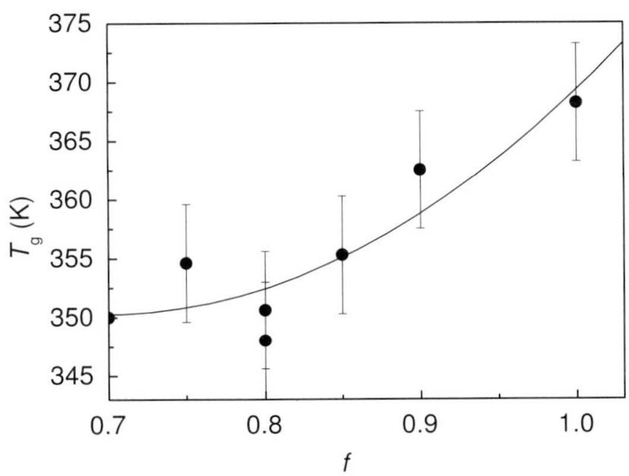

Fig. 8. The measured T_g of 33 nm thick PS films spin-coated on P(S-*r*-MMA) brushes as a function of the styrene fraction, *f*, of the brush (solid symbols). The solid line is a guide-to-the-eye.

energy, $N_\xi \varepsilon_a$, required to activate all the molecules in a cooperative region to move simultaneously. By using the published data of PS with M_w similar to ours,[63] we estimate that ε_a ~ 130 Jcm^{-3}.[25] To the first-order approximation, an interfacial energy of γ_{sf} should produce a change in the glass transition temperature, $\delta T_g \sim \gamma_{sf} T_g/(\varepsilon_a \xi)$, where ξ is the size of a cooperative region. Based on previous measurements of γ_{sf} *vs.* f,[64,65] the change in γ_{sf} when f is changed by 0.3 is 0.43 erg/cm^2. Hence, to obtain the observed change of 20 K in T_g as f was decreased from 1 to 0.7, ξ would have to be 0.6 Å, which is unphysically small. This simple estimate shows that the value of γ_{sf} alone does not provide sufficient ground to understand the effect of interfacial interaction on the T_g of polymer films.

At the interface between the homopolymer and the brush, the specific interactions between monomers and packing constraints would produce perturbations to the chain conformations. Consider $(\delta T_g/\delta \rho)$, the reduction in T_g due to a decrease in the mass density, ρ, because of the perturbation. $(\delta T_g/\delta \rho) = (\delta T_g/\delta P)(\delta P/\delta \rho) = (\delta T_g/\delta P)(1/\rho\kappa)$, where, for PS, the pressure coefficient of T_g is 3.09×10^{-7} K/Pa,[65] the isothermal compressibility, $\kappa = 2.2 \times 10^{-10}$ Pa^{-1} [66] and $\rho = 1.04$ g/cm^3, yielding $(\delta T_g/\delta \rho) = 1.35 \times 10^3$ K/(g cm^{-3}). To produce a 20 K drop in the T_g for a 10 nm thick film would require a density decrease of ~1.4%. Such a large change in the average density of a thin film polymer has not been observed in experiment.[53,67] Thus, a reduced *average* film density cannot be used to explain the present observations. A mechanism focusing on changes at the interface between the polymer and the brush is more likely to be operative. Consider the following bilayer model for a film with thickness h: At the polymer-brush interface, the density of the polymer is ρ_i over a distance ζ from the interface, but is ρ, the bulk density, for the remaining part, $h - \zeta$ of the film. The T_g of the film as a whole would be given by $T_g = T_g^{bulk} + (\delta T_g/\delta \rho)(\zeta/h)(\Delta\rho)$, where $\Delta\rho = \rho_i - \rho$. For a 10 nm thick film, a 20 K reduction in T_g would be found when $\zeta\Delta\rho \sim -1.48 \times 10^{-8}$ g/cm^2. If we take $\zeta = 3$ nm, several Kuhn lengths, then $\Delta\rho/\rho = -0.05$, i.e., a 5% density decrease at the interface. It should be remarked that $\Delta\rho\,\zeta$ represents the surface excess (or surface deficit if $\Delta\rho < 0$) of polymer segments at the substrate wall due to the polymer-substrate interactions. As h is increased, this effect diminishes according to $\Delta\rho\,\zeta/h$ and explains the decreasing T_g reduction with increasing film thickness.

The foregoing discussion has focused on cases where the T_g decreases with decreasing film thickness. By considering favorable interactions that make $\Delta\rho$ positive, the same model can also describe systems with enhanced T_g with decreasing film thickness. Under the special condition where $\Delta\rho = 0$, the T_g of a film should remain the same as the bulk. This is the most closely fulfilled in this experiment with the $f = 1$ brush. Previous theoretical estimate[68] shows that the interfacial energy for the $f = 1$ sample would be small but finite ~0.1 erg/cm^2. The observed T_g with the $f = 1$ brush (Fig. 8), being ~5 K below the bulk T_g, is consistent with this estimate. This result shows that the effect of interfacial interaction on the T_g of polymer films cannot be formulated simply in terms of the polymer-substrate interfacial energy, but rather the surface excess, $\Delta\rho\,\zeta$, is more suitable. Recently, McCoy and Curro[2,69] estimated the surface excess of polystyrene sandwiched between confining walls that are non-wetting, neutral and strongly wetting, respectively, to the polymer, and found that the surface excess was negative in the first two cases, but positive in the last one. These estimates are in keeping with the present experimental and the ideas proposed thus far about the effect of the surface excess on the sign of the T_g shift.

In comparing our model with the $T_g(h)$ reported by Keddie *et al.*,[18,19] $\zeta\Delta\rho$ should be $\sim 1/h^{1.8}$. Confinement effect may give rise to strong perturbations to chain conformations.[70,71] However, the thickness at which changes in T_g start to occur (~50 nm for PS on SiO_2), being bigger than the gyration radius of the polymer, R_g (~15 nm, here), would be too large to produce any noticeable effect. Computer simulation studies[70,71] show that perturbations to the segment density due to the interface persist over a distance, $\xi_p > \sim R_g$. Insofar, the estimates of the surface excess had been performed on polymer slabs with $h >> \xi_p$. It would be interesting to estimate the surface excess for polymer slabs where this does not hold and see if it will demonstrate any thickness dependence. In particular, one would like to see if it will reproduce the $1/h^{1.8}$ dependence deduced above.

3. A Major Issue

The fact that the T_g of polymer films deviates from the bulk value when the film thickness is decreased below a certain value suggests that finite size effects of some sort must be involved. The biggest mystery about the problem of thin film T_g is the origin of such finite size effects. The onset of the T_g anomaly begins at a thickness that is surprisingly large (~50 nm for supported films and ~100 nm for low-M_w freely standing films) compared to the cooperativity size of glass transition, which is ≈ 1–3 nm.[72,73] At the same time, the molecular interactions at the interfaces – expected to be of the Lennard-Jones type because the polymers are apolar – are not long-range enough to provide a direct explanation either. While chain connectivity may account for the chain confinement effect noticed in the high-M_w freely standing films (Fig. 4), it must not be relevant for the low-M_w freely-standing films nor supported films since no M_w dependence was found in the $T_g(h)$ dependence of these films. In particular, for supported PS films on Si, the absence of M_w dependence in $T_g(h)$ was found for $2.3 \leq M_w \leq 3000$ kDa,[18,19,24,58] corresponding to $3.5 \leq R_{EE} \leq 128$ nm[55] that embraces very well the onset thickness of 50 nm for the finite size effect.

4. Theoretical Models

There have been only a few models proposed to explain the T_g anomaly in polymer films, attributable to the fact that the physics of the glass transition is not very well understood.[74] In this section, I shall outline the ideas of these models. The interested readers are referred to the original papers for more details.

4.1. *Layered Model*

The layered model discussed in Sec. 2 is the most popular and can be used to explain essentially all experimental observations. However, it is mostly phenomenological at this time and lacks any predictive power. For instance, the model by itself cannot predict the observed functional form of $T_g(h)$ nor can it predict *a priori* whether the T_g of a polymer supported film should increase or decrease with respect to the bulk T_g. Nonetheless, it is often a good starting point for constructing further details to describe the data.[18,19,25,54]

4.2. *Sliding Chain Model*

De Gennes[75] proposed the sliding chain model to explain the T_g reduction in freely-standing films. He suggested that it arose from the sliding motion of loops of polymer chains along their own contour. Only those loops with the two end points in contact with the polymer-air surfaces as shown in Fig. 9 were considered so that the barrier to the motion at the end points could be ignored. By taking into account the cooperativity size for such motion and arguing that it is those loops that are long enough to reach the mid-plane of the film that have the most influence on the T_g of the film, he arrived at a glass transition temperature, T_g(slide), that is linear in h. The glass transition temperature of the film would be determined by the sliding motion if $T_g(\text{slide}) < T_g(\infty)$. This condition occurs when the film thickness is smaller than a threshold value that is of the order of the end-to-end distance, R_{EE}, of the polymers. The model prediction provides a good description for the experimental result of the high-M_w freely-standing films (Fig. 4). However, it does not predict a M_w-dependence in the slope of $T_g(h)$ as observed in experiment nor does it predict a crossover from a high-M_w to low-M_w behavior (Fig. 4). The latter implies there to be an additional mechanism for the enhanced molecular mobility that competes with the sliding chain mechanism.

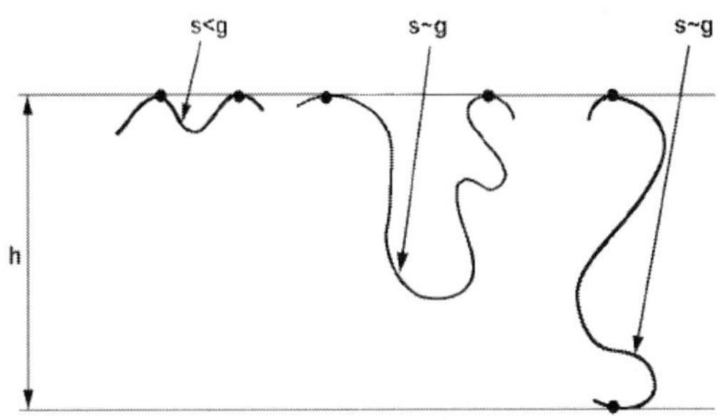

Fig. 9. Illustration of polymer loops with various loop lengths, s, sliding about the polymer-air surface. Only those loops with s $\sim g \sim (h/a)^2$ and thus extend to the mid-plane of the film can have an impact on the T_g of the film. Figure was reproduced from Ref. 73, copyright EDP Sciences.

4.3. *Percolation Model*

Long and co-worker[76,77] proposed a mechanism for the T_g reduction for apolar liquid films based on a percolation model for the glass transition. The idea is founded on spatial dynamic heterogeneity, *i.e.*, the presence of spatially distributed domains with slow dynamics near the glass transition. In the percolation model, the glass transition occurs when the slow domains percolate through the sample. This condition defines the critical value ρ_c for the local mass density of a region above which the region is reckoned "slow":

$$\int_{\rho_c}^{\infty} P(T,\rho)d\rho = p_c^{3D}, \tag{6}$$

where $P(T,\rho)$ is the probability density distribution of the mass density ρ of the liquid at temperature T; p_c^{3D} is the three-dimensional (3D) percolation threshold. (The exact value of p_c^{3D} depends on the form of the percolation network.) If the liquid is made to form a film, and the interactions between the liquid and the substrate are weak, the condition for glass transition would be modified to one requiring the slow domains to percolate in the plane. In the limit where the film thickness is smaller than the 3D percolation correlation length, the film is 2D and the percolation threshold on the RHS of Eq. 6 should be the 2D percolation threshold, p_c^{2D}. For an intermediate thickness h, the percolation threshold $p_c(h)$ lies between p_c^{3D} and p_c^{2D}. Since $p_c^{2D} > p_c^{3D}$ in general, $p_c(h)$ increases and thereby $T_g(h)$ decreases as h decreases. It can be shown that[76]

$$p_c(h) - p_c^{3D} \sim h^{-1/\mu}, \tag{7}$$

where μ (≈ 0.88 universally) is the critical exponent for the 3D percolation correlation length. By using Eq. 7 and assuming that $P(T,\rho)$ is Gaussian and sharply peaked at the average mass density of the glass-forming liquid at temperature T, the authors[76] derived an expression for the T_g of the liquid confined in a film with thickness h:

$$T_g(h) \approx T_g(bulk)[1-(a/h)^{1/\mu}], \tag{8}$$

where a is, within a factor of order unity, equal to the monomer length. Clearly, the exponent given in Eq. 8, $= 1/\mu = 1/0.88 = 1.136$ deviates from the value of $\nu = 1.8$ (in Eq. 1) quoted above.[18,19] But as pointed out by several researchers,[2] the range of film thickness studied in experiment, which is usually 10 to 50 nm, is probably too small for the functional form of $T_g(h)$ to be determined precisely. In fact, our group had found ν = 1.4 for the same polymer film system.[24] Kim *et al.*[78,79] and Herminghaus *et al.*[58,80] even suggested a different function to describe $T_g(h)$, which is pertinent to the capillary wave model to be discussed next.

4.4. *Capillary Wave Model*

In the capillary wave model proposed by Herminghaus,[58,80] a reduction in the T_g of polymer films can arise from the coupling of the viscoelastic surface capillary modes to the bulk of the film. Correspondingly, the thickness dependence of T_g is caused by the cut-off in the wavevector, q, of the surface capillary modes that varies as $1/h$ – because only those modes with wavelengths longer than h can penetrate deep enough into the film to affect the T_g. With additional physical arguments,[58] Herminghaus simplified the relaxation rate of a film to:

$$\omega(q) \approx \frac{E}{\eta} + \frac{F(1)\gamma}{\eta} q, \tag{9}$$

where γ and E are the surface tension and elastic modulus of the polymer, respectively, and $F(1)$ is a constant of order unity. By arguing that the memory kernel, which accounts for memory effects during relaxations of strains in the polymer, scales as $1/T$ (where T is temperature), and applying the same criterion used in the mode-coupling framework for the transition of a system to be frozen into a non-ergodic state, Herminghaus arrived at the following relation for the T_g of thin films:

$$T_g(h) = T_g(\infty)\frac{\omega(0)}{\omega(h^{-1})} \tag{10a}$$

$$= T_g(\infty)\left(1 + \frac{h_0}{h}\right)^{-1}, \tag{10b}$$

where $h_0 = F(1)\gamma/E \approx \gamma/E$ by Eq. 9. Kim *et al.*[78] showed that Eq. 10b could in fact also describe the data of Keddie *et al.* in Fig. 1 if

h_0 = 0.68 nm. Herminghaus *et al.*[58] found that Eq. 10b could describe their own data better than Eq. 1 if h_0 was taken to be 0.82 nm. By putting γ = 31 mN/m and $h_0 = \gamma/E$ = 0.82, they obtained $E \approx$ 44 MPa, which corresponds to the value of E for PS in the middle of the glass transition on log scale.[58]

5. Dynamical Measurements of Polymer Films

The majority of dynamical studies of polymer films have measured the thickness dependence of T_g. Since all the models discussed above were constructed to give the $T_g(h)$ dependence observed in experiment, more dynamical information about the films is needed to discriminate these models. Below I review the dynamical measurements made in the past on polymer films, and discuss any insights one may draw from those outcomes about the anomalous dynamics of polymer films.

5.1. *Diffusion Experiments*

Soon after the reduction of T_g was observed in polymer films, experiments were carried out to measure the diffusion rate (including tracer and inter-diffusion) in both supported polymer films[81,82] and later also freely-standing films.[83,84] In films showing a reduction of T_g with decreasing film thickness, only bulk-like or slower diffusion rates had been found. For example, in PS films supported by Si, the in-plane diffusion rate decreased monotonically with decreasing film thickness below h = 150 nm.[81] In freely-standing PS films with M_w = 6900 kDa, no thickness dependence was found for the out-of-plane diffusion rate down to a film thickness of 69 nm,[84] where there should be a T_g reduction of ~40 K according to Fig. 4. This finding corresponds well with that of a holes growth measurement in freely-standing PS films (for $51 \leq h \leq 91$ nm) where no significant growth of holes was observed until the temperature reached the bulk T_g.[3] Simple inference from these results would expect anything but a reduction of thin film T_g, which has caused some to query the interpretation of T_g reduction in those systems. It is recently pointed that the results on the thin film diffusion and T_g

measurements can be reconciled: The higher mobility implied by the T_g reduction in polymer films is applicable only to motions on the segmental length scale; whole chain motions relevant to diffusion and growth of holes may not occur until temperatures reach the bulk T_g.[3,4] This picture was suggested[3] to be consistent with Semenov's analysis.[85]

5.2. *Viscosity or Dynamical Mechanical Measurements*

While the measured diffusion rate can be dominated by motions with length scales irrelevant to the glass transition, viscosity has traditionally been the dynamical quantity used to define T_g. In particular, glass transition is commonly described as a kinetic transition whereat the viscosity of a liquid increases by 10^{13} times within a few degrees upon cooled across the glass transition temperature.[86] The information most often extracted from viscosity (η) measurements is the functional dependence of η or relaxation time, τ, on the temperature, T. For glass-formers in general, such dependences have been found to display the Vogel-Tammann-Fucher (VTF) scaling, which reads:

$$\tau(T) = \tau_0 \exp\left(\frac{U}{T - T_0}\right), \qquad (11)$$

where τ_0 is the reciprocal attempt rate of molecular motion, U is related to the activation energy of the motion, and T_0 is the Kauzmann or Vogel temperature.[62] From Eq. 11, it is clear that either a reduction in U or T_0 would lead to a decrease in τ and hence T_g. There have been numerous efforts to measure the $\tau(T)$ relation for polymer films at different film thicknesses. But they are relatively scarce, attributable to the difficulty generally involved in detecting responses from a small sample. (*c.f.* For a 1 $cm^2 \times 10$ nm film, the mass of the polymer is only ~1 μg.) Three main kinds of techniques have been used for the measurement of $\tau(T)$ although other techniques have been used as well.[84,87,88] They are atomic force microscopy (AFM),[37–42] ac dielectric (and capacitance) spectroscopy,[89-95] and x-ray photon correlation spectroscopy (XPCS).[96-98] I briefly discuss the status quo of each of them.

Atomic Force Microscopy: Dynamical studies employing AFM techniques generally probe the surface dynamics of the polymer.[37–42] Because of the large number of factors that can influence the measurements, the conclusion can often depend on the interpretation adopted for the data, which may explain the vastly different AFM results reported by different groups on polystyrene (PS). By using forced modulation microscopy (FMM), Kajiyama *et al.*[36–38] showed that the surface of PS was already in a viscoelastic state even at room temperature, provided $M_n <$ ~30K Da or the polymer contains some $M_n <$ 30K Da components. With monodispersed PS with M_n = 40K Da, Hammerschmidt *et al.*[39] observed a modest reduction in the surface T_g (< 10 K) by friction force microscopy (FFM), which is consistent with Kajiyama's observations. On the other hand, Ge *et al.*[40] measured no change or a slight increase in the T_g of monodispersed supported PS with 3K $\leq M_w \leq$ 6.7M Da and freely-standing PS films with M_w = 697K Da and $32 \leq h \leq 140$ nm by using shear modulation force microscopy (SMFM). To understand these different results, one should take a closer look at the experimental details. In the FMM studies of Refs. 36–38 the AFM cantilever was driven into vertical vibration at a frequency of 4 kHz, the in-phase and quadrature responses of the tip (under an average load of 25 nN against the sample) were measured at room temperature and used to deduce the storage modulus and loss tangent of the sample. The researchers concluded the surface of a polymer to be rubbery at room temperature on the basis that the loss tangent was an order of magnitude larger than the corresponding values of the bulk. In SMFM,[40] which is similar to FMM in spirit, the cantilever was driven in sideway oscillations at 1.4 kHz; the resulting amplitude of the tip, under a constant load of ~12 nN in Ref. 39 and 25 nN in Ref. 40, was monitored as the temperature was swept from room temperature to above the bulk T_g. The T_g of the polymer was determined by drawing linear extrapolations from the low- and high-temperature asymptotes of the data and seeking the intersection. Since the response amplitude increases as the tip penetrates into the sample (as demonstrated by the creep data in Fig. 2 of Ref. 40), if a surface rubbery existed and expanded with increasing temperature, the T_g so measured could be dictated by the expansion of this surface layer. We have estimated[41,42] the indentation of an AFM tip into the surface of a polymer near its glass-to-rubber transition by using the Johnson-Kendall-Roberts (JKR) model.[99] The estimated indentation at loading, δ_{load}, is represented by the dashed line in Fig. 10. Assuming the surface rubbery layer to be 10 (or 20 nm) thick, the creep compliance of this layer only needs to be 3 (or 6) times of $J(0)$ in order for the tip to penetrate through this rubbery layer. (Note that if we had used the more popular Hertz model[100] to estimate δ_{load}, which ignores the tip-sample adhesion interaction, we would have overestimated the needed increase of the creep compliance to be ~150 (or ~400) times.) Since the creep compliance usually changes by several orders of magnitude across the glass transition, this estimate shows that we may expect an AFM tip to penetrate through the surface rubbery layer if it exists. In Fig. 11 is shown the T dependence of the thickness of the rubbery layer, $\delta(T)$, modeled by Keddie *et al.*,[18,19] *i.e.*, = $\delta(1-T/T_g(\infty))^{-1/\nu}$ (see Sec. 2). The intercept obtained from asymptotic linear extrapolations at the two ends of this curve is very near $T_g(\infty)$ (Fig. 11). Although the AFM response *vs.* T curve is strictly speaking different from this curve due to the finite creep rate for the tip to penetrate the rubbery layer, one may envision that the position of the intercept would still be very close to $T_g(\infty)$ due to the theoretical divergence of the rubbery layer thereat. Given the vast number of recent experiments showing the surface dynamics of PS to be enhanced,[29,44,45,49] the scenario portrayed in Fig. 11 is probable.

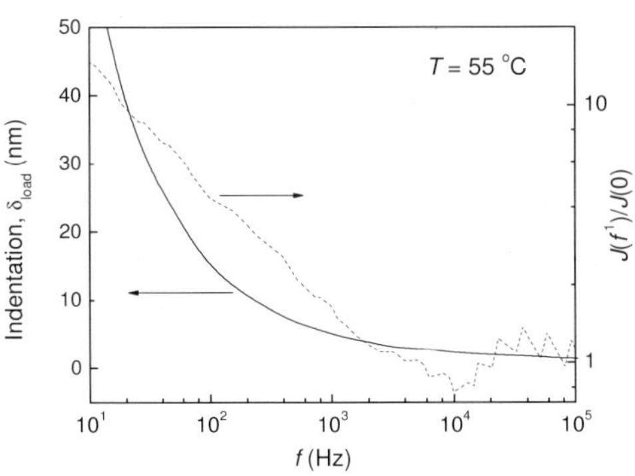

Fig. 10. Calculated tip indentation into a flat poly(*tert*-butyl acrylate) (P*t*BuA) (M_w = 148 kDa, M_w/M_n = 17, T_g = 50°C) plotted as a function of measurement probe rate, f. The data was obtained assuming the JKR model, and the radius of the AFM tip to be 50 nm, the applied force at loading to be 2.5 nN, and the adhesion data published in Refs. 41 and 42. The normalized creep compliance, $J(f^{-1})/J(0)$ (dashed line; left scale) is also shown.

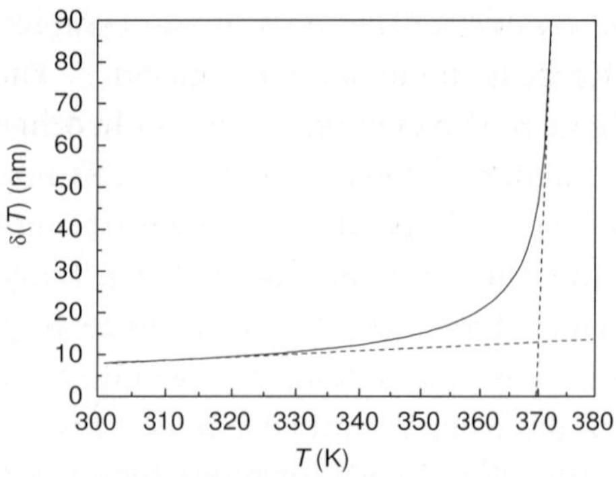

Fig. 11. Simulated temperature dependence of the thickness of the surface rubbery layer according to the model of Keddie *et al.*[18,19] discussed in Sec. 2. $T_g(\infty)$ is 373 K here.

Dielectric Spectroscopy: Like AFM techniques, dielectric spectroscopy has been used in a large fraction of the dynamical measurements of polymer films, and produced some of the most elaborate data for $\tau(T)$. In this technique, the $\tau(T)$ of a film is typically obtained by measuring the dielectric loss peak from a frequency (or temperature) scan at different temperatures (or frequencies). For PS films, it was found that when the film thickness was decreased, the dielectric loss peak broadened and at the same time shifted to higher frequencies or lower temperatures.[89,90] These results suggest that the dynamics in the film become more heterogeneous and at the same time more mobile on average. From the measured $\tau(T)$, the T_g reduction was found to be largely caused by a reduction of the Vogel temperature with decreasing film thickness.[89,90] But it is noteworthy that, except for Ref. 99, all dielectric measurements have been performed on polymer films sandwiched between metal electrodes as illustrated in Fig. 3(c). To make these films, the upper electrode was deposited onto the polymer film by thermal evaporation. Concerns have been raised on whether the evaporation process might alter the properties of the polymer film. In a recent experiment, Sharp *et al.*[101] fused two $h/2$ thick supported films to make one h thick capped film. They found that the T_g of these films remained constant equal to $T_g(\infty)$ for all thicknesses down to 8 nm, contrary to the results of previous dielectric measurements.[89,90] This result reaffirms the significance of a free surface to the reduction of T_g in polymer films, but at the same time casts doubts to the existing dielectric results, which had mostly been obtained from capped films.

X-ray Photon Correlation Spectroscopy: This is a state-of-the-art x-ray scattering technique in which a synchrotron x-ray beam is cut down to ~5^2 to ~ 20^2 μm^2 sizes to gain sufficient coherence in the beam so that a speckle pattern is produced upon scattering from a sample.[96,102,103] The relaxation time τ of a polymer film is determined from the exponential decay of the intensity autocorrelation of the x-ray beam that is totally externally reflected from the film surface. According to Kim *et al.*,[96] this technique measures the dynamics in the top 10 nm of the film from the free surface. For PS films on Si,[96] the result from XPCS was found to display the relation, $\tau(q_{//}) \sim Q(q_{//}h)/q_{//}$ (where $q_{//}$ is the wave vector of the fluctuations on the film surface and $Q(q_{//}h)$ is a function of $q_{//}h$ only), which is consistent with the dynamic capillary wave theory and hence confirms that XPCS monitors the time variations in the surface structure of the film due to motions of the surface capillary waves. The smallest film thickness studied was 84 nm. The authors found no difference for the viscosity *vs.* temperature relation, $\eta(T)$, between the studied films and bulk PS.[96] By analyzing the dynamics of the surface capillary modes in a dynamically stratified film, the authors estimated that if the polymer film were composed of a 10 nm thick high-mobility surface layer with a viscosity 10 times smaller than the bulk value, resting on the remainder of the film that is "normal", the total thickness of the film needs to be ~20 nm in order for the effect to be reflected in the dynamics of the capillary waves.[96]

Several other studies have measured the viscosity of polymer films by analyzing the rate of growth of holes in the films.[104-106] In this method, the number density of holes increases rapidly with decreasing film thickness.[107] A high number density of holes causes the coalescence of holes at early times and thus prevents the opening of holes to be followed for an adequate amount of time.[105,107]

From the above brief overview of dynamical measurements of polymer thin films, we observe that the smallest thickness of films that have been studied for the viscosity is 27 nm by the hole-growth method.[105] As for $\eta(T)$ measurements, the thinnest film that have been studied with no ambiguity is 84 nm by XPCS.[96] Clean measurements of $\eta(T)$ in thin films over the full thickness range of 5 nm < h < ~50 nm are still lacking, and this is necessary for making connections with the anomalous T_g observed in polymer film.

6. Concluding Remarks and Outlook

In conclusion, the problem concerning the dynamics of polymers confined in thin films with thicknesses in the nanometer range is more than two decades old, and has become one of the most challenging current problems of polymer physics. Due to the widespread application of polymer films and the drastic thickness dependence of T_g that can sometimes take place, the issue is practical and has significant technological implications. One difficulty of the problem stems from the fact that most experiments had measured the glass transition temperature only, which, however, only represents one façade of the glass transition. To fully characterize the dynamical behavior would require the dynamical relaxation curve, $\tau(T)$ to be measured for films with thicknesses where the T_g anomaly was observed, *i.e.*, 5 nm $< h <$ ~50 nm (Fig. 1). However, dynamical measurements of polymer films in the < 20 nm thickness range are scarce, attributable to the small amount of polymer material (~ 1 μg) contained in these films. With the recent theoretical development, having the knowledge about the $\tau(T)$ dependence of polymer films with thickness as such can be important: The capillary wave model discussed in Sec. 4.4 requires the free surface of the polymer film to be more mobile than the bulk since the fitted value of $E \approx 44$ MPa corresponds to the value of a rubber. From the estimate of Kim *et al.*,[96] in order to see a reduced viscosity in the surface capillary modes due to a 10 nm thick mobile surface layer, one must examine films with thicknesses ≤ 20 nm. If the capillary wave model is operative, the $\tau(T)$ function probed in a sufficiently thin film would be that of the surface layer and hence should demonstrate a much weaker temperature dependence than that of the bulk as in a rubber, but the absolute value of the relaxation time may depend on the film thickness as the coupling between the surface layer and the rest of the film may vary. On the other hand, if the percolation model is correct, the functional form of $\tau(T)$ should vary continuously with decreasing film thickness as the system approaches the 2D limit with the high-temperature limit of the relaxation time remaining the same for all thicknesses. The qualitatively different behaviors of $\tau(T)$ of the two models should enable one to distinguish which model is at work. One should also mention the recent result of Fakhraai *et al.*[108] who measured the T_g of PS films on Si at different cooling rates, and found that the cooling rate *vs.* T_g followed the VTF scaling above bulk T_g, but became Arrhenius below it with the activation energy decreasing with decreasing film thickness. The former is consistent with a recent viscosity measurement of 18 nm PS films by time-resolved AFM.[109] If this behavior indeed reflects $\tau^{-1}(T)$, as tentatively suggested by the authors,[108] the physics of thin film T_g may require ideas different from the ones suggested so far. On the basis of the above discussions, obtaining a measurement of $\tau(T)$ for films much thinner than 20 nm would be important for understanding the anomalous dynamics of polymer films. With the active pursuit of the problem by researchers around the globe, and new experiments with better approach and higher accuracy are designed, it is likely that new breakthroughs in the understanding of the problem will soon occur.

Acknowledgements

The author would like to thank all the students, postdocs and colleagues who have worked and explored with her on this exciting area of polymer thin film dynamics. She also wants to thank the Polymer Division of NSF for supporting her research through the project DMR-0706096.

References

1. M. Alcoutlabi and G. B. McKenna, *J. Phys. Condens. Matter*, 17, R461 (2005).
2. J. Baschnagel and F. Varnik, *J. Phys. Condens. Matter*, 17, R851 (2005).
3. C. B. Roth and J. R. Dutcher, *Soft Condensed Matters: Structure and Dynamics*. (Dekker, New York, 2004).
4. J. A. Forrest and K. Dalnoki-Veress, *Adv. Colloid Interface Sci.*, 94, 167 (2001).
5. G. Reiter, M. Hamieh, P. Damman, S. Sclavons, S. Gabriele, T. Vilmin and E. Raphael, *Nat. Mater.*, 4, 754 (2005).
6. G. Reiter and P.-G. de Gennes, *Eur. Phys. J. E*, 6, 25 (2001).
7. T. Kanaya, T. Miyazaki, H. Watanabe, K. Nishida, H. Yamano, S. Tasaki and B. Bucknall, *Polymer*, 44, 3769 (2003).
8. M. D. Morariu, E. Schaffer and U. Steiner, *Eur. Phys. J. E*, 12, 375 (2003).
9. C. Bollinne, S. Cuenot, B. Nysten and A. M. Jonas, *Eur. Phys. J. E*, 12, 389 (2003).

10. H. Zhao, Y. J. Wang and O. K. C. Tsui, *Langmuir*, 21, 5817 (2005).
11. O. K. C. Tsui, Y. J. Wang, H. Zhao and B. Du, *Eur. Phys. J. E*, 12, 417 (2003).
12. Y. J. Wang and O. K. C. Tsui, *Langmuir*, 22, 1959 (2006).
13. T. Miyazaki, K. Nishida and T. Kanaya, *Phys. Rev. E*, 69, 022801 (2004).
14. C. L. Jackson and G. B. McKenna, *J. Non-Cryst. Solids*, 131-133, 221 (1991).
15. G. Reiter, *Europhys. Lett.*, 23, 579 (1993).
16. G. Reiter, *Macromolecules*, 27, 3046 (1994).
17. G. Reiter, *Eur. Phys. J. E*, 8, 251 (2002).
18. J. L. Keddie, R. A. L. Jones and R. A. Cory, *Europhys. Lett.*, 27, 59 (1994).
19. J. L. Keddie, R. A. L. Jones and R. A. Cory, *Faraday Discuss.*, 98, 219 (1994).
20. J. H. van Zanten, W. E. Wallace and W.-L. Wu, *Phys. Rev. E*, 53, R2053 (1996).
21. W. E. Wallace, J. H. van Zanten and W.-L. Wu, *Phys. Rev. E*, 52, R3329 (1995).
22. L. Xie, G. B. DeMaggio, W. E. Frieze, J. DeVries, D. W. Gidley, H. A. Hristov and A. F. Yee, *Phys. Rev. Lett.*, 74, 4947 (1995).
23. J. A. Forrest and J. Mattsson, *Phys. Rev. E*, 61, R53 (2000).
24. O. K. C. Tsui and H. F. Zhang, *Macromolecules*, 34, 9139 (2001).
25. O. K. C. Tsui, T. P. Russell and C. Hawker, *Macromolecules*, 34, 5535 (2001).
26. A. M. Mayes, *Macromolecules*, 27, 3114 (1994).
27. H. Brown and T. P. Russell, *Macromolecules*, 29, 798 (1996).
28. T. Kajiyama, K. Tanaka and A. Takahara, *Macromolecules*, 28, 3482 (1995).
29. C. J. Ellison and M. Torkelson, *Nat. Mater.*, 2, 695 (2003).
30. F. Varnik, J. Baschnagel and K. Binder, *Phys. Rev. E*, 65, 021507 (2002).
31. P. Scheidler, W. Kob and K. Binder, *Eur. Phys. J. E*, 59, 701 (2002).
32. P. Scheidler, W. Kob and K. Binder, *J. Phys. Chem. B*, 108, 6673 (2004).
33. T. R. Bohme and J. J. de Pablo, *J. Chem. Phys.*, 116, 9939 (2002).
34. J. A. Torres, P. F. Nealey and J. J. De Pablo, *Phys. Rev. Lett.*, 85, 3221 (2000).
35. A. R. C. Baljon, M. H. M. van Weert, R. B. DeGraaff and R. Khare, *Macromolecules*, 38, 2391 (2005).
36. T. Kajiyama, K. Tanaka and A. Takahara, *Macromolecules*, 30, 280 (1997).
37. K. Tanaka, A. Taura, S.-R. Ge, A. Takahara and T. Kajiyama, *Macromolecules*, 29, 3040 (1996).
38. K. Tanaka, A. Takahara and T. Kajiyama, *Macromolecules*, 30, 6626 (1997).
39. J. A. Hammerschmidt, W. L. Gladfelter and G. Haugstade, *Macromolecules*, 32, 3360 (1999).
40. S. Ge, Y. Pu, W. Zhang, M. Rafailovich, J. Sokolov, C. Buenviaje, R. Buckmaster and R. M. Overney, *Phys. Rev. Lett.*, 85, 2340 (2000).
41. O. K. C. Tsui, X. P. Wang, J. Y. L. Ho, T. K. Ng and X. Xiao, *Macromolecules*, 33, 4198 (2000).
42. X. P. Wang, X. Xiao and O. K. C. Tsui, *Macromolecules*, 34, 4180 (2001).
43. Y. Liu, T. P. Russell, M. G. Samant, J. Stohr and H. Brown, *Macromolecules*, 30, 7768 (1997).
44. W. E. Wallace, D. A. Fischer, K. Efimenko, W.-L. Wu and J. Genzer, *Macromolecules*, 34, 5081 (2001).
45. Y. C. Jean, R. Zhang, H. Cao, J.-P. Yuan, C.-M. Huang, B. Nielsen and P. Asoka-Kumar, *Phys. Rev. B*, 56, R8459 (1997).
46. A. D. Schwab, D. M. G. Agra, J.-H. Kim, S. Kumar and A. Dhinojwala, *Macromolecules*, 33, 4903 (2000).

47. O. C. Tsang, F. Xie, O. K. C. Tsui, Z. Yang, J. Zhang, D. Shen and X. Yang, *J. Polym. Sci. B*, 39, 2906 (2001).
48. O. C. Tsang, O. K. C. Tsui and Z. Yang, *Phys. Rev. E*, 63, 061603 (2001).
49. T. Kerle, Z. Q. Lin, H. C. Kim and T. P. Russell, *Macromolecules*, 34, 3484 (2001).
50. M. Hamdorf and D. Johannsmann, *J. Chem. Phys.*, 112, 4262 (2000).
51. J. A. Forrest, K. Dalnoki-Veress, J. R. Stevens and J. R. Dutcher, *Phys. Rev. Lett.*, 77, 2002 (1996).
52. J. A. Forrest, K. Dalnoki-Veress and J. R. Dutcher, *Phys. Rev. E*, 56, 5705 (1997).
53. J. A. Forrest, K. Dalnoki-Veress and J. R. Dutcher, *Phys. Rev. E*, 58, 6109 (1998).
54. J. Mattsson, J. A. Forrest and L. Borjesson, *Phys. Rev. E*, 62, 5187 (2000).
55. K. Dalnoki-Veress, J. A. Forrest, C. Murray, C. Gigault and J. R. Dutcher, *Phys. Rev. E*, 62, 5187 (2001).
56. C. B. Roth and J. R. Dutcher, *Eur. Phys. J. E*, 12, 103 (2003).
57. C. B. Roth, A. Pound, S. W. Kamp and C. A. Murray, *Eur. Phys. J. E*, 20, 441 (2006).
58. S. Herminghaus, K. Jacobs and R. Seemann, *Eur. Phys. J. E*, 5, 531 (2001).
59. Pemra Doruker and Wayne L. Mattice, *J. Phys. Chem. B*, 103, 178 (1999).
60. T. G. Fox and P. J. Flory, *J. Appl. Phys.*, 21, 581 (1950).
61. A. Hariharan, S. Kumar and T. P. Russell, *J. Chem. Phys.*, 99, 4041 (1993).
62. G. Adam and J. H. Gibbs, *J. Chem. Phys.*, 43, 139 (1965).
63. P. G. Santangelo and C. M. Roland, *Macromolecules*, 31, 4581 (1998).
64. P. Mansky, Y. Liu, E. Huang, T. P. Russell and C. Hawker, *Science*, 275, 1458 (1997).
65. A. Eisenberg, *J. Phys. Chem.*, 67, 1333 (1963).
66. J. Brandrup and E. H. Immergut eds., *Polymer Handbook*, 3rd edn. (Wiley, New York, 1989).
67. W. E. Wallace, N. C. Beck-Tan and W.-L. Wu, *J. Chem. Phys.*, 108, 3798 (1998).
68. K. R. Shull, *Faraday Discuss.*, 98, 203 (1994).
69. John D. McCoy and John G. Curro, *J. Chem. Phys.*, 116, 9154 (2002).
70. I. Bitsanis and G. Hadziioannou, *J. Chem. Phys.*, 92, 3827 (1990).
71. J. Baschnagel and K. Binder, *Macromolecules*, 28, 6808 (1995).
72. U. Tracht, M. Wilhelm, A. Heuer, H. Feng, K. Schmidt-Rohr and H. W. Spiess, *Phys. Rev. Lett.*, 81, 2727 (1998).
73. M. T. Cicerone, F. R. Blackburn and M. D. Ediger, *Macromolecules*, 28, 8224 (1995).
74. P. W. Anderson, *Science*, 267, 1615 (1995).
75. P.-G. De Gennes, *Eur. Phys. J. E*, 2, 201 (2000).
76. D. Long and F. Lequeuz, *Eur. Phys. J. E*, 4, 371 (2001).
77. S. Merabia and D. Long, *Eur. Phys. J. E*, 9, 195 (2002).
78. J.-H. Kim, J. Jang and W.-C. Zin, *Langmuir*, 16, 4064 (2000).
79. J.-H. Kim, J. Jang and W.-C. Zin, *Langmuir*, 17, 2703 (2001).
80. S. Herminghaus, *Eur. Phys. J. E*, 8, 237 (2002).
81. B. Frank, A. P. Gast, T. P. Russell, H. Brown and C. Hawker, *Macromolecules*, 29, 6531 (1996).
82. X. Zheng, M. Rafailovich, J. Sokolov, Y. Strzhemechy and S. A. Schwarz, *Phys. Rev. Lett.*, 79, 241 (1997).
83. E. K. Lin, W.-L. Wu and S. K. Satija, *Macromolecules*, 30, 7224 (1997).

84. Y. Pu, H. White, M. Rafailovich, J. Sokolov, A. Patel, C. White, W.-L. Wu, V. Zaitsev and S. A. Schwarz, *Macromolecules*, 34, 8518 (2001).
85. A. N Semenov, *Phys. Rev. Lett.*, 80, 1908 (1998).
86. E. Donth, *The glass transition: relaxation dynamics in liquids and disordered materials*. (Springer-Verlag, Berlin, 2001).
87. C. Carelli, A. M. Higgins, R. A. L. Jones and M. Sferrazza, *Phys. Rev. E*, 73, 061804 (2006).
88. P. A. O'Connell and G. B. McKenna, *Science*, 307, 1760 (2005).
89. K. Fukao and Y. Miyamoto, *Phys. Rev. E*, 61, 1743 (2000).
90. K. Fukao and Y. Miyamoto, *Phys. Rev. E*, 64, 011803 (2001).
91. K. Fukao, S. Uno, Y. Miyamoto, A. Hoshino and H. Miyaji, *Phys. Rev. E*, 64, 051807 (2001).
92. K. Fukao, *Eur. Phys. J. E*, 12, 119 (2003).
93. L. Hatmann, W. Gorbatschow, J. Hauwede and F. Kremer, *Eur. Phys. J. E*, 8, 145 (2002).
94. A. Serghei and F. Kremer, *Phys. Rev. Lett.*, 91, 165702 (2003).
95. M. Wubbenhorst, C. A. Murray and J. R. Dutcher, *Eur. Phys. J. E*, 12, S109 (2003).
96. H. Kim, A. Ruhm, L. B. Lurio, J. K. Basu, J. Lal, D. Lumma, S. G. J. Mochrie and S. K. Sinha, *Phys. Rev. Lett.*, 90, 068302 (2003).
97. X. Hu, Z. Jiang, S. Narayanan, X. Jiao, A. R. Sandy, S. K. Sinha, L. B. Lurio and J. Lal, *Phys. Rev. E*, 74, R010602 (2006).
98. Z. Jiang, H. Kim, S. G. J. Mochrie, L. B. Lurio and S. K. Sinha, *Phys. Rev. E*, 74, 011603 (2006).
99. K. L. Johnson, K. Kendall and A. D. Roberts, *Proc. R. Soc. London A*, 324, 301 (1971).
100. H. Hertz, *J. Reine Angew. Math.*, 92, 156 (1882).
101. J. S. Sharp and J. A. Forrest, *Phys. Rev. Lett.*, 91, 235701 (2003).
102. O. K. C. Tsui and S. G. J. Mochrie, *Phys. Rev. E*, 57, 2030 (1998).
103. O. K. C. Tsui, S. G. J. Mochrie and L. E. Berman, *J. Synchrotron Rad.*, 5, 30 (1998).
104. K. Dalnoki-Veress, B. G. Nickel, C. Roth and J. R. Dutcher, *Phys. Rev. E*, 59, 2153 (1999).
105. J.-L. Masson and P. F. Green, *Phys. Rev. E*, 65, 031806 (2002).
106. C. Li, T. Koga, C. Li, J. Jiang, S. Sharma, S. Narayanan, L. B. Lurio, X. Hu, X. Jiao, S. K. Sinha, S. Billet, D. Sosnowik, H. Kim, J. Sokolov and M. Rafailovich, *Macromolecules*, 38, 5144 (2005).
107. G. Reiter, *Phys. Rev. Lett.*, 87, 186101 (2001).
108. Z. Fakhraai and J. A. Forrest, *Phys. Rev. Lett.*, 95, 025701 (2005).
109. O. K. C. Tsui, Y. J. Wang, F. K. Lee, C.-H. Lam and Z. Yang, *Macromolecules*, 41, 1465 (2008).

INDEX